THE CELL MEMBRANE

Its Role in Interaction with the Outside World

Herman M. Kalckar

THE CELL MEMBRANE

Its Role in Interaction with the Outside World

A Volume in Honor of Professor Herman Kalckar on His Seventy-fifth Birthday

Edited by

Edgar Haber

Massachusetts General Hospital
Harvard Medical School
Boston, Massachusetts

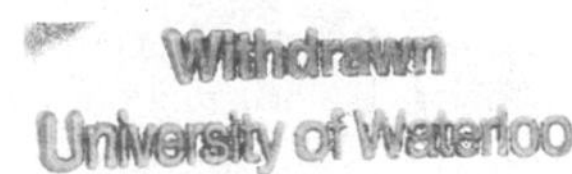

PLENUM PRESS • NEW YORK AND LONDON

Library of Congress Cataloging in Publication Data

Main entry under title:

The Cell membrane.

"Proceedings of a meeting in honor of Professor Herman Kalckar, held December 6, 1983, at Harvard Medical School, Boston, Massachusetts, under the sponsorship of Harvard Medical School"—T.p. verso.
Bibliography: p.
Includes index.
1. Cell membranes—Congresses. 2. Kalckar, Herman M. (Herman Moritz), 1908–
I. Kalckar, Herman M. (Herman Moritz), 1908– . II. Haber, Edgar. III. Harvard Medical School.
QH601.C34 1984 574.87′5 84-18099
ISBN 0-306-41827-4

Proceedings of a meeting in honor of Professor Herman Kalckar, held December 6, 1983, at Harvard Medical School, Boston, Massachusetts, under the sponsorship of Harvard Medical School

Supported in part by the Upjohn Company; Burroughs-Wellcome; Merck, Sharp and Dohme; and Schering-Plough

A Division of Plenum Publishing Corporation
233 Spring Street, New York, N.Y. 10013

Printed in the United States of America

PREFACE

The contents of this book reflect a symposium held in honor of Professor Herman Kalckar's seventy-fifth birthday. His impact on the history of biochemistry is reflected by the diversity of the contributions of his former students and friends. Speakers came from Asia, Europe, and the United States to discuss both procaryotes and eukaryotes. The unifying theme was the cell membrane, both its organization and its function. Ektobiology, a topic that has held the attention of Professor Kalckar for many years, was clearly defined as a central topic in biology. This subject deals with the key structure whereby the cell interacts with the outside world and which, in a sense, defines the boundary between what is the cell and what is not.

Topics discussed include the biogenesis of membrane proteins, sugars and lipids, the role of membrane components in osmoregulation, and mechanisms of nutrient transport. Of great interest is the system for surface recognition evolved in vertebrates, exemplified by the HLA system of man. Neoplasia causes changes in the cell membrane that may be of significant future potential in the diagnosis and treatment of malignancies as well as in the understanding of the process of transformation. The changes in glycosphingolipids and carbohydrate antigens in relation to oncogenesis are detailed. I should like to recognize Doctors Kurt J. Isselbacher, Phillips W. Robbins, Victor Ginsburg, and Hiroshi Nikaido for their assistance in organizing the symposium. Ms. Jean Brumbaugh deserves special thanks for putting this book together.

Edgar Haber

CONTENTS

HERMAN KALCKAR: AN APPRECIATION

Edgar Haber

Cardiac Unit
Massachusetts General Hospital
Boston, MA 02114

A Brief Biographical Sketch

Herman Moritz Kalckar was born of Ludvig Kalckar and Bertha Melchior in Copenhagen, Denmark, on March 26, 1908. He received his M.D. degree from the University of Copenhagen and later served as Associate Professor in the University's Institute of Medical Physiology. In January 1939 he left for the United States where he travelled under the auspices of the Rockefeller Foundation. He then worked successively at the Public Health Research Institute of New York and at the National Institutes of Health in Bethesda. In 1958 he was appointed Professor of Biology at Johns Hopkins University, and in 1961 as Professor of Biological Chemistry at Harvard Medical School and Wellcome Biochemist and Chief of Biochemical Research at the Massachusetts General Hospital. At present he is Emeritus Professor at Harvard University and Distinguished Research Professor of Biochemisty at Boston University.

Professor Kalckar is the recipient of many honors including membership in the National Academy of Sciences, and the American Academy of Arts and Sciences and foreign membership in the Royal Danish Academy. He has been a Harvey Lecturer and a Fogarty Scholar of the National Institutes of Health and recipient of honorary degrees from Washington University, the University of Chicago and the University of Copenhagen.

It is not my intent to write even a short biography of Herman Kalckar, but simply to point to a few highlights of his career, particularly the early and formative years during which a revolution in fundamental biochemistry was revealed, and the period when he had a major influence on the development of a relationship between medicine and biochemistry with the discovery of the mechanism of galactosemia. His interest in Ektobiology that characterized the Massachusetts General Hospital-Harvard years are the subject of this volume. The most recent decade is discussed here in a chapter by C. William Christopher.

Oxidative Phosphorylation

Herman Kalckar's career was shaped by notable biochemists, many of whom became his close personal friends. During the Copenhagen days, his mentors were Ejnar Lundsgaard and Fritz Lipman. Seminal to the beginning of Kalckar's research was the major discovery by Lundsgaard that lactic acid formation alone could not explain the source of the energetics needed for muscle contraction. As a consequence of showing that iodoacetate-treated muscle could carry out a considerable amount of mechanical work anaerobically without any lactic acid being formed, the Meyerhof-Hill hypothesis was shattered. Phosphocreatine was shown to be essential by Lundsgaard's demonstration that there was a linear relationship between the amount of phosphate released from phosphocreatine and the amount of mechanical work performed by the muscle. Kalckar, working with kidney tissues, began to appreciate the respiration as a critical factor in active phosphorylation. He varied the nature of the electron donors and demonstrated that glutamic acid, dicarboxylic and tricarboxylic acids markedly increase Qo_2 as well as phosphorylation. Thus, the concept of oxidative phosphorylation was born. In 1937, Kalckar was able to show that in thoroughly dialyzed kidney extracts or washed kidney mince where glucose was absent, malate underwent oxidation and phosphopyruvate accumulated. Thus one of the main pathways of gluconeogenesis from pyruvate and CO_2 underwent its initial definition.

By the time Kalckar left for the United States as a Rockefeller Research Fellow in January 1939, his work on oxidative phosphorylation was already well recognized and of interest to a number of investigators, among them Robert Robison at the Lister Institute in London and Carl and Gerty Cori at Washington University in St. Louis. He stopped to visit each of these laboratories and in the latter demonstrated the methods that were needed to detect oxidative phosphorylation. By adequate aeration and the use of Warburg vessels, the Coris were able to confirm the phenomenon.

At the end of February of 1939, he had arrived at the California Institute of Technology, which was then, as today, a center of important research. He came in contact with Henry Borsook, Max Delbruck, Norman Horowitz, Thomas H. Morgan, and Alfred Sturtevant among others. In discussions with Charles Coryell, he was encouraged to write about energetic coupling via phosphorylation in biological syntheses. Linus Pauling effected the publication of this work in Chemical Reviews in 1941. This was a monumental paper that had an exceptional impact on the field of biochemistry.

Hereditary Human Galactosemia

Galactosemia is a disease of infants characterized by severe vomiting, diarrhea, and jaundice. If ingestion of mother's or cow's milk is not stopped, permanent changes including cataracts, cirrhosis of the liver, and mental retardation occur. The consequences of the disease may be effectively prevented if milk (containing galactose) is omitted from the diet and other foods substituted. It is of particular interest that in the absence of lactose or galactose, normal development and growth is not impaired. With renewed lactose administration, galactosemia and galactosuria resume. How does the organism manage to grow normally without the ability to synthesize essential galactosyl compounds, such as galactolipids of the brain?

In 1956, Kalckar, working with Isselbacher and Anderson, designed a sensitive and specific methods that pinpointed the enzymatic defect. It had been demonstrated by other investigators earlier that the administration of galactose to the affected children gave rise to an accumulation of alpha-galactose-1-phosphate in red cells. This excluded a galactokinase defect as the basis of the disease. Kalckar and his coworkers did, however, identify a clear deficiency of galactose-1-phosphate uridylyl transferase. It is the conversion of Gal-1-P to UDP-Gal that is blocked, resulting in the accumulation of toxic level of of Gal-1-P. Since the conversion of UDP-Glucose to UDP-Gal is unimpaired, there is adequate substrate for the synthesis of essential galactose containing compounds from glucose compounds.

Aside from the interest in defining the mechanism responsible for a disease, this discovery had wider importance. It was a graphic demonstration to physicians of the value of biochemical investigation in understanding pathophysiology and a catalyst to further research on inborn errors of metabolism, of which many have now been defined. On a more personal level, it was the beginning of the distinguished research career, largely based on the biochemical investigation of disease, of the young postdoctoral fellow participating in the research, Kurt Isselbacher.

These vignettes are but a narrow window on the many contributions to science by Herman Kalckar. I hope that a biographer takes up the challenge of revealing Kalckar's entire career, which will certainly be an inspiration to all of us. As is clear in Christopher's chapter, it is a career that is still in full development.

THE CELL MEMBRANE OF PROCARYOTES

OUTER MEMBRANE PERMEABILITY OF GRAM-NEGATIVE BACTERIA IN THEIR INTERACTION WITH THE OUTSIDE WORLD

Hiroshi Nikaido

Department of Microbiology and Immunology
University of California, Berkeley
Berkeley, California 94720

ABSTRACT

Gram-negative bacteria including Esherichia coli and Salmonella typhimurium produce outer membranes that serve as an effective permeability barrier. In order to shut out noxious substances as much as possible, which include the bile salts in the case of enteric organisms, yet allow the rapid influx of nutrients, these bacteria have evolved porins, proteins with channels that are ideally suited for this purpose. Furthermore, the enteric organisms contain multiple species of porins, and have developed an elaborate mechanism of regulating the production of different species of porins so that their chances of survival would be maximized under different environmental conditions, providing us with a remarkable example of ecological adaptation at the molecular level. These porins are also of interest because analogous proteins occur in the outer membranes of organelles of eukaryotic cells, such as mitochondria.

INTRODUCTION

Kalckar (1965) has always emphasized that the phenomena we encounter in the world of higher animals and plants, for example, the phenotypes of Escherichia coli mutants defective in their galactose metabolisms, were very instructive in our understanding of the potential and real consequences of such defects in humans. This idea has always sustained me in my studies of the ways the bacteria adapted to various environments, especially during the years in which the "relevance" of basic research has been questioned, sometimes rather severely, by the society in general and by the administrations in power.

It has been known for a long time that Gram-positive bacteria and Gram-negative bacteria are quite different in the organization of their surface layers, and that much higher amounts of lipids and proteins were found in the "cell wall" of Gram-negative bacteria (Salton, 1964). Kellenberger and Ryter (1958) also noted clearly the presence of a triple-layered, membrane-like structure on the surface of E. coli cells. However, to my knowledge, Bladen and Mergenhagen (1964) were the first to coin the word "outer membrane" for the extra membrane structure that was found in a Gram-negative species. Glauert and Thornley (1969) also contributed significantly to our understanding by emphasizing, in their review, the ubiquitous presence of this membrane layer in all Gram-negative bacteria so far examined, and its absence in all Gram-positive species.

STRUCTURE AND PROPERTIES OF THE LIPID BILAYER REGION

In spite of our knowledge on the presence of this extra membrane layer, its functions are poorly understood. We were led to the study of the structure-function relationship of this unusual membrane through our interest in the lipopolysaccharide (LPS), a characteristic component of the outer membrane. Toward the end of the 1960's both Westphal and his associates in Germany and the Roantree-Stocker group at Stanford found that mutants of E. coli and Salmonella producing severely truncated LPS molecules were strikingly more sensitive than the wild type organisms toward various antibiotics and other inhibitors (Schlecht and Westphal, 1969; Roantree et al., 1969). The inspection of the structures of the agents to which the mutants became hypersensitive suggested to me that they were mostly hydrophobic or large molecules. Assay of the rate of diffusion of a hydrophobic semisynthetic penicillin, nafcillin, indeed confirmed that the cells of wild type S. typhimurium were completely impermeable to this drug, but those mutants with defective LPS had an outer membrane permeable to this agent (Nikaido, 1976). These experiments, then, showed that the outer membrane of E. coli or S. typhimurium is normally a very effective barrier against the penetration by hydrophobic molecules. This was surprising because lipid bilayers are usually quite permeable to any hydrophobic molecules, and the search for the basis of this low permeability led to the discovery that the bilayer in the bacterial outer membrane is unusual in its asymmetric construction, containing only LPS in its outer leaflet and only phospholipids in the inner leaflet (Smit et al., 1975; Kamio and Nikaido, 1976). We do not know why the leaflet composed entirely of LPS would inhibit the diffusion of hydrophobic molecules, but we can certainly speculate about it. Unlike phospholipids, which contain only two hydrocarbon chains linked to the glycerol backbone, LPS has seven fatty acid chains connected to the glucosaminyl-glycosamine backbone (Qureshi et al., 1982). If, as is commonly believed, the diffusion of hydrophobic molecules involves the pushing aside of the lipid molecules to create a cavity for the solute, such an act will be much more difficult with the LPS monolayer than with the phospholipid bilayer. Furthermore, the fatty acid residues in LPS are all saturated. This will reduce the fluidity of the hydrocarbon domain of LPS, as indeed found by the insertion of spin-labeled probe into LPS bilayers (Nikaido et al., 1977). It is well known that bilayers with less fluid hydrocarbon regions show lower permeability to hydrophobic solutes (de Gier et al., 1971). Regardless of the precise molecular mechanism of this low permeability, clearly it is an ecological advantage, or necessity, for many Gram-negative organisms to have an outer membrane with these properties. For example, E. coli and other enteric organisms normally live in the upper part of the intestinal tract of higher animals, where one encounters high concentrations of bile salts, powerful detergents that would dissolve the cytoplasmic membrane of these bacteria if the outer membrane barrier were absent. The classical selective media for enteric bacteria use this barrier property, in order to inhibit the growth of Gram-positive and some other Gram-negative bacteria by the inclusion of hydrophobic inhibitors such as dyes, bile salts, and other detergents. This concept also explains the fact that only a few Gram-positive species are found in the intestinal tract, and that either they are unusual in their resistance to bile salts (enterococci) or found predominantly in the lower part of the tract (clostridia).

PORINS AND PROPERTIES OF THE PORIN CHANNEL

The finding of the low permeability of the lipid bilayer matrix still did not explain how various nutrients crossed the outer membrane, because most of the nutrients used by bacteria are very hydrophilic, and the lipid matrix would be even less permeable for these solutes. When we examined the permeability of intact outer membrane by studying the influx of radioactive solutes into the enlarged periplasmic space, i.e., the space between the outer and cytoplasmic membranes, of plasmolyzed E. coli and S. typhimurium cells, we found that solutes with less than about 600 molecular weight were able to cross the outer membrane extremely rapidly in the time scale of our experimental set-up (Decad and Nikaido, 1976). What then produces this apparently nonspecific permeability? We have studied this problem through a reconstitution approach. The first few tries involving the reconstitution of LPS/phospholipid liposomes produced no encouraging results, but then Taiji Nakae in the laboratory had the first promising results by reconstituting vesicles with phospholipids, LPS, and a crude mixture of outer membrane proteins (Nakae, 1975). The results were not the strongest, in the sense that even when 200 μg protein was added to 1 μmol phospholipids only 75 percent of the [^{14}C]-sucrose diffused out of the vesicles during gel filtration. However, Nakae quickly found out that the most active proteins were located in the fraction that remained insoluble after detergent extraction, and could purify a group of proteins, porins, that allowed the efflux of more than 90 percent of the entrapped [^{14}C]-sucrose when added only in small amounts (10 μg per 1 μmol phospholipids) (Nakae, 1976a). Interestingly, the porin turned out to be the protein (Nakae, 1976b) that was previously purified by Rosenbusch (1974), and was erroneously believed to be the shape-forming component of the outer membrane (see Henning and Haller, 1975).

Porins are very interesting proteins. In terms of mass, it usually represents a few percent of the total E. coli protein, and thus is one of the most abundant proteins in these bacteria. In addition to the fact that they were the first proteins identified to produce large, nonspecific channels, they are also unusual in their structure. Three subunits of this protein, each with a molecular weight around 36,000, produce a trimer that is very stable, even to the denaturing action of sodium dodecylsulfate, and is apparently the functional unit that produces the pores. The protein also contains hardly any detectable α-helix, but an unusually large amount of β-sheet structure (Rosenbusch, 1974). The sequence of E. coli porin has been determined in the laboratory of U. Henning (Chen et al., 1982), and it was surprising to see no long (20-30 residues) stretches of hydrophobic amino acids, and to see that the overall hydrophobicity of the protein was so low. However, when one considers that the backbone of β-sheet structure is much more extended than that of α-helix, and could cross the thickness of the membrane in a stretch of 12-15 residues, then the primary sequence becomes remarkably suggestive, as the Chou-Fasman prediction of its secondary sequence (Chou and Fasman, 1978) shows that 12-16 residue stretches containing few charged amino acid residues, predicted as β-sheets, alternate with regions predicted as β-turns and containing a number of charged, and otherwise hydrophilic, amino acid residues.

The radioisotope efflux assay we used in our early studies had a rather poor time resolution, and was not suitable for studying the kinetics of solute diffusion through the porin channels. A liposome swelling assay was therefore developed by incorporating small amounts of porins into the phospholipid bilayers of liposomes. When these liposomes were diluted into isotonic solutions of various hydrophilic solutes, the solutes diffused through the porin channels into the intravesicular space following their concentration gradient, and the accompanying influx of water produced the swelling of liposomes which could be followed continuously by recording the turbidity of the liposome suspension in a spectrophotometer (Luckey and Nikaido, 1980). This is a modification of the method used by workers who studied the permeability of phospholipid bilayers (for example, see de Gier et al, 1971), and was very fast and sensitive. In order to measure the permeability of E. coli porin channel for relatively small molecules such as L-arabinose, the liposomes used in one assay had to contain only about 20 ng of purified porin.

Studies of the properties of the porin channel by this method produced the following results.

1. With any porin channel, the size of the solute had a very strong influence on its rate of diffusion through the channel. With E. coli porin channels, for example, N-acetyl-D-glucosamine (221 daltons) penetrated at a rate two- to threefold lower than that of L-arabinose (150 daltons), and the rates of diffusion of hexose disaccharides (342 daltons) were about 50-500 times lower than the rate for arabinose (Nikaido and Rosenberg, 1983). This is exactly the behavior of the type predicted for narrow water-filled channels (Renkin, 1954), and calculations show that the E. coli pores can be approximated by hollow cylinders of about 1.1 to 1.2 nm diameter.

2. With any porin channel, more hydrophobic solutes penetrated the pore more slowly than the more hydrophilic solutes of the same size (Nikaido et al., 1983). It is possible that the water molecules within the pore are more strongly hydrogen-bonded to each other and also to the groups on the channel wall, and that the penetration of hydrophobic molecules is not favored because many of these have to be broken.

3. The electrical charges on the solutes have strong influences on the rates of penetration. With the porins produced by E. coli under the usual growth conditions (OmpF and OmpC), the presence of negative charges strongly reduce the rates of penetration (Nikaido and Rosenberg, 1983). In contrast, with the porin originally discovered in the "pseudorevertants" of porin-deficient mutants, PhoE, the negative charge did not hinder the penetration and in fact accelerated it (Nikaido et al., 1983; Korteland et al., 1982).

4. In spite of early claims from several other laboratories, no evidence for the configurational specificity was observed. Because the effect of the general physicochemical properties of the solutes mentioned above can be quite pronounced, it is possible that limited comparisons of only a few compounds could have created false impressions of "specificity."

PORINS IN THE INTERACTION OF BACTERIA WITH THE EXTERNAL WORLD

Because porins constitute one of the first sites of contact between the molecules in the outside world and the bacterial cell, one would expect that evolution must have produced, in the porin channel, features that are best adapted for the survival of the particular bacterial species in its particular ecological niche. This is seen to some extent in comparisons of porins from various species.

Thus the normal E. coli or S. typhimurium porin channel is narrow and tends to reject hydrophobic and negatively charged molecules, all these properties presumably being ideal for preventing the influx of bile salts. In contrast, saprophytic bacteria living in nature do not have the luxury of our intestinal enzymes breaking down macromolecules into small constituents, and thus they tend to produce porins with larger channels presumably so that they can take up larger peptides and oligosaccharides as nutrients (L.S. Zalman and H. Nikaido, to be published). An interesting anomaly, though, is Pseudomonas aeruginosa. Its porin channel is quite large, with an estimated diameter of about 2.0 nm, which allows the penetration of polysaccharides of several thousand daltons (Benz and Hancock, 1981; Yoshimura et al., 1983), yet the rate of penetration of solutes through liposome membranes containing P. aeurginosa porin as well as through intact outer membranes of this organism is quite slow (Yoshimura and Nikaido, 1982). Possibly this organism avoids the excessive vulnerability to various inhibitors, which this large pore size could produce, by closing down most of the pores. The well-known intrinsic antibiotic resistance of this organism is largely caused by this poor permeability of its outer membrane.

Each strain of E. coli or S. typhimurium produces multiple species of porins, and this is a phenomenon that seems to be restricted to enteric bacteria. The production of each species of porin appears to be controlled very carefully, to maximize the chances of survival of these bacteria.

Firstly, it was mentioned above that PhoE porin, which is not produced under the usual growth conditions, favors the diffusion of negatively charged solutes. We could not imagine the benefit this porin would give to the E. coli cell, but Boos, as well as Lugtenberg, found that PhoE protein is a part of the alkaline phosphatase regulon and is derepressed only under phosphate starvation (Argast and Boos, 1980; Tommassen and Lugtenberg, 1980). Thus, obviously the main function of this porin is to bring in phosphate and phosphorylated compounds more efficiently across the membrane; it is produced under phosphate limitation conditions only, presumably because normally the enteric organism tries to shut out bile salts by having a poor permeability to negatively charged compounds.

Secondly, the production of OmpF porin is repressed by high osmotic pressure, such as that caused by the presence of 10 percent sucrose in the medium (Nakamura and Mizushima, 1976). It seems to be repressed also by high growth temperature (Lugtenberg et al., 1976). We could not understand the physiological significance of these regulatory changes, but recently we found a very suggestive piece of evidence. Dr. A.A. Medeiros of Brown University isolated from a patient suffering from a renal abscess and bacteremia, two strains of S. typhimurium, one before the antibiotic therapy, and the other after several days of cephalexin treatment. Both strains contained the same R-factor, but the "mutant" isolated after the therapy was more resistant to a wide variety of β-lactams, and it was shown, by the use of the method proposed by Zimmermann and Rosselet (1977), that the mutant produced an outer membrane with a lowered permeability. When we examined the porin composition of these two strains in cells grown in Trypticase-Soy medium (a very low osmolarity medium), we found that the parent strain produced OmpF and OmpC porins, whereas the mutant produced only OmpF porin. The loss of OmpC porin, however, did not explain the low permeability seen by Medeiros, because (1) OmpC channel has a slightly smaller diameter than the OmpF channel, and therefore the loss of OmpC does not visibly decrease the permeability toward large, negatively charged, somewhat hydrohobic solutes such as β-lactams (Nikaido and Rosenberg, 1983), and (2) the outer membranes of the two strains grown in Trypticase-Soy broth indeed showed very similar permeability to β-lactams. However, when the strains were grown in L broth, which contains 1 percent NaCl, the OmpF porin was almost completely repressed and the mutant strain produced no visible porin, and its outer membrane permeability was indeed very low. The very fact that this mutant was selected in the body of the patient through the β-lactam therapy shows us that indeed the OmpF porin must have been repressed almost entirely in the normal habitat of Salmonella, i.e., in the body of animals and humans. "10% sucrose" sounds like a very high osmolarity medium, yet its osmolarity is similar to that of 1% NaCl, which of course is close to the osmotic activity in our body. These results then suggest that the enteric bacteria live in their normal habitat by repressing the wider OmpF porin; this would be beneficial to these bacteria as animal body is full of inhibitory substances for bacteria. What useful purpose could the OmpF porin serve for these bacteria? M.J. Osborn (personal communication) suggested that the bacteria would derepress the production of OmpF porin when they are in natural waters, presumably through the influence of the low osmolarity and low temperature; the wider OmpF channel would help them in the uptake of nutrients that exist in very low concentrations in the environment of this type (see also Nikaido, 1979).

CHANNEL-FORMING PROTEINS IN THE EUKARYOTIC CELL

Do proteins similar to porins occur in eukaryotic cells? It seems unlikely to find a permeability barrier outside the cytoplasmic membrane of the cells of multicellular organisms, because there is no need for such a barrier. However, in view of the probable endosymbiotic origin of mitochondria and chloroplasts, and of the presence of morphologically recognizable outer membrane in these organelles, it was of interest to see whether their outer membrane contained porins. Such a study showed clearly that the outer membrane of mitochondria from diverse sources did contain porin (Zalman et al., 1980). Although the mitochrondrial porin has a molecular weight similar to that of the bacterial porin, it is not clear whether they share a common evolutionary origin. However, like the bacterial porin, the mitochrondrial porin appears to occur as trimers (Mannella et al., 1983). It is also interesting that the mitochrondrial porin channel prefers anionic solutes (Colombini, 1980), a feature obviously suited for the passage of metabolic intermediates, practically all of which are negatively charged.

If cells of multicellular organisms incorporated porin-like molecules in their cytoplasmic membrane, they would lose metabolite molecules into extracellular space, and obviously this is impossible. However, if the channels of the porin-like molecule opened into the cytoplasm of another cells, this will merely result in the connecting together of two cells. In fact, this type of connection has been known for many years as "gap junctions," and many truly ingenious experiments have been performed to study the properties of the presumed channel (see Loewenstein, 1979). However, to our knowledge no direct reconstitution has been performed with this system, and thus we know very little about the properties of the channel. More recently, we have been able to apply the techniques we developed with the porin system to the reconstitution of gap junction proteins (H. Nikaido and E.Y. Rosenberg, to be published). It is hoped that the studies of the bacterial system will prove useful in studying what happens in much more complex cells of eukaryotes.

ACKNOWLEDGEMENT

It is an honor for the author to dedicate this article for the celebration of Prof. Herman Kalckar's 75th birthday, because the author learned practically everything he knows now about science from Prof. Kalckar. The studies described in this article were supported in part by grants from American Cancer Society (grant BS-20) and U.S. Public Health Service (research grant AI-09644).

REFERENCES

Argast, M., and Boos, W., 1980. Co-regulation in Escherichia coli of a novel transport system for sn-glycerol-3-phosphate and outer membrane protein Ic (e,E) with alkaline phosphatase and phosphate-binding protein, J. Bacteriol., 143:142.

Benz, R., and Hancock, R.E.W., 1981, Properties of the large ion-permeable pores formed from protein F of Pseudomonas aeruginosa in lipid bilayer membranes, Biochim. Biophys. Acta, 646:298.

Bladen, H.A., and Mergenhagen, S.E., 1964, Ultrastructure of Veillonella and morphological correlation of an outer membrane with particles associated with endotoxic activity, J. Bacteriol., 88:1482.

Chen, R., Krämer, C., Schmidmayr, W., Chen-Schmeisser, U., and Henning, U., 1982, Primary structure of major outer-membrane protein I (ompF protein, porin) of Escherichia coli, Biochem. J., 203:33.

Chou, P.Y., and Fasman, G.D., 1978, Prediction of secondary structure of proteins from their amino acid sequence, Advan. Enzymol., 47:45.

Colombini, M., 1980, Structure and mode of action of a voltage-dependent anion-selective channel (VDAC) located in the outer mitohondrial membrane, Ann. N.Y. Acad. Sci., 341:552.

Decad, G.M., and Nikaido, H., 1976, Outer membrane of Gram-negative bacteria. XII. Molecular-sieving function of cell wall, J. Bacteriol., 128:325.

de Gier, J., Mandersloot, J.G., Hupkes, J.V., McElhaney, R.N., and van Beek, W.P., 1971, On the mechanism of non-electrolyte permeation through lipid bilayers through biomembranes, Biochim. Biophys. Acta, 233:610.

Glauert, A.M., and Thornley, M.J., 1969, The topography of the bacterial cell wall, Ann. Rev. Microbiol. 23:159.

Henning, U., and Haller, I., 1975, Mutants of Escherichia coli K12 lacking all the "major" proteins of the outer cell envelope membrane, FEBS Lett., 55:161.

Kalckar, H.M., 1965, Galactose metabolism and cell "sociology", Science, 150:305.

Kamio, Y., and Nikaido, H., 1976, Outer membrane of Salmonella typhimurium: Accessibility of phospholipid head groups to phospholipase C and cyanogen bromide activated dextran in the external medium, Biochemistry, 15:2561.

Kellenberger, E., and Ryter, A., 1958, Cell wall and cytoplasmic-membrane of Escherichia coli, J. Biophys. Biochem. Cytol., 4:323.

Korteland, J., Tommassen, J., and Lugtenberg, B., 1982, PhoE protein pore of the outer membrane of Escherichia coli K12 is a particularly efficient channel for organic and inorganic phosphate, Biochim. Biophys. Acta, 690:282.

Loewenstein, W.R., 1979, Junctional intercellular communication and the control of growth, Biochim. Biophys. Acta, 560:1.

Luckey, M., and Nikaido, H., 1980, Specificity of diffusion channels produced by phage receptor protein of Escherichia coli, Proc. Natl. Acad. Sci. USA, 77:167.

Lugtenberg, B., Peters, R., Bernheimer, H., and Berendsen, W., 1976, Influence of cultural conditions and mutations on the composition of the outer membrane proteins of Escherichia coli, Molec. Gen. Genet., 147:251.

Mannella, C., Colombini, M., and Frank, J., 1983, Structural and functional evidence for multiple channel complexes in the outer membrane of Neurospora crassa mitochondria, Proc. Natl. Acad. Sci. USA, 80:2243.

Nakae, T., 1975, Outer membrane of Salmonella typhimurium: Reconstitution of sucrose-permeable membrane vesicles, Biochem. Biophys. Res. Commun., 64:1224.

Nakae, T., 1976a, Outer membrane of Salmonella. Isolation of protein complex that produces transmembrane channels, J. Biol. Chem., 251:2176.

Nakae, T., 1976b, Identification of the outer membrane protein of E. coli that produces transmembrane channels in reconstituted vesicle membranes, Biochem. Biophys. Res. Commun., 71:877.

Nakamura, K., and Mizushima, S., 1976, Effects of heating in dodecyl sulfate solution on the conformation and electrophoretic mobility of isolated major outer membrane proteins from Escherichia coli K-12, J. Biochem. (Tokyo), 80:1411.

Nikaido, H., 1976, Outer membrane of Salmonella typhimurium: Transmembrane diffusion of some hydrophobic substances, Biochim. Biophys. Acta, 433:118.

Nikaido, H., 1979, Nonspecific transport through the outer membrane, in: "Bacterial Outer Membranes," M. Inouye, ed., John Wiley and Sons, New York.

Nikaido, H., and Rosenberg, E.Y., 1983, Porin channels in Escherichia coli: Studies with liposomes reconstituted from purified proteins, J. Bacteriol. 153:241.

Nikaido, H., Rosenberg, E.Y., and Foulds, J., 1983, Porin channels in Escherichia coli: Studies with β-lactams in intact cells, J. Bacteriol., 153:232.

Nikaido, H., Takeuchi, Y., Ohnishi, S., and Nakae, T., 1977, Outer membrane of Salmonella typhimurium. Electron spin resonance studies, Biochim. Biophys. Acta, 465:152.

Qureshi, N., Takayama, K., and Ribi, E., 1982, Purification and structural determination of non-toxic lipid A obtained from the lipopolysaccharide of Salmonella typhimurium, J. Biol. Chem., 257:11808.

Renkin, E.M., 1954, Filtration, diffusion, and molecular sieving through porous cellulose membranes, J. Gen. Physiol., 38:225.

Roantree, R.J., Kuo, T., MacPhee, D.G., and Stocker, B.A.D., 1969, The effects of various rough lesions in Salmonella typhimurium upon sensitivity to penicilins, Clin. Res., 17:157.

Rosenbusch, J.P., 1974, Characterization of the major envelope protein from Escherichia coli. Regular arrangement on the peptidoglycan and unusual dodecylsulfate binding, J. Biol. Chem., 249:8019.

Salton, M.R.J., 1964, "The Bacterial Cell Wall," Elsevier, Amsterdam.

Schlecht, S., and Westphal, O., 1970, Untersuchungen zur Typisierung von Salmonella-R-Formen. 4. Mitteilung: Typisierung von S. minnesota-T-Mutanten mittels Antibiotica, Zentralbl. Bakteriol. I. Orig., 213:354.

Smit, J., Kamio, Y., and Nikaido, H., 1975, Outer membrane of Salmonella typhimurium: Chemical analysis and freeze-fracture studies with lipopolysacharide mutants. J. Bacteriol., 124:942.

Tommassen, J., and Lugtenberg, B., 1980, Outer membrane protein e of Escherichia coli K-12 is coregulated with alkaline phosphatase, J. Bacteriol., 143:151.

Yoshimura, F., and Nikaido, H., 1982, Permeability of Pseudomonas aeruginosa outer membrane to hydrophilic solutes, J. Bacteriol., 152:636.

Yoshimura, F., Zalman, L.S., and Nikaido, H., 1983, Purification and properties of Pseudomonas aeruginosa porin, J. Biol. Chem., 258:2308.

Zalman, L.S., Nikaido, H., and Kagawa, Y., 1980, Mitochondrial outer membrane contains a protein producing nonspecific diffusion channels, J. Biol. Chem., 255:1771.

Zimmermann, W., and Rosselet, A., 1977, The function of the outer membrane of Escherichia coli as a permeability barrier to the β-lactam antibiotics, Antimicrob. Ag. Chemother., 12:368.

BIOGENESIS OF MEMBRANE LIPOPROTEINS IN BACTERIA

Henry C. Wu

Department of Microbiology
Uniformed Services University of the Health Sciences
Bethesda, Maryland 20814

ABSTRACT

The major outer membrane lipoprotein in *Escherichia coli* has provided a useful model for the studies of the biogenesis of bacterial lipoproteins in general. Lipoproteins with covalently linked glycerides and amide-linked fatty acids are present in both gram-negative and gram-positive bacteria. The primary structures and functions of these lipoproteins differ greatly but they share common features with respect to their modes of biosynthesis. They are first synthesized as precursor proteins, and in their precursor forms they are modified to become glyceride-modified prolipoproteins which are proteolytically processed by a unique prolipoprotein signal peptidase to form mature lipoproteins. This unique signal peptidase (signal peptidase II) is distinct from the signal peptidase for the processing of nonlipoprotein precursors (signal peptidase I) in its exquisite sensitivity to the cyclic antibiotic globomycin. Recent studies on the mechanism of export of prolipoprotein in *E. coli* and the specificities of the modification and processing enzymes are reviewed.

The outer membrane of gram-negative bacteria contains a few major proteins, one of which is the free-form of murein lipoprotein (Osborn and Wu, 1980). In the past 15 years since the discovery of the murein-bound lipoprotein by Braun and his coworkers, this major outer membrane lipoprotein has been one of the most thoroughly studied outer membrane proteins with respect to its structure, biosynthesis and assembly (Braun and Rehn, 1969; Braun, 1975). In 1977, Inouye and his coworkers identified a precursor form of the lipoprotein, the prolipoprotein, and determined the structure of the NH_2-terminal signal sequence of prolipoprotein (Inouye et al., 1977).

This timely discovery has provided a new dimension for the study of outer membrane lipoprotein, i.e., the mechanism of lipoprotein export in bacteria.

Studies in the last five years have revealed both unique and common features in the biogenesis of membrane lipoproteins in bacteria as compared to the export of proteins without covalently attached lipids (Silhavy et al., 1983). This paper summarizes recent advances in the studies of biogenesis of membrane lipoproteins in bacteria.

The biosynthesis and assembly of membrane lipoproteins in bacteria can be divided into discrete steps: synthesis of a prolipoprotein, insertion of the newly synthesized prolipoprotein into the cytoplasmic membrane, subsequent translocation across the cytoplasmic membrane, modifications and processing of the prolipoprotein, and in the case of the murein lipoprotein the intermembrane translocation and assembly of mature lipoprotein into the outer membrane of the cell envelope.

INITIATION OF LIPOPROTEIN EXPORT

The overall rate of assembly of Braun's lipoprotein into the outer membrane is an extremely rapid process under physiological conditions (Lin et al., 1980a). However, the rate of the initial step of the lipoprotein export process can be decreased by any of the following experimental manipulations: abolishment of membrane potential (Russell and Model, 1982), abortive export of structurally abnormal hybrid proteins which results in jamming of the export machinery (Ito et al., 1981; Tokunaga et al., 1982a), alterations in the signal sequences of prolipoprotein (Vlasuk et al., 1983), and mutations in genes the products of which are required for protein secretion (Hayashi and Wu, 1984). Under each of these conditions, unmodified prolipoprotein was detected in the cell, presumably due to the failure in the export of unmodified prolipoprotein through the cytoplasmic membrane. These results taken together suggest that the initial step in the export of prolipoprotein through the cytoplasmic membrane requires: (a) certain structural or conformational features of the prolipoprotein signal sequence which are recognized by a putative secretory machinery; (b) proteins coded by genes required for protein export in general, e.g., secA, secB, secY (prlA) (Michaelis and Beckwith, 1982; Silhavy et al., 1983); and (c) membrane potential. All these requirements appear to be shared by other exported proteins (Silhavy et al., 1983). The fact that export of prolipoprotein is affected by mutations in the secA, secB, or secY gene and by the accummulation of hybrid proteins coded by lamB-lacZ or malE-lacZ fused genes, strongly suggests that the export of prolipoprotein in E. coli shares a common secretory machinery used for the export of nonlipoprotein precursors.

STRUCTURES OF PROLIPOPROTEIN SIGNAL SEQUENCES REQUIRED FOR THE MODIFICATION AND PROCESSING OF PROLIPOPROTEIN

Both in vitro and in vivo experiments have shown that the modification of prolipoprotein by sequential transfer of glyceryl and O-acyl moieties, precedes the processing of modified prolipoprotein by the prolipoprotein signal peptidase (Tokunaga et al., 1982a; Tokunaga, H., and Wu, H.C., unpublished data). Thus the first enzyme in the pathway for lipoprotein biogenesis is the glyceryltransferase. Comparison of known sequences of prolipoproteins in bacteria has revealed a consensus sequence of leu-ala-gly-cys-x-ser-asn at the modification and cleavage site (Table 1).

If one examines the sequences immediately preceding the modifiable cysteine residues in the prolipoproteins, a consensus sequence of leu-ala-gly-cys or leu-ser-gly-cys can be readily recognized; leu at -3 can be replaced by ser, and gly at -1 can be substituted with ala. The sequences immediately following the cys at +1 are more difficult to evaluate. Combining the sequences of murein prolipoproteins from various gram-negative bacteria into a single group, the +2 residues are highly variable (ser, ala, asn, or gly). Ser is slightly favored at +3 and +5 while ala and asn are acceptable at both positions. Asn is favored at +4 position.

Mutations in the signal sequence of prolipoprotein affect its modification and/or processing. The following examples will illustrate the subtle specificity in the recognition of prolipoprotein by glyceryltransferase. Substitution of gly_{14} by asp_{14} (the <u>mlp</u>A allele) almost completely abolishes the modification, and consequently the processing as well, of the <u>mlp</u>A mutant prolipoprotein (Lin et al., 1978). Second-site reversion of asp_{14} to asn_{14} restores the prolipoprotein structure in such a way that the rate of modification of the revertant prolipoprotein is greatly reduced, as compared to that for the wild-type prolipoprotein. In contrast, the rate of processing of modified revertant prolipoprotein (containing asn_{14} instead of gly_{14}) is 1.5 times as rapid as that of the modified wild-type prolipoprotein (Tokunaga, H., and Wu, H.C., unpublished data). Another pseudorevertant of the <u>mlp</u>A allele has been isolated and characterized; i.e., a double mutant with asp_{14} ile_{16}, instead of gly_{14} thr_{16}, in the signal sequence of prolipoprotein. This pseudorevertant synthesizes both the unmodified prolipoprotein (two-third) and mature lipoprotein (one-third). Thus a conformational change in the prolipoprotein of this pseudorevertant allows partial modification of prolipoprotein to form mature lipoprotein (Giam, 1983) (Table 2).

A more systematic approach to ascertaining the structural requirements of signal sequences for the modification and processing of prolipoprotein has been taken by Inouye and his coworkers using site-directed mutagenesis in vitro (Inouye et al., 1983; Vlasuk et al.,

Table 1: Concensus Sequences for Prolipoprotein Modification/Processing Site

Protein	Amino Acid Sequence of Modification/Processing Site	References
Murein Lipoprotein		
E. coli	Gly Ser Thr Leu Leu Ala Gly Cys Ser Ser Asn Ala	Nakamura and Inouye, 1979
S. marcesens	Gly Ser His Ser Ala Gly Cys Ser Ser Asn Ala	Nakamura and Inouye, 1980
E. amylovora	Gly Ser Thr Leu Leu Ala Gly Cys Ser Ser Asn Ala	Yamagata et al., 1981
M. morganii	Ala Ser Ala Leu Leu Ala Gly Cys Ser Ser Asn Ala	Huang et al., 1983
Penicillinase		
B. licheniformis 749/C	Leu Phe Ser Cys Val Ala Leu Ala Gly Cys Ala Asn Asn Gln	Lai et al., 1981
S. aureus PC1	Ile Ala Leu Val Leu Ser Ala Cys Asn Ser Asn Ser	Nielsen and Lampen, 1982
PraT protein (pR100)	Ser Thr Leu Ala Leu Ser Gly Cys Gly Ala Met Ser	Ogata et al., 1982; Perumal and Minkley, 1982
Concensus Sequence	Leu (Ser) Ala (Ser) Gly[-1] (Ala) Cys[+1] - Ser (Ala) (Asn) Asn (Met) Ser (Ala) (Gln)	

Table 2: Modification and Processing of Outer Membrane Prolipoprotein in Wild-type and *lpp* Mutants.

	Prolipoprotein Structure	Prolipoprotein Modification	Prolipoprotein Processing
Wild-type	Gly_9 Ala Val Ile Leu Gly_{14} Ser Thr_{16} Leu Leu Ala Gly_{20} Cys_{21}	+	+
mlpA	Asp_{14}	-	-
Pseudorevertant 1	Asn_{14}	(rate decreased)	(rate increased)
pseudorevertant 2	Asp_{14} Ile_{16}		

Table 3: Comparison of Two Distinct Peptidases in E. coli.

	SPase I[a]		SPase II[b]	
Structural gene (map position)	lep (55 min)		lsp (0.5 min)	
Molecular weight	37,000		18,000	
Subcellular localization	Inner and outer membrane		Inner membrane	
Globomycin	Insensitive		Sensitive	
Anti-SPase I serum	Reactive		Nonreactive	
Substrates and cleavage sites	M13 procoat protein	Ala-Ala	Prolipoprotein	Gly-(Glyceride-Cys)
	LamB precursor	Ala-Asp	Prepenicillinase	Gly-(Glyceride-Cys)
	LSBP precursor	Als-Asp	TraT precursor	Gly-(Glyceride-Cys)
	OmpA precursor	Ala-Ala		
	MalE precursor	Ala-Lys		

[a]Silver and Wickner, 1983; Wolfe et al., 1983; Zwizinski et al., 1981.
[b]Regue et al., in press; Tokunaga et al., 1982b; Tokunaga et al., in press.

1984). Of particular interest are the results obtained from two mutants: a mutant with the deletion of a single amino acid (∇gly_{14}) in the signal sequence of prolipoprotein synthesizes modified but unprocessed prolipoprotein. This defect in processing, due to the deletion of gly_{14}, can be suppressed by a second deletion of gly_9. Another mutant prolipoprotein with the deletion of gly_{20} is unmodified and therefore not processed, whereas the replacement of gly_{20} by an ala_{20} does not affect the modification nor the processing. Thus the modification and processing of prolipoprotein are novel examples of covalent modifications of proteins which include glycosylation, phosphorylation, proteolytic processing and methylation, etc.

A UNIQUE SIGNAL PEPTIDASE FOR PROLIPOPROTEIN

The identification of a distinct prolipoprotein signal peptidase (signal peptidase II) resulted from the discovery of a novel cyclic peptide antibiotic globomycin (Nakajima et al., 1978). Globomycin specifically inhibits the processing of modified prolipoprotein by signal peptidase II without affecting the processing of nonlipoprotein precursors by signal peptidase I (Inukai et al., 1978) while the precise mode of action of globomycin remains unknown, its specificity in inhibiting signal peptidase II may be related to the unique requirement of this enzyme for glyceride-modified cysteine as part of its recognition site.

Signal peptidase I and II (Table 3) appears to be antigenically unrelated and differ in their subunit molecular weights, their substrate specificities, amino acid sequences of their cleavage sites sensitivity to globomycin and possibly the regulation of their gene expression. They are coded by genes half-genome apart on the E. coli chromosome.

While the identification of two distinct signal peptidases in E. coli has provided a partial answer to the question of substrate specificities of the processing enzymes, the precise structural

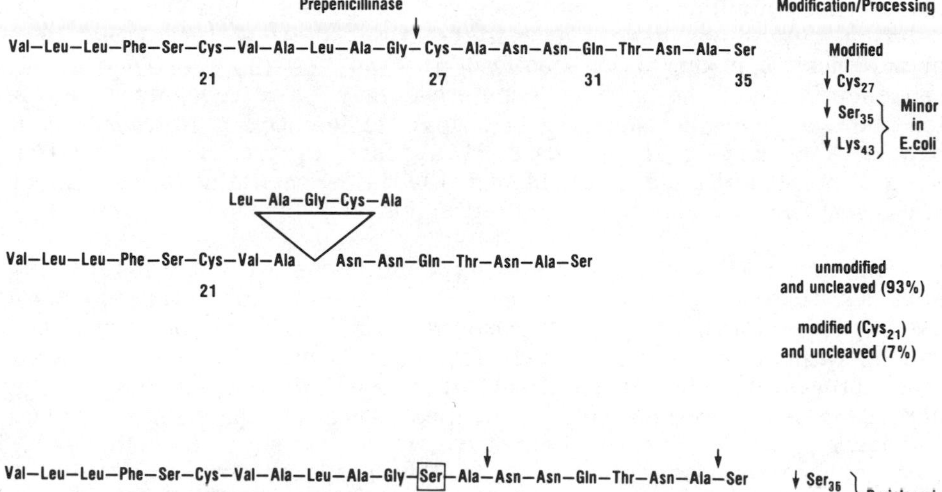

Fig. 1. Modification and processing of B. licheniformis wild-type and mutant prepenicillinases in E. coli.

features recognized by these two enzymes remain obscure. On the one hand, there is no clear-cut consensus sequence for signal peptidase I cleavage site, ala-x; the nature of x appears to be highly variable with respect to the size and charge of the side chains of this residue. On the other hand, lpp mutant prolipoproteins which are not modified by glyceryltransferase (e.g., $gly_{14} \rightarrow asp_{14}$, ∇gly_{20}, or ∇cys_{21}) remain uncleaved in vivo, even though there are ala-x sequences in the mutant prolipoproteins which are potential cleavage sites for signal peptidase I. The subtle specificity of signal peptidase I is further demonstrated by the studies of two mutant forms of B. licheniformis penicillinase genes in E. coli (Hayashi et al., 1983; Hayashi, S., and Wu, H.C., unpublished data). It has been shown previously that penicillinase exists in three forms in B. licheniformis: a membrane-bound form and two exo-forms (exo-large and exo-small). When expressed in E. coli, the penicillinase (pen) gene product is largely present in the outer membrane of E. coli cell envelope. The membrane-bound form of B. licheniformis penicillinase has been shown to correspond to a processed lipoprotein (Lai et al., 1980). This finding provides strong support for the hypothesis that the sequence of leu_{24}-ala-gly-cys_{27} in the signal sequence of prepenicillinase constitutes a modification and processing site for the formation of membrane lipoproteins in bacteria. We have studied the fate of two mutant penicillinases in which the modifiable cys_{27} is deleted by site-directed mutagenesis in vitro. In the first mutant, the pentapeptide leu_{24}-ala-gly-cys-ala_{28} is deleted. The deleted mutant prepenicillinase is not processed by E. coli cells harboring the deletion mutant pen gene; a small fraction (7 %) of the mutant pre-penicillinase appears to be modified at cys_{21} of the shortened signal sequence. Thus the glyceryltransferase may have recognized leu_{18}-leu-phe-cys_{21}-val-ala, albeit much less efficiently. Neither signal peptidase I nor signal peptidase II appears to process the deletion mutant prepenicillinase even though the latter contains a number of ala-x sequences within the shortened signal sequence.

In contrast, a point mutant prepenicillinase with a single amino acid substitution of cys_{27} by ser_{27} is efficiently processed to form two soluble forms of penicillinase with NH_2-termini of ser_{35} and asn_{29}, respectively. These two forms of penicillinase would result from processing of the point-mutant prepenicillinase at ala_{34}-ser_{35} and ala_{28}-asn_{29}, respectively. The processing at ala_{34}-ser_{35} appears to occur more efficiently than that at ala_{28}-asn_{29}, and the ser_{35} form of penicillinase corresponds to the exo-large form known to be present as one of the secreted penicillinases in B. licheniformis. Thus alternative processing of this mutant prepenicillinase takes place, resulting in the secretion of penicillinase into the periplasmic space. These results are summarized in Figure 1.

TRANSLOCATION AND ASSEMBLY OF MATURE LIPOPROTEIN INTO THE OUTER MEMBRANE OF E. COLI CELL ENVELOPE

Kinetic studies of the assembly of lipoprotein have revealed an extremely rapid overall process so that the detection of biosynthetic intermediates in the assembly pathway requires experimental manipulations such as lowering of growth temperature, instantaneous termination of reactions by the addition of trichloracetic acid and perturbation of relative rates in discrete steps. Pulse chase experiments have revealed the transient appearance of mature lipoprotein in the cytoplasmic membrane. This observation strongly suggests that modification and processing of prolipoprotein takes place in the cytoplasmic membrane prior to its translocation to the outer membrane. The localization of signal peptidase II activity in the cytoplasmic membrane of the cell envelope has since been confirmed by direct determination of the distribution of this enzyme activity in subcellular fractions (Tokunaga, M., and Wu, H.C., unpublished data).

Neither the modification nor the processing of prolipoprotein is required for the translocation of prolipoprotein into the outer membrane. Thus unmodified (and uncleaved) prolipoprotein is translocated into the outer membrane in lpp mutants which synthesize structurally altered prolipoproteins (Lin et al., 1980b). Likewise, modified but uncleaved prolipoprotein in globomycin-treated E. coli cells is also translocated to the outer membrane (Ichihara et al., 1982; Inukai and Inouye, 1983). It is likely that the signal peptides in the uncleaved prolipoproteins, regardless whether they are modified or not, remain anchored to the cytoplasmic membrane as envisioned by the loop model (Inouye and Halegoua, 1980). The bulk polypeptides corresponding to the mature lipoproteins may contain the topogenic sequences targeting their assembly into the outer membrane (Blobel, 1980).

EPILOGUE

Our current understanding of the biosynthesis and assembly of outer membrane lipoproteins can be summarized in Figure 2 and Table 4. It is clear that many details are lacking. The biggest gaps in our knowledge in the processes of protein secretion and assembly of outer membrane proteins are the mechanisms of translocation of polypeptides across the lipid bilayer of the cytoplasmic membrane in prokaryotic cell (or the rough endoplasmic reticulum membrane in eukaryotic cell) and the mechanism of intermembrane translocation of processed proteins from the inner cytoplasmic membrane to the outer membrane. In my efforts to learn more about these fascinating biological processes in all living cells, how much I wish I might have benefited from epigenetic processes which would have endowed me with a small fraction of the infectious enthusiasm and fertile mind so characteristic of my mentor, Dr. Herman M. Kalckar, whom we honor on this occasion.

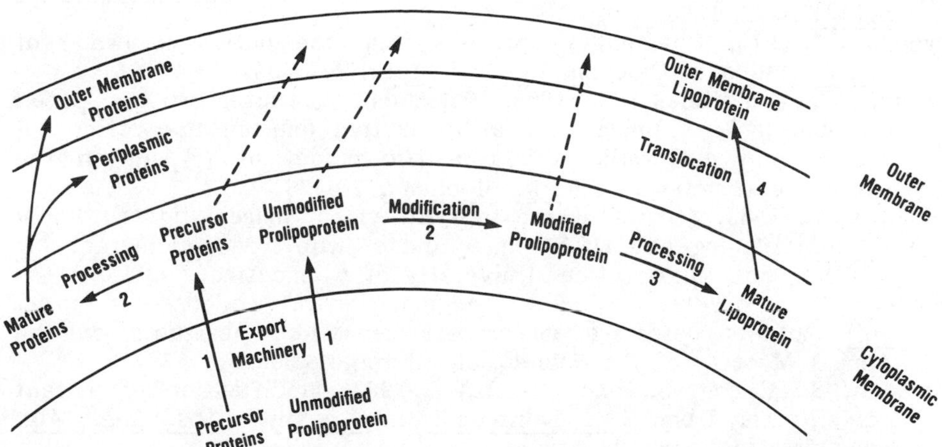

Fig. 2. Schematic pathway for the export of lipoproteins and non-lipoproteins in E. coli.

Table 4: Biogenesis of Lipoproteins in Bacteria

Step	Affected By
1. Initiation of protein export	Mutations in lpp gene Mutations in secA, secB, or secY (prlA) gene Accumulation of hybrid proteins (malE-lacZ; lamB-lacZ)
2. Modification	Mutations in lpp genes Accumulation of hybrid proteins
3. Processing	Mutations in lpp gene Mutations in lsp gene Antibiotic (globomycin)
4. Translocation	

ACKNOWLEDGEMENTS

The work described in this article was supported by a United States Public Health Service Grant GM-28811 and American Heart Association Grant 81-663.

REFERENCES

Blobel, G., 1980, Intracellular protein topogenesis, Proc. Natl. Acad. Sci. USA, 77:1496.

Braun, V., 1975, Covalent lipoprotein from the outer membrane of Escherichia coli, Biochim. Biophys. Acta, 415:335.

Braun, V., and Rehn, K., 1969, Chemical characterization, spatial distribution and function of a lipoprotien (murein lipoprotein) of the E. coli cell wall. The specific effect of trypsin on the membrane structure. Eur. J. Biochem., 10:426.

Giam, C.Z., 1983, Studies of the biosynthesis of murein lipoprotein in E. coli: cloning and DNA sequencing of wild-type and mutant lpp alleles. PhD Dissertation, University of Connecticut, p. 79.

Hayashi, S., and Wu, H.C., 1984, Accumulation of prolipoprotein in E. coli mutants defective in protein secretion, Abstracts of the Annual Meeting of the Society of Microbiologists, p. 157.

Hayashi, S., Chang, S., and Wu, H.C., 1983, Modification of mutant penicillinase from B. licheniformis in E. coli. Abs. Ann. Mtg. Soc. Microbiol., p. 199.

Huang, Y.X., Ching, G., and Inouye, M., 1983, Comparison of the lipoprotein gene among the Enterobacteriacea. DNA sequence of Morganella morganii lipoprotein gene and its expression in

Escherichia coli, J. Biol. Chem., 258:8139.

Ichihara, S., Hussain, M., and Mizushima, S., 1982, Mechanism of export of outer membrane lipoproteins through the cytoplasmic membrane in Escherichia coli. Binding of lipoprotein precursors to the peptidoglycan layer, J. Biol. Chem., 257:495.

Inouye, M., and Halegoua, S., 1980, Secretion and membrane localization of proteins in Escherichia coli, CRC Crit. Rev. Biochem., 7:339.

Inouye, S., Hsu, C.P.S., Itakura, K., and Inouye, M., 1983, Requirement for signal peptide cleavage of Escherichia coli prolipoprotein, Science, 221:59.

Inouye, S., Wang, S., Sekizawa, J., Halegoua, S., and Inouye, M., 1977, Amino acid sequence for the peptide extension on the prolipoprotein of the Escherichia coli outer membrane. Proc. Natl. Acad. Sci. USA, 74:1004.

Inukai, M., and Inouye, M., 1983, Association of the prolipoprotein accumulated in the presence of globomycin with the Escherichia coli outer membrane, Eur. J. Biochem., 130:27.

Inukai, M., Takeuchi, M., Shimizu, K., and Arai, M., 1978, Mechanism of action of globomycin, J. Antibiotics, 31:1203.

Ito, K., Bassford, P.J., Jr., and Beckwith, J., 1981, Protein localization in E. coli: Is there a common step in the secretion of periplasmic and outermembrane protein? Cell, 24:707.

Lai, J.S., Sarvas, M., Brammar, W.J., Neugebauer, K., and Wu, H.C., 1981, Bacillus licheniformis penicillinase synthesized in Escherichia coli contains covalently linked fatty acid and glyceride. Proc. Natl. Acad. Sci. USA, 78:3506.

Lin, J.J.C., Giam, C.Z., and Wu, H.C., 1980a, Assembly of the outer membrane lipoprotein in Escherichia coli. J. Biol. Chem., 255:807.

Lin, J.J.C., Kanazawa, H., and Wu, H.C., 1980b, Assembly of outer membrane lipoprotein in an Escherichia coli mutant with an amino acid replacement within the signal sequence of prolipoprotein. J. Bacteriol., 141:550.

Lin, J.J.C., Kanazawa, H., Ozols, J., and Wu, H.C., 1978, An Escherichia coli mutant with an amino acid alteration within the signal sequence of outer membrane prolipoprotein. Proc. Natl. Acad. Sci. USA, 75:4891.

Michaelis, S., and Beckwith, J., 1982, Mechanism of incorporation of cell envelope proteins in Escherichia coli, Ann. Rev. Microbiol., 36:435.

Nakajima, M., Inukai, M., Haneishi, T., Terahara, A., and Arai, M., 1978, Globomycin, a new peptide antibiotic with spheroplast-forming activity III. Structural determination of globomycin. J. Antibiotics, 31:426.

Nakamura, K., and Inouye, M., 1979, DNA sequence of the gene for the outer membrane lipoprotein of E. coli: an extremely AT-rich promoter, Cell, 18:1109.

Nakamura, K., and Inouye, M., 1980, DNA sequence of the Serratia marcescens lipoprotein gene, Proc. Natl. Acad. Sci. USA, 77:1369.

Nielsen, J.B.K., and Lampen, J.O., 1982, Nucleotide sequence analysis of the complement resistance gene from plasmid R100, J. Bacteriol., 151:819.

Ogata, R.T., Winters, C., and Levine, R.P., 1982, Nucleotide sequence analysis of the complement resistance gene from plasmid R100, J. Bacteriol., 151:819.

Osborn, M.J., and Wu, H.C.P., 1980, Proteins of the outer membrane of gram-negative bacteria. Ann. Rev. Microbiol., 34:369.

Perumal, N.B., and Minkley, E.G., Jr., 1982, Processing of the TraT protein of the Escherichia coli sex factor F: Signal sequence cleavage and attachment of glyceride and fatty acids. Abst. Ann. Mtg. Am. Soc. Microbiol., p. 167.

Regue, M., Remenick, J., Tokunaga, M., and Wu, H.C., in press, Mapping of the lipoprotein signal peptidase gene in Escherichia coli. Abst. Ann. Mtg. Am. Soc. Microbiol.

Russell, M., and Model, P., 1982, Filamentous phage pre-coat is an integral membrane protein: analysis by a new method of membrane preparation. Cell, 28:177.

Silhavy, T.J., Benson, S.A., and Emr, S.D., 1983, Mechanism of protein localization. Microbiol. Rev., 47:313.

Silver, P., and Wickner, W., 1983, Genetic mapping of the Escherichia coli leader (signal) peptidase gene (lep): a new approach for determining the map position of a cloned gene, J. Bacteriol., 154:569.

Tokunaga, M., Tokunaga, H., and Wu, H.C., 1982a, Post-translational modification and processing of Escherichia coli prolipoprotein in vitro, Proc. Natl. Acad. Sci. USA, 79:2255.

Tokunaga, M., Loranger, J.M., Wolfe, P.B., and Wu, H.C., 1982b, Prolipoprotein signal peptidase in Escherichia coli is distinct from the M13 procoat protein signal peptidase. J. Biol. Chem., 257:9922.

Tokunaga, M., Loranger, J.M., Chang, S.Y., Chang, S., and Wu, H.C., in press, Identification and genomic organization of prolipoprotein signal peptidase in Escherichia coli. Abst. Ann. Mtg. Am. Soc. Microbiol.

Vlasuk, G.P., Ghrayeb, J., and Inouye, M., in press, The major outer membrane lipoprotein of Escherichia coli: secretion, modification and processing, in: "The Enzymes of Biological Membranes," A. Martonosi, ed., Plenum Publishing Co., New York.

Vlasuk, G.P., Inouye, S., Ito, H., Itakura, K., and Inouye, M., 1983, Effects of the complete removal of basic amino acid residues from the signal peptide on secretion of lipoprotein in Escherichia coli. J. Biol. Chem., 258:7141.

Wolfe, P.B., Wickner, W., and Goodman, J.M., 1983, Sequence of the leader peptidase gene of Escherichia coli and the orientation of leader peptidase in the bacterial envelope, J. Biol. Chem., 258:12073.

Yamagata, H., Nakamura, K., and Inouye, M., 1981, Comparison of lipoprotein gene amoung the Enterobacteriacae. DNA sequence of Erwina amylovora lipoprotein gene, J. Biol. Chem., 256:2194.

Zwizinski, C., Date, T., and Wickner, W., 1981, Leader peptidase is found in both the inner and outer membrane of *Escherichia coli*. *J. Biol. Chem.*, 256:3593.

PERIPLASMIC MEMBRANE-DERIVED OLIGOSACCHARIDES AND OSMOREGULATION IN *ESCHERICHIA COLI**

Eugene P. Kennedy

Department of Biological Chemistry
Harvard Medical School
Boston, Massachusetts 02115

ABSTRACT

Membrane-derived oligosaccharides (MDO), found in the periplasmic space of *Escherichia coli* and other Gram-negative bacteria, are composed of 8-10 glucose united in a branched structure, joined by β1→2 and β1→6 linkages. They are a family of compounds, variously substituted with *sn*-1-phosphoglycerol residues, derived from the head-groups of phosphatidylglycerol, and with succinate in 0-ester linkage, as well as smaller amounts of phosphoethanolamine. A study of the turnover of membrane lipids led to the discovery of this novel class of cell constituents. The synthesis of MDO is strictly regulated by the osmolarity of the medium in which the cells are growing. A study of the enzymes catalyzing the biosynthesis of MDO may therefore shed light on the fundamental mechanisms by which cells sense the osmolarity of their surroundings and adapt to it. Progress in the elucidation of the biosynthetic pattern is reviewed.

It is a pleasure and a privilege for me to participate in this symposium in honor of Professor Herman Kalckar. Herman Kalckar is one of the few who can be said to have truly shaped modern biochemistry. For his many students and friends, however, the force of his example, and the grace of his personal style have been as important as his scientific contributions.

---

*This work was supported by Grants GM19822 and GM22057 from the National Institute of General Medical Sciences and GM26625 from the National Institutes of Health

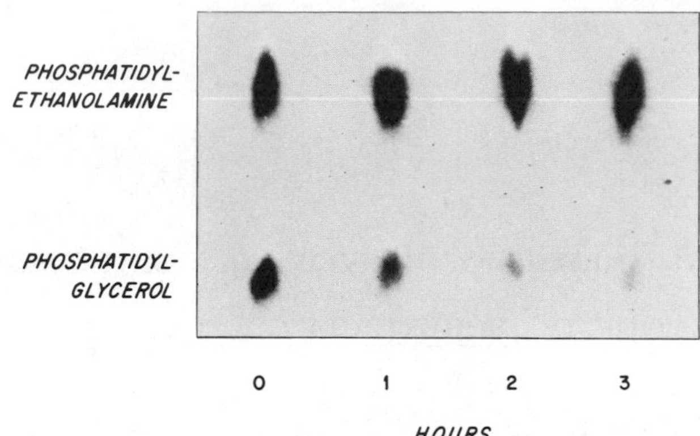

Fig. 1. Turnover of the hydrophilic head-group of phosphatidylglycerol in log-growing cells of E. coli B. Cells were labeled with ^{32}P and then transfered to an unlabeled medium for the hours of growth indicated. From Kanfer and Kennedy (1963) with permission of the publishers.

DISCOVERY OF MEMBRANE-DERIVED OLIGOSACCHARIDES IN E. COLI

In 1963, I began a systematic investigation of the biosynthesis of membrane lipids in Escherichia coli. Our previous work on the biosynthesis of membrane lipids had been carried out on animal tissues, and the decision to turn to an examination of bacterial systems was based on the conviction that problems of membrane function could more readily be approached in an organism like E. coli, for which an extremely rich background of genetic information was already available.

Kanfer and Kennedy (1963) measured the turnover of the principal membrane phospholipids of E. coli under conditions of steady-state, logarithmic growth. It was found (Fig. 1) in a pulse-chase experiment that ^{32}P incorporated into the head group of phosphatidylethanolamine was metabolically stable under these conditions. In contrast, radioactivity was steadily lost from the hydrophilic head group of phosphatidylglycerol. The results suggested some essential function of phosphatidylglycerol that required the continuous renewal of its hydrophilic head group. This interesting observation was set aside for a period of 10 years, during which we worked out the complete pattern of the biosynthesis of membrane phospholipids in E. coli in considerable detail.

In 1973, we turned once again to an examination of the meaning of the turnover of phosphatidylglycerol. We discovered that the phosphoglycerol head group of phosphatidylglycerol was continuously transferred to a novel type of water-soluble oligosaccharides which, from their relation to membrane phospholipids, were called membrane-derived oligosaccharides (MDO) (van Golde et al., 1973).

STRUCTURE OF MEMBRANE-DERIVED OLIGOSACCHARIDES

The MDO of E. coli are a heterogeneous family of closely related oligosaccharides containing glucose as the sole sugar. They range in size from 6 to 12 D-glucopyranoside residues per mole, linked solely by $\beta 1 \rightarrow 2$ and $\beta 1 \rightarrow 6$ bonds. The principal species appear to contain 7 to 9 glucose residues per mole. They are multiply branched structures with glucose units at the branch points doubly substituted at positions 2 and 6. Most species of MDO are multiply substituted with sn-1-phosphoglycerol residues linked to the 6 position of certain of the glucose residues (Kennedy et al., 1976). MDO may also contain smaller amounts of phosphoethanolamine residues linked in phosphodiester bonds, also to the 6 position of glucose residues. The finding of phosphoethanolamine residues in MDO was surprising, because our first experiments (Fig. 1) did not detect a turnover of phosphatidylethanolamine under the particular conditions of that experiment. MDO may also contain succinic acid in 0-ester linkage, adding to the net negative charge of these molecules. Distinct families of MDO may be subfractionated on DEAE cellulose as indicated in Figure 2. Fractions designated A, B, and C may be further subfractionated by chromatography on Dowex-1 acetate ion exchange resin at pH 3.7 into subfamilies such as A1, A2, etc. (van Golde, et al., 1973). A tentative structure for MDO A2, which contains three residues of phosphoglycerol per mole (Schneider and Kennedy, 1978) is indicated in Figure 3. It should be emphasized that the exact arrangement of the glucose residues in this structure is arbitrarily presented.

PERIPLASMIC LOCALIZATION AND DISTRIBUTION OF MDO IN GRAM-NEGATIVE BACTERIA

Schulman and Kennedy (1979) reported that the MDO of E. coli are localized in the periplasmic compartment of that organism from which they can be readily released by treatment of the cells with EDTA under conditions that avoid osmotic shock. About 84 percent of the MDO were released, while low molecular weight constituents of the cytoplasm were retained by the cells. Schulman and Kennedy (1979) also examined a group of representative Gram-negative bacteria for the presence of MDO. These oligosaccharides were found in all of the E. coli strains examined, as well as in Salmonella anatum, Pseudomonas aeruginosa, Proteus mirabilis, and Enterobacter aerogenes. A few Gram-positive organisms were also examined and found to be devoid of soluble oligosaccharides of the size and general properties of MDO.

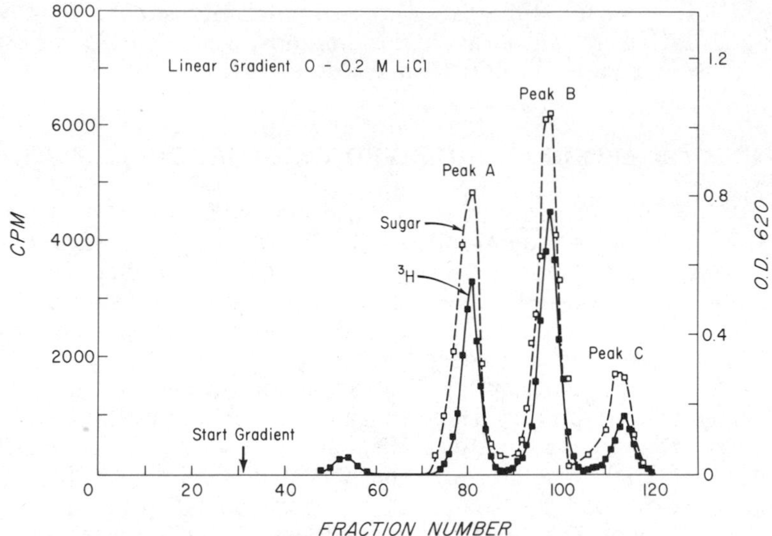

Fig. 2. Separation of families of MDO by chromatography on DEAE-cellulose. The MDO fraction of cells labeled by growth on [2-^{3}H]glycerol was mixed with carrier material from a large-scale preparation of frozen cells and chromatographed on a column on DEAE-cellulose (31 x 2.4 cm) that had been previously equilibrated with 10 mM Tris HCl, pH 7.4. The column was washed with several bed volumes of water and then eluted with a linear gradient of LiCl (0 - 0.2 M in 2 liters). Fractions of 10 ml each were collected and analyzed for radioactivity (left-hand ordinate) and anthrone-reactive sugar (shown in arbitrary units as A_{620} on the right-hand ordinate). From van Golde et al (1973) with permission of the publishers.

The fact that MDO are rather generally distributed in Gram-negative bacteria suggests that they are carrying out some function, as yet undetermined, of general importance.

OSMOREGULATION AND THE SYNTHESIS OF MDO

Because of the localization in the periplasmic space of E. coli and their highly anionic character, MDO must contribute to the osmolarity of the periplasmic compartment and to the Donnan potential shown by Stock et al. (1977) to exist across the outer membrane of this organism. Stock et al. (1977) also offered evidence that the osmolarity of the periplasmic space is, in fact, regulated and is maintained at a value closely similar to that of the cytoplasmic contents of the cell. Kennedy (1982) therefore examined the synthesis of MDO as a function of the osmolarity of the medium. It was found (Fig. 4) that the synthesis of MDO in cells grown in low osmolarity was 16 times higher than in cells grown in medium with added 0.4 M NaCl. Control experiments revealed that the effect was not a specific one of NaCl, but was a property of other osmotically active solutes in the medium such as ammonium sulfate or sucrose. Under conditions of maximum synthesis in medium of low osmolarity, MDO may constitute as much as 7 percent of the cell substance produced in the culture. This fact, together with the complex pattern of enzymic reactions needed for MDO biosynthesis, further suggests an important, although still unidentified, function for MDO.

BIOSYNTHESIS OF MDO

The enzymes of MDO biosynthesis, particularly those involved in the first stages of the process, are of special interest, because it should be possible to determine at what stage MDO assembly is regulated and what are the effectors that are modulated by the osmolarity of the medium.

A partial working model for the biosynthesis of MDO in E. coli is shown in Figure 5.

Schulman and Kennedy (1977) found evidence that UDP-glucose is an essential intermediate in the biosynthesis of the polysaccharide chains of MDO. Thus, galU mutants blocked in the formation of UDP-glucose are unable to synthesize MDO, while revertants that regained UDP-glucose pyrophosphorylase activity simultaneously regained the ability to make MDO. In confirmation of these genetic studies, pulse-label isotope tracer studies were carried out with glucose of high specific activity, under conditions in which UDP-glucose comprised a large fraction of the total radioactivity in the

```
GLC-1 → 2-GLC-1 → 2-GLC-1 → 2-GLC-1 → 2-GLC-1 → 2-GLC
|                  |                     |
6                  6                     6
↑                  ↑                     ↑
1       O          1       O             1       O
|       ||         |       ||            |       ||
GLC-6-O-P-O-GRL    GLC-6-O-P-O-GRL       GLC-6-O-P-O-GRL
        |                  |                     |
        O-                 O-                    O-
```

Fig. 3. Tentative structure of MDO fraction A-2. This fraction contains 3 phosphoglycerol residues linked to the 6 positions of glucose units in a polysaccharide of about 9 glucose residues/mole (Schneider and Kennedy, 1978). The pattern of branching and the exact position of the phosphoglycerol residues have not yet been determined and are shown arbitrarily in the tentative structure.

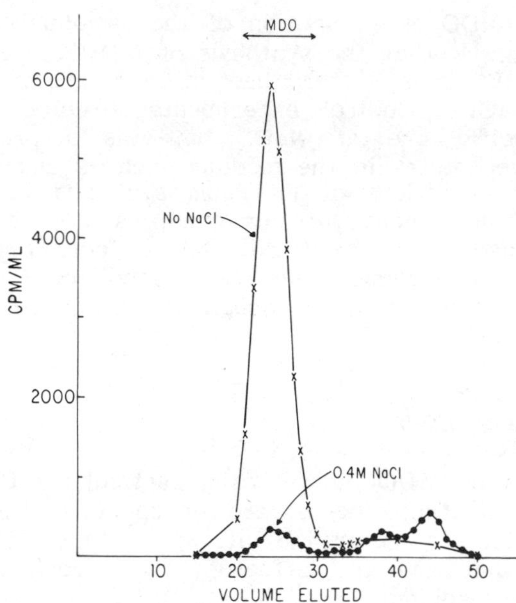

Fig. 4. Osmotic regulation of the biosynthesis of MDO. Cells of strain BB26-36 plsB were grown in the presence of [2-^{3}H]-glycerol, in medium of low osmolarity, with or without the addition of 0.4 M NaCl. The MDO fraction was extracted and chromatographed on a column of Sephadex G-25. From Kennedy (1982) with permission of the publishers.

lower molecular weight pool of cells of strain DF214. Subsequent chase experiments in the presence of unlabeled glucose clearly revealed the conversion of labeled UDP-glucose to the higher molecular weight, membrane-derived oligosaccharides.

Recently, Weissborn and Kennedy (1983) have discovered a novel glucosyl transferase system in E. coli that is thought to catalyze the elongation of MDO chains. This system will elongate suitable "primers" but has not yet been observed to initiate the formation of new chains. The enzyme system requires the cell-membrane fraction and also a heat-stable protein derived from the soluble supernatant fraction of the cell. The system required UDP-glucose, in confirmation of the previous studies of Schulman and Kennedy (1977), and also requires a primer. Study of the system has been greatly facilitated by the discovery that simple β-glucosides, such as 2-0-β-D-glucosyl-D-glucose or octyl-β-D-glucoside, are effective primers in this system. The added β-glucoside primers are thought to substitute for the physiological acceptors, the lipid-linked polyglucose chains indicated in Figure 5. As yet there is no direct evidence for the role of lipid carrier in the biosynthetic process, but it is known that

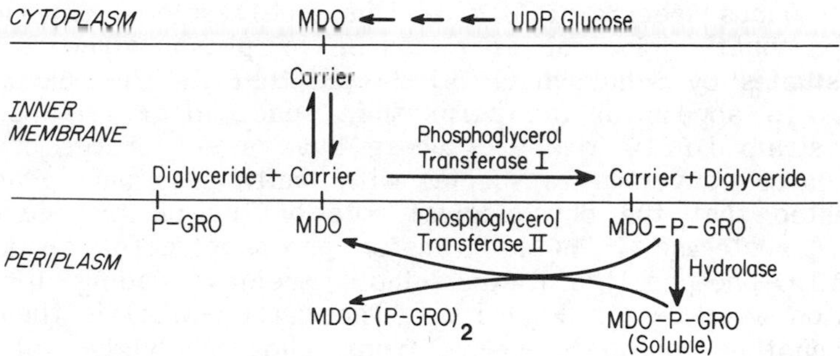

Fig. 5. Working model for the biosynthesis of membrane-derived oligosaccharides.

free glucose cannot act as a primer, and therefore the initiation of MDO chains must involve some kind of activation of a glucose unit. Based on analogy with other systems, it is tentatively suggested that the initiation step involves the attachment of a glucose unit derived from UDP-glucose to a bactoprenol derivative. The products synthesized by the glucosyl transferase system are glucose oligomers joined by β1-2 linkages and the principal products are approximately the same size as MDO molecules. They appear to represent the backbone structure of MDO at a stage prior to branching.

Assay of mdoA mutants (Bohin and Kennedy, in press) revealed that these strains are defective in the membrane component of the glucosyl transferase system of Weissborn and Kennedy (1983). The heat-stable protein component, on the other hand, is present in the mdoA mutants at the same level as in the wild-type. These findings offer strong evidence that the glucosyl transferase system catalyzes an essential step in MDO synthesis since the mdoA mutation leads to a block in MDO synthesis in vivo at an early stage (Bohin and Kennedy, in press).

Phosphoglycerol Transferase I: En Enzyme of the Inner Membrane

Jackson and Kennedy (1983) reported the initial characterization of an enzyme catalyzing the transfer of phosphoglycerol residues from phosphatidylglycerol to membrane-derived oligosaccharides or to synthetic β-glucoside acceptors. The products are sn-1,2-diglyceride and β-glucoside-6-phosphoglycerol. The enzyme was found to be localized in the inner membrane of E. coli but with its active site directed outwards.

Phosphoglycerol Transferase II: A Periplasmic Enzyme

Goldberg et al. (1981) reported the discovery of a periplasmic enzyme that catalyzes the interchange of phosphoglycerol residues between various species of MDO. This freely soluble enzyme cannot utilize phosphatidylglycerol as a source of phosphoglycerol residues. Earlier studies by Schulman (1976) showed that the first soluble MDO molecules to appear in the periplasmic space after brief pulsing of cells of strain DF214 with ^{3}H-glucose have a net charge of only -1, but are later converted to species with multiple anionic charges. It is suggested that the physiological role of the periplasmic phosphoglycerol transferase II is to transfer phosphoglycerol residues from carrier-MDO-phosphoglycerol to soluble species leading to multiple substitution as shown in Figure 5. The carrier-MDO is then free to accept another phosphoglycerol from phosphatidylglycerol in the reaction catalyzed by the membrane-bound phosphoglycerol transferase I.

OSMOREGULATION AND THE BIOSYNTHESIS OF MDO: PROBLEMS AND PERSPECTIVES

Osmotic regulation is a fundamental problem faced by all living cells, arising from the fact that the plasma membrane of cells is freely permeable to water. The total concentration of osmotically active solutes in cells is of the order of 300 milliosmolar (Stock et al., 1977; Epstein and Laimins, 1980). A high proportion of these solutes consist of potassium salts; a concentration of potassium of about 0.1 M is needed for the activity of intracellular enzymes and for the biosynthesis of proteins.

The vital importance of osmoregulation is clearly shown by the intricate and elaborate mechanisms developed by mammals to balance the osmolarity of the extracellular fluids with that of cellular contents. The highly efficient functioning of these systems for the close regulation of the osmolarity of the extracellular fluids is absolutely essential for life.

Plants and bacteria, lacking the highly developed regulation of the extracellular fluid that has been developed by mammals, face two distinct types of osmotic challenge. The first of these is encountered when cells are growing in a medium of osmolarity markedly higher than 300 milliosmolar. Under these conditions water flows from the cytoplasm, through the membrane leading to loss of cell volume. Bacterial and plant cells respond to this type of osmotic stress by increasing their intracellular content of potassium (Epstein and Laimins, 1980; Measures, 1975) and of glucose and its derivatives (Roller and Anagnostopoulous, 1982). In this response, the higher level of osmolarity of the cytoplasm balances that of the medium, preventing further constriction of cell volume. Osmotic stress of this type is particularly important for plants growing in surroundings of high salinity, or during periods of drought. The enormous economic importance for agriculture of an understanding of osmoregulation from this point of view has been stressed in the recent review by Le Rudulier and Valentine (1982).

Cells of E. coli and other enteric bacteria may at first glance appear to be living in a highly protected environment in the gut. However, passage from one host to another very often requires adaptation to media of very low osmolarity for growth or survival in sewage or water supplies. During the long course of evolution, adaptation to conditions of low osmolarity must have been absolutely essential for the survival of the bacterial species.

If cells of E. coli, with an internal osmolarity of about 300 milliosmolar, are growing in medium containing solutes at about 50 milliosmolar, water must flow into the cell across the plasma membrane down the gradient of activity of water until, at equilibrium, its flow is resisted by a hydrostatic pressure of about 6.4 atmospheres. The peptidoglycan component of the E. coli cell envelope must play an essential role in maintaining the structural integrity of the cell against such pressures because it is well known that lesions in the peptidoglycan structure caused by treatment with lysozyme plus EDTA, or by treatment of growing cells with penicillin, leads to the swelling of cells and their lysis. This role of peptidoglycan is even more striking and obvious in Gram-positive bacteria in which the peptidoglycan of the wall is much more prominent, and which lack the outer membrane characteristic of Gram-negative bacteria.

It was therefore rather surprising when Stock et al. (1977) discovered that the contents of the periplasmic space of E. coli appeared to have an osmolarity approximately equal to that of the cytoplasm. For cells growing in medium of low osmolarity with an internal osmolarity of 300 milliosmolar, this means that the periplasmic space must contain solutes totaling approximately 300 milliosmolar that are impermeable to the outer membrane. If this is the case, the outer membrane must be capable of withstanding the resultant high hydrostatic pressure, in part, perhaps, due to the bonding of outer membrane proteins such as the major lipoprotein to the murein structures of the cell. Stock and Roseman (1977) also demonstrated that there is a Donna potential across the outer membrane of E. coli with the periplasmic space negative with respect to the medium. These facts pointed to the existence of a high concentration of fixed anions in the periplasmic compartment, amounting to about 150-120 milliquivalents/gram (dry weight) of cells. The authors attempted to account for this fixed negative charge in terms of the known negatively charged components of the cell envelope, such as periplasmic proteins, peptidoglycan and lipopolysaccharide, but concluded that a large fraction remained unidentified. It has now been discovered (Kennedy, 1982) that the principal osmotically active solutes in the periplasmic space of cells grown in medium of low osmolarity are MDO molecules which on the average have a net negative charge of about 5 units/mole.

Because MDO may comprise as much as 7 percent of the total cell substance when grown in medium of low osmolarity and because of the rather general occurrence of the MDO in Gram-negative bacteria, it seems certain that they play an important function in cellular adaptation to low osmolarity. The first rather naive postulate was that the presence of these molecules as a kind of osmotic buffer may be essential for the structural integrity of the cells. This has proved not be be the case. Mutants have been isolated that are defective in the *mdoA* locus described by Bohin and Kennedy (in press). These mutants are blocked in an early stage of MDO assembly since they are defective in the membrane component of the glucosyl transferase system of Weissborn and Kennedy (1983), described above. In certain genetic backgrounds, at least, *mdoA* mutants continue to grow in medium of low osmolarity. Thus, the presence of MDO is not needed for the structural integrity of the cell envelope. Indeed to date no striking phenotype can be associated with their absence.

Difficulty in observing the phenotypic expression of the *mdoA* mutation may not be so surprising as may first appear. There appears to be considerable redundancy in certain cell envelope localized processes such as the multiple transport systems for the uptake of the essential cations, potassium and magnesium. It is possible that the synthesis of MDO is only one of several mechanisms for coping with medium of low osmolarity, any single one of which is dispensible. We are currently pursuing investigations along these lines by inserting the *mdoA* mutation into strains of various genetic backgrounds thought to affect osmotic adaptation.

The fundamental mechanisms by which cells of *E. coli* recognize the osmolarity of the medium in which they are growing and adapt to it by appropriate signaling systems are completely unknown. From this point of view the study of the enzymology of MDO biosynthesis, particularly in its early stages is especially attractive. This pathway now appears to be open to direct biochemical investigation. A careful study of the effectors that modulate the activity of enzymes required in the initial stages of MDO synthesis should shed light on the detailed mechanisms of osmotic regulation.

REFERENCES

Bohin, J.-P., and Kennedy, E.P., in press, *J. Bacteriol.*

Epstein, W., and Laimins, L.A., 1980, *Trends Biochem. Sci.*, 5:21.

Goldberg, D.E., Runley, M.K., and Kennedy, E.P., 1981, *Proc. Natl. Acad. Sci. USA*, 78:5513.

Jackson, B., and Kennedy, E.P., 1983, *J. Biol. Chem.*, 258:2394.

Kanfer, J., and Kennedy, E.P., 1963, *J. Biol. Chem.*, 238:2919.

Kennedy, E.P., Runley, M.K., Schulman, H., and van Golde, L.M.G., 1976, *J. Biol. Chem.*, 251:4208.

Kennedy, E.P., 1982, Proc. Natl. Acad. Sci. USA, 79:1092.
Le Rudulier, D., and Valentine, R.C., 1982, Trends Biochem. Sci., 7:431.
Measures, J.C., 1975, Nature, 257:398.
Roller, S.D., and Anagnostopoulos, G.D., 1982, J. Applied Bacteriol., 52:425.
Schneider, J.E., and Kennedy, E.P., 1978, J. Biol. Chem., 253:7738.
Schulman, H., 1976, Dissertation, Harvard University.
Schulman, H., and Kennedy, E.P., 1979, J. Bacteriol., 137:686.
Stock, J.B., Rauch, B., and Roseman, S., 1977, J. Biol. Chem., 252:7850.
van Golde, L.M.G., Schulman, H., and Kennedy, E.P., 1973, Proc. Natl. Acad. Sci. USA, 70:1368.
Weissborn, A., and Kennedy, E.P., 1983, Fed. Proc., 42:2122.

APPROACHES TO THE BIOCHEMISTRY OF PROTEIN SECRETION IN BACTERIA*

Bernard D. Davis, Stephen Lory, Michael Caulfield, and P.C. Tai

Bacterial Physiology Unit
Harvard Medical School
Boston, Massachusetts 02115

ABSTRACT

Studies in this laboratory on the mechanism of protein secretion in bacteria are reviewed. In Bacillus subtilis the membrane complexed with ribosomes and the free membrane were found to differ extensively in protein composition. A 64 Kd protein in the former is almost certainly involved in secretion, since it is protected by the ribosomes from interacting with protease or with antibody. This protein is also present in the cytoplasm as a complex with three additional proteins, which sediments with the 70S ribosome fraction; its relation to the ribosomes, and to the initiation of protein secretion, is under investigation. In addition, the membrane fraction associated with polysomes contains a 38 Kd protein that appears to be involved in secretion, since it remains with the polysomes after solubilization of the membrane with Triton.

A related problem is how gram-negative bacteria (with a double membrane) excrete proteins to the exterior. We found that with exotoxin A of Pseudomonas aeruginosa, which is normally excreted as rapidly as it is completed, ethanol inhibits processing of a larger precursor and causes it to accumulate on the outer surface of the outer membrane. This and other findings suggest that the protein is not secreted through either membrane but moves laterally, via

*This work has been supported by grants from the American Cancer Society and the National Institutes of Health.

intermembrane junctions, from the inner to the outer membrane, from which the cleaved form is released.

INTRODUCTION

By and large, bacteria have been the most rewarding organisms for the study of universal molecular properties of cells because they are simpler than eukaryotes and their genes are more easily manipulated. However, in the study of membrane functions, including the transport of proteins as well as of small molecules, many key discoveries were made first in animals cells. Among these, Palade (see review, 1975) demonstrated ribosomes attached to the endoplasmic reticulum and provided indirect evidence that they secrete proteins; Milstein et al. (1972) discovered that a secreted protein was made as a precursor with a cleavable N-terminal signal sequence; similar cleavable signals were also found on precursors of proteins incorporated into membranes; and Blobel and Dobberstein (1975) developed a system for studying sequestration of growing proteins into vesicles in vitro.

In recent years the potential advantages of bacteria for studying protein translocation have begun to be realized. In highly successful genetic studies Beckwith's group has isolated mutants altered in various components of the secretory process, and through gene fusion they have incorporated soluble proteins into proteins attached to the membrane (reviewed in Michaelis and Beckwith, 1982; Silhavy et al., 1983). This approach is now being extended in many laboratories. Our group has undertaken a more traditional biochemical approach, even though one of us (BDD) was proselytizing for the use of bacterial mutants in biochemistry 35 years ago.

MECHANISM OF SECRETION ACROSS A MEMBRANE

Extracellular Labeling of Growing Chains

Our initial studies took advantage of the fact that the far side of the membrane in bacteria, from which the secreted protein emerges, is exposed to the external medium rather than being buried in the endoplasmic reticulum. Using nonpenetrating radioactive reagents to label proteins exposed on the external surface of protoplasts and building on our experience in isolating active polysomes, we were able to show that the extracellularly labeled material included nascent chains of secreted proteins, still attached to membrane-associated ribosomes in the cell (Smith et al., 1977; Davis and Tai, 1980). We could also show that treatment of intact cells with Pronase, to digest polypeptides exposed on the surface, shortened the chains attached to polysomes (Smith et al., 1978).

These unequivocal demonstrations of cotranslational secretion have turned out to be more than a confirmation of an already known process. Work in several laboratories has since shown that secretion is not always cotranslational: some proteins containing a cleavable signal peptide are synthesized on free rather than on membrane-bound polysomes, and so they evidently enter the membrane post-translationally. Hence, while the discovery of hydrophobic signal peptides had suggested a spontaneous attachment of these N-terminal sequences to the membrane, this finding was not sufficient to prove cotranslational secretion.

The Proteins of Complexed and Free Membrane

With cotranslational secretion established, the key question is its mechanism. Does the signal peptide bind to a specific complex of proteins in the membrane? Does the membrane provide a machinery of active secretion, dependent on energy from the membrane or from ATP or does it provide a passive channel through which the energy of protein synthesis forces the chain? Or does the penetration of the protein depend (see Wickner, 1980) on a conformational change triggered by cleavage of the signal peptide? We decided to pursue these problems by building on an adventitious observation: in the course of separating the membrane-ribosome complexes from the denser noncomplexed ribosomes in the experiments on extracellular labeling, we had also encountered a third fraction, free membrane (without ribosomes), at the top of the two-step sucrose gradient. Since Kreibich et al. (1978) had observed that the proteins of the rough and those of the smooth microsomes differed in only two major bands in SDS-polyacrylamide gel electrophoresis, we thought that a comparison of free and complexed membrane, from Bacillus subtilis, might reveal a set of proteins associated with protein secretion.

In fact, a large number of differences were observed, including not only many bands unique to the complexed membrane but also two bands unique to the free membrane (Marty-Mazars et al., 1983). These differences therefore cannot all be related to secretion: when the membrane of the lysing cell breaks up into vesicles additional functions evidently separate more or less cleanly. For example, the penicillin-binding proteins, concerned with peptidoglycan formation and morphogenesis, are present only in the free membrane fraction (Caulfield et al., 1983). It is thus possible to separate various functional domains in the fragmented bacterial membrane. Such fractionation, including the use of additional methods, should eventually shed light on the topography of the membrane.

To study the topography of those proteins that might be involved in secretion we prepared rabbit antibodies to six of the proteins unique to the complexed membrane. The protein of 64 Kd proved to be of great interest. It was the only one found in the cytoplasm as well as in the membrane fraction, suggesting a cyclical attachment (Horiuchi et al., 1983a). Moreover, in the membrane it had an interesting location: unlike the other proteins, it was not digested by trypsin either on the external surface (accessible in intact protoplasts) or on the cytoplasmic surface (in complexed vesicles); but it could be completely digested if the ribosomes were released from the vesicles by lowering the Mg^{++} concentration (Horiuchi et al., 1983b). Another test for accessibility, binding of anti-64 Kd antibody (labeled with ^{125}I) by vesicles, also required that the ribosomes be released. It thus appears that the 64 Kd protein is closely covered, in the membrane, by the attached ribosomes; hence it is almost certainly involved in protein secretion.

The 64 Complex

The 64 Kd protein has in turn provided a handle for identifying additional proteins that are presumably concerned with secretion. We will briefly summarize these findings, whose documentation will appear elsewhere.

As a first approach the proteins in complexed vesicles were crosslinked with dithiobis-(succinimidyl propionate), solubilized with SDS, precipitated with anti-64 Kd antibody, cleaved at the crosslinks with mercaptoethanol, and then analyzed by SDS-polyacrylamide gel electrophoresis. The immunoprecipitate yielded heavy bands not only at 64 Kd but also at 60, 41, and 36 Kd. Moreover, the association appears to be physiological since it could also be demonstrated without crosslinking: the same complex of four proteins (called the 64 complex) remained associated with the ribosomal fraction after the membrane was gently dissolved with 1% Triton X-100. It could be recovered by lowering the Mg^{2+} (to release the nascent chain and associated proteins) and then precipitating with anti-64 Kd antibody.

At first we assumed that this complex was the machinery of secretion, or part of it, embedded in the membrane. However, further analysis revealed a more complex picture. The bulk of the complex is present in the cytoplasm and sediments with the monosome fraction, but its relation to the ribosomes is not clear. In eukaryotic cells initiation of protein translocation has been shown to involve binding to the ribosome of a signal recognition particle (Walter and Blobel, 1980, 1981, 1982). This complex of several proteins and 7S RNA halts protein synthesis when a signal sequence emerges, and it allows resumption of synthesis on contacting a membrane receptor (Meyer et al., 1982; Gilmore et al., 1982). How closely our 64 complex resembles the signal recognition particle is under investigation.

The Machinery of Translocation after Initiation

What began as a search for a machinery of secretion in the membrane has thus far revealed what appears to be a complex, or a portion of a complex, involved in the initiation of protein translocation into membrane. Meanwhile, we applied the same Triton treatment to the membrane-bound polysomes and found that they retained only 1 major band of 38 Kd (along with a small amount of the 64 complex). This protein may represent a true membrane machinery involved in the later stages of translocation. It is, of course, not certain whether that machinery is simpler in composition than the initiating particle or whether it contains additional proteins which are not firmly bound to the secreting ribosome.

MECHANISN OF EXCRETION PAST A DOUBLE MEMBRANE

Gram-negative bacteria have an outer membrane (OM) outside their rigid peptidoglycan wall in addition to the inner, cytoplasmic membrane (IM), and they secrete proteins into the intervening periplasmic space. In addition, they often excrete proteins to the exterior. This excretion presents a special problem, for secretion across the IM requires energy from the protonmotive force, as has been shown in cells (Daniels et al., 1981; Enequist et al., 1981) and also in vesicles in vitro (D. Rhoads, P.C. Tai, and B.D. Davis, manuscript submitted for publication). If excreted proteins were secreted successively through IM and OM they would presumably also require energy for passage through the latter. Yet no source for such energy is apparent. In addition, the folding of the protein as it emerged in the periplasmic space might also be expected to impair its further passage.

We have therefore undertaken to test for alternative excretion mechanisms that bypass the periplasm. Our results have led us to propose the model schematized in Figure 1. In this mechanism the protein would be attached to the IM and move laterally to the OM through the Bayer junctions (regions of fusion between IM and OM), and cleavage at some stage in its journey would weaken its attachment and allow release to the exterior.

We investigated this problem (Lory et al., 1983) with exotoxin A of Pseudomonas aeruginosa, which is freely excreted by growing cells. This toxin, of 66 Kd, is very similar in structure and action to diphtheria toxin. It is excreted as soon as it is completed, since antibodies to it failed to reveal any detectable precursor in the culture or any toxin in the cells, even after very brief pulse-labeling. However, when the growing culture was treated with 10% EtOH, which perturbs the membrane and blocks the secretion of periplasmic proteins (Palva et al., 1981), the cells immediately ceased excreting toxin, and instead they accumulated, for several minutes, a precursor about 3 Kd larger. Moreover, when the cells were fractionated this precursor was found entirely in the OM fraction. In intact cells it could be digested by Pronase, leaving no protected residue large enough to detect by immunoprecipitation. The bulk of the precursor molecule is thus exposed on the outer surface of the OM.

These findings all fit the model of Figure 1. That model is also supported by the effects of removing the EtOH after accumulation of the precursor: the cells rapidly resume excretion of mature toxin, but precursor already accumulated in the OM remains there during further growth without EtOH. These results suggest that the EtOH reversibly inhibits the specific peptidase for processing the exotoxin precursor, and after this enzyme has regained activity it can act on new precursor but the already accumulated precursor can no longer reach it. Our model would predict this irreversibility, for lipopolysaccharide, the lipid of the outer leaflet of the OM, is known not to flow back through the intermembrane junction.

The location of this signal peptidase is not known. Efforts to detect it, or to cleave the precursor with purified signal peptidase from E. coli (kindly provided by W. Wickner), have not been successful. If the Pseudomonas processing enzyme were located at the far side of the intermembrane junction, its action on the precursor might immediately release the mature exotoxin to the exterior. However, other findings suggest that the signal peptidase is located in the IM, and after it has acted the polysomes remain attached to the membrane. Thus when the membrane-complexed polysomes that form the toxin were freed of membrane by detergent before chain completion, they continued to yield almost all mature toxin, rather than precursor. Presumably some hydophobic segment beyond the signal holds the growing chain to the IM (and to the vesicles in vitro). This post-signal segment would also provide information directing the protein to its destination (as is widely assumed for other proteins that have different destinations). In this model the outer leaflet of the OM would have a lower affinity than the IM for the postulated residual hydrophobic sequence, and so the mature toxin would be released rapidly after passing the junction. In EtOH-treated cells, however, the uncleaved signal, retained on the precursor, would strengthen its attachment and prevent release from the OM.

The mechanism of lateral flow from IM to OM via junctions extends a concept that is generally accepted for OM lipids. It has also been suggested for integral proteins of the OM (Bayer, 1979; Osborn and Wu, 1980; Silhavy et al., 1979). Since this mechanism of excretion would not involve secretion through either membrane, we would recommend avoiding terminological bias by restricting the term secretion to translocation that unequivocally crosses a membrane.

The mechanism of lateral flow through a junction is an attractive possibility for other transfer processes besides that of proteins and lipids in gram-negative bacteria. One example is the transfer of peptidoglycan precursor from its carrier lipid (polyisoprenol phosphate) to the growing wall, in either gram-positive or gram-negative bacteria. For it seems unlikely that this carrier cyclically reverses its orientation across the membrane, as assumed earlier,

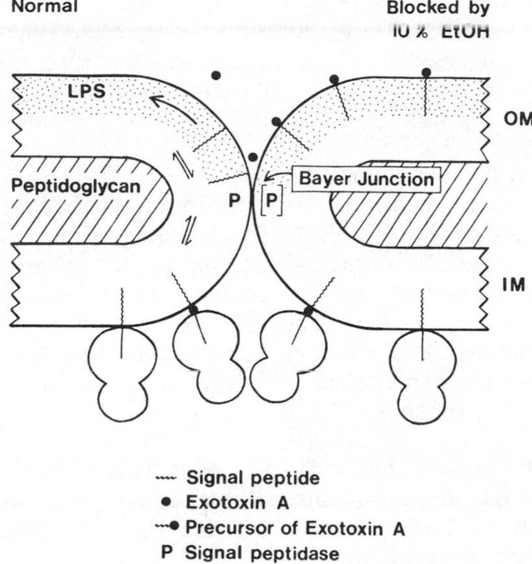

Fig. 1. Model of protein excretion via the Bayer junction in gram-negative bacteria. In this model, the growing precursor of exotoxin A is held to the inner face of the IM by its signal peptide, together with a subsequent region. After chain completion, the product is transported laterally via a Bayer junction to the outer surface of the outer membrane, and in transit it is cleaved by a signal peptidase, P, resulting in release to the medium. Whether P is located in the IM or in the Bayer junction is not certain; if it is the former, the residual attachment after cleavage would be stable in the IM, but not in the OM. In the presence of EtOH (right half of diagram) the peptidase [P] fails to cleave the precursor, which then ends up in the OM surrounded by lipopolysaccharide (LPS). Removal of EtOH restores normal processing and excretion of newly synthesized exotoxin, but not of the accumulated precursor. The diagram also illustrates the reversible flow of phospholipid and the irreversible flow of lipopolysaccharide and exotoxin precursor. (From Lory, Tai, and Davis, 1983; reprinted by permission of the publisher)

since we now know that phospholipids flip only very slowly between the two leaflets of a membrane. The large, highly polar complex attached to the polyisoprenol carrier would impair its flipping even more. A reasonable alternative is flow (without change of orientation) along a membranous junction, in this case connecting an edge in the IM and the adjacent edge of the growing wall. Indeed, one must consider the possibility that proteins are also excreted in gram-positive bacteria at such sites, rather than being secreted through the cytoplasmic membrane and diffusing through the intact, thick peptidoglycan.

The mechanism proposed here should also be considered for transport of proteins across other double membranes, as in mitochondria. To be sure, the contacts between the two mitochondrial membranes are generally depicted as parallel. However, since there is no rigid layer to permit one to observe their behavior when the IM is osmotically contracted from the OM (as one can do in plasmolyzed bacteria), one cannot exclude the possibility that the membranes are fused at the points of contact. If so, the parallel to gram-negative bacteria would seem close.

As Herman Kalckar has often emphasized, advances in science depend not only on logic but also on intuition and on serendipity. I hope Herman will find our recent wanderings sufficiently tortuous and serendipitous to be entertaining.

REFERENCES

Bayer, M.E., 1979, The fusion sites between outer membrane and cytoplasmic membrane of bacteria: their role in membrane assembly and virus infection, in: "Bacterial Outer Membranes: Biogenesis and Function," M. Inouye (ed.), p. 167, John Wiley and Sons, Inc., New York.

Blobel, G., and Dobberstein, B., 1975, Transfer of proteins across membranes. I. Presence of proteolytically processed nascent immunoglobulin light chains on membrane-bound ribosomes of murine myeloma, J. Cell Biol., 67:835.

Caulfield, M.P., Tai, P.C., and Davis, B.D., 1983, Association of penicillin-binding proteins and other enzymes with the ribosome-free membrane fraction of Bacillus subtilis, J. Bacteriol., 156:1.

Daniels, C.J., Bole, D.G., Quay, S.C., and Oxender, D.L., 1981, Role of membrane potential in the secretion of protein into the periplasm of Escherichia coli, Proc. Natl. Acad. Sci. USA, 78:4396.

Davis, B.D., and Tai, P.C., 1980, Mechanism of protein secretion across membranes, Nature (London), 283:433.

Enequist, H.G., Hirst, T.R., Harayama, S., Hardy, S.J.S., and Randall, L.L., 1981, Energy is required for maturation of exported proteins in Escherichia coli, Eur. J. Biochem., 116:227.

Gilmore, R., Walter, P., and Blobel, G., 1982, Protein translocation across the endoplasmic reticulum. II. Isolation and characterization of the signal recognition particle receptor, J. Cell Biol., 95:470.

Horiuchi, S., Marty-Mazars, D., Tai, P.C., and Davis, B.D., 1983a,

Localization and quantitation of proteins characteristic of the complexed membrane of Bacillus subtilis, J. Bacteriol., 154:1215.

Horiuchi, S., Tai, P.C., and Davis, B.D., 1983b, A 64 kilodalton membrane protein of Bacillus subtilis covered by secreting ribosomes, Proc. Natl. Acad. Sci. USA, 80:3287.

Kreibich, G., Czako-Graham, M., Grebenau, R., Mok, W., Rodriguez-Boulan, E., and Sabatini, D., 1978, Characterization of the ribosomal binding site in rat liver rough microsomes: ribophorins I and II, two integral membrane proteins related to ribosome binding, J. Supramol. Struct., 8:279.

Lory, S., Tai, P.C., and Davis, B.D., 1983, Mechanism of protein excretion by gram-negative bacteria: Pseudomonas aeruginosa exotoxin A, J. Bacteriol., 156:695.

Marty-Mazars, D., Horiuchi, S., Tai, P.C., and Davis, B.D., 1983, Proteins of ribosome-bearing and free membrane domains in Bacillus subtilis, J. Bacteriol., 154:1381.

Meyer, D.I., Krause, E., and Dobberstein, B., 1982, Secretory protein translocating across membranes—the role of the "docking protein," Nature (London) 297:647.

Michaelis, S., and Beckwith, J., 1982, Mechanism of incorporation of cell envelope proteins in Escherichia coli, Ann. Rev. Microbiol., 36:435.

Milstein, C., Brownlee, G.G., Harrison, T.M., and Mathews, M.B., 1972, A possible precursor of immunoglobulin light chains, Nature (London) New Biol., 239:117.

Osborn, M.J., and Wu, H.C.P., 1980, Proteins of the outer membrane of gram-negative bacteria, Ann. Rev. Microbiol., 34:369.

Palade, G.E., 1975, Intracellular aspects of the process of protein synthesis, Science, 189:347.

Palva, E.T., Mirst, R.T., Hardy, S.J.S., Holmgren, T., and Randall, L., 1981, Synthesis of a precursor to the B subunit of heat-labile enterotoxin in Escherichia coli, J. Bacteriol., 146:325.

Silhavy, T.J., Bassford, P.J., Jr., and Beckwith, J.R., 1979. A genetic approach to the study of protein localization in Escherichia coli, in: "Bacterial Outer Membranes: Biogenesis and Function," M. Inouye (ed.), p. 203, John Wiley and Sons, Inc., New York.

Silhavy, T.J., Benson, S.A., and Emr, S.D., 1983, Mechanisms of protein localization, Microbiol. Rev., 47:313.

Smith, W.P., Tai, P.C., Thompson, R.T., and Davis, B.D., 1977, Extracellular labeling of nascent polypeptides traversing the membrane of E. coli, Proc. Natl. Acad. Sci. USA, 74:2830.

Smith, W.P., Tai, P.C., and Davis, B.D., 1978, Interaction of secreted nascent chains with surrounding membrane in Bacillus subtilis, Proc. Natl. Acad. Sci. USA, 75:5922.

Walter, P., and Blobel, G., 1980, Purification of a membrane-associated protein complex required for protein translocation across the endoplasmic reticulum, Proc. Natl. Acad. Sci. USA, 77:7112.

Walter, P., and Blobel, G., 1981, Translocation of proteins across the endoplasmic reticulum. III. Signal recognition protein (SRP) causes signal sequence-dependent and site-specific arrest of chain elongation that is released by microsomal membranes, J. Cell Biol., 91:557.

Walter, P., and Blobel, G., 1982, 7S small cytoplasmic RNA is an integral component of the signal recognition particle, Nature (London) 299:691.

Wickner, W., 1980, Assembly of proteins into membranes, Science, 210:861.

TWO MECHANISMS OF BIOSYNTHESES OF ANTIBIOTIC PEPTIDES

Kiyoshi Kurahashi and Chika Nishio

Institute for Protein Research
Osaka University
3-2, Yamadaoka, Suita
Osada 565, Japan

ABSTRACT

The biosynthesis of antibiotic peptides such as gramicidin S, tyrocidines, bacitracins, gramicidin A, and polymyxin E, is carried out by the multienzyme thiotemplate mechanism without the involvement of either ribosomes or mRNAs. However, a larger antibiotic peptide produced by Bacillus subtilis ATCC 6633, subtilin, with 32 amino acid residues, was found to be formed by processing of its precursor protein. There are two different mechanisms for the synthesis of antibiotic oligopeptides by microorganisms.

INTRODUCTION

The biosynthetic mechanism of antibiotic peptides, such as gramicidin S, tyrocidines, bacitracins, gramicidin A and polymyxin E, by the multienzyme thiotemplate mechanism has been studied extensively in recent years (cf. Kurahashi, 1981; Kurahashi et al., 1982, Kleinkauf and Koischwitz, 1978, 1980; Frøyshov et al., 1978; Katz and Demain, 1977) and is considered to be fairly well understood, though the details of the mechanism still remain ambiguous.

I illustrate here the present-day knowledge of the multienzyme thiotemplate mechanism of antibiotic peptide biosyntheses, using gramicidin S synthesis as an example. Figure 1 shows the structure

---

This work was supported by Grants-in-Aid for Scientific Research from the Ministry of Education, Science and Culture of Japan.

```
             L-Leu --> D-Phe
            ↗   10  |  1   ↘
       L-Orn 9      |      2 L-Pro
        ↑           |           ↓
      L-Val 8       |        3 L-Val
        ↑           |           ↓
       L-Pro 7      |      4 L-Orn
            ↖    6  |  5   ↙
             D-Phe <-- L-Leu
```

Fig. 1. Structure of gramicidin S.

of gramicidin S. It is a cyclic decapeptide made up of two identical pentapeptides. Gramicidin S synthetase consists of a light enzyme, Enzyme II, and a heavy enzyme, Enzyme I, as shown in Figure 2A (Tomino et al., 1967; Itoh et al., 1968; Kurahashi et al., 1969; Kleinkauf and Gevers, 1969; Kleinkauf and Koischwitz, 1978; Gilhuus-Moe et al., 1970). The light enzyme has a molecular weight of 100,000 and activates L-phenylalanine to L-phenylalanyl adenylate and binds it as a thioester. Then, the L configuration is converted to the D configuration (Yamada and Kurahashi, 1969; Takahashi et al., 1971; Gevers et al., 1969). The heavy enzyme activates the rest of the constituent amino acids, L-proline, L-valine, L-ornithine and L-leucine as adenylates and binds them as thioesters. This enzyme contains a phosphopantetheine arm which participates in the elongation of the peptide chain as revealed by the work of Laland's group (Gilhuus-Moe et al., 1970; Laland et al., 1972) and Lipmann's group (Kleinkauf et al., 1971; Lipmann, 1971). The interaction of the light enzyme and the heavy enzyme brings about the synthesis of D-phenylalanyl-L-proline dipeptide thioesterified to the heavy enzyme as shown in Figure 2A and B. A point under discussion was whether the condensation is a process of a direct transfer or mediated by phosphopantetheine, but recently Hori et al. (1981) showed that a mutant enzyme lacking pantetheine is unable to form the dipeptide. The dipeptide is then transferred onto L-valine thioesterified to the enzyme by the mediation of the phosphopantetheine arm to form a tripeptide, D-phenylalanyl-L-prolyl-L-valine, as shown in Figure 2C and D. By repeating the sequence of reactions a pentapeptide thioesterified to the enzyme is formed as shown in Figure 2E (Lipmann, 1973; Laland and Zimmer, 1973). The mechanisms of condensation of the two pentapeptides formed is not completely elucidated. Stoll et al. (1970) presented evidence that there is a waiting site, Site 6, on the heavy enzyme where the first pentapeptide sits and waits for the completion of the second pentapeptide as shown in Figure 2E. Then a head-to-tail condensation of the two pentapeptides takes place. Results by Roskoski et al. (1971) favor intermolecular cyclization between the two pentapeptides. The heavy enzyme has a molecular weight 280,000, and studies by Koischwitz and Kleinkåuf (1976) and Hori et al. (1982) indicate that it is a single polypeptide chain in contrast to tyrocidine synthetase. Intermediate and heavy enzyme of tyrocidine synthetase were dissociated into three and six 70,000 dalton amino acid activating subunits, respectively, by SDS-polyacrylamide gel electrophoresis (Lee et al., 1973).

The features of the multienzyme thiotemplate mechanisms of antibiotic oligopeptide synthesis are summarized as follows:

1. The enzymes that carry out peptide synthesis consist of one or several protein components, each of which, in general, contains several subunits.

2. The constituent amino acids are activated by each subunit as aminoacyl adenylates and transferred to the respective subunit to form amino-acyl-thioesters.

3. The peptide bond formation is mediated by a phosphopantetheine arm present on the multifunctional enzyme components.

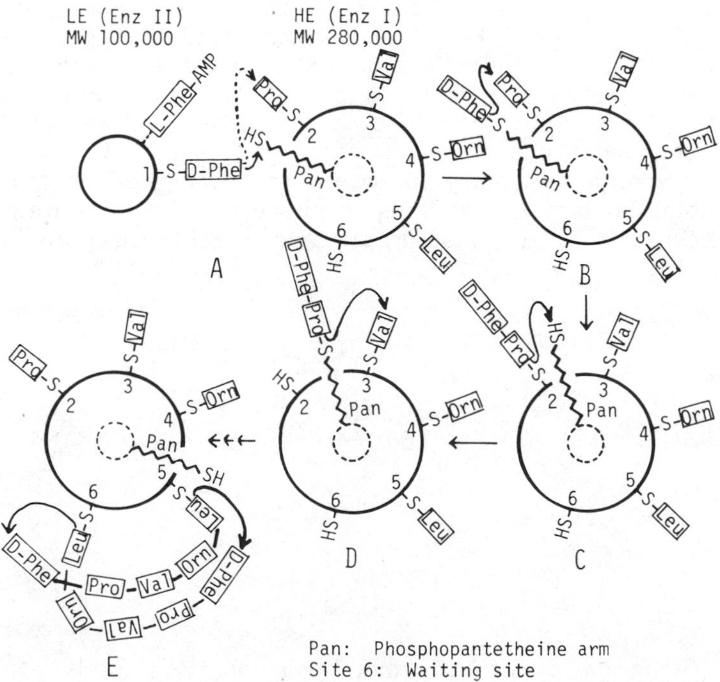

Fig. 2. Schematic model of gramicidin S synthesis.

4. Peptide synthesis initiates generally with either unnatural or modified amino acids and the direction of elongation is towards the C-terminus.

5. The composition and the sequence of the constituent amino acids in peptides are determined by the substrate specificity and the spatial arrangement of the active thiol groups on the enzyme.

6. The fidelity of peptide synthesis therefore is not as strict as that of the ribosome-mRNA system of protein synthesis.

How large a peptide can be synthesized by this kind of mechanism is an interesting question. If we divide the molecular weight of the heavy enzyme of gramicidin S synthetase (280,000) by 4, that is the number of amino acids activated, we get 70,000 as an average molecular weight for each subunit like the subunits of tyrocidine synthetase. Thus, for the synthesis of a peptide consisting of 30 amino acid residues a multienzyme with a molecular weight of 2.1 million is required. The largest peptide which is reported to be synthesized by the multienzyme thiotemplate mechanism is alamethicin which consists of 19 amino acid residues (Mohr and Kleinkauf, 1978). Suzukacillin with 23 amino acid residues is a higher homolog of alamethicin. This may also be synthesized by the similar mechanism. However, there are two larger well-known antibiotic peptides, called nisin and subtilin. Figures 3 and 4 show their structures elucidated by Gross et al. (1973). Nisin consists of 34 amino acid residues and is produced by Streptococcus lactis. Subtilin consists of 32 amino acid residues and produced by Bacillus subtilis ATCC 6633. Despite the difference in taxonomy between the two producing organisms, both antibiotics have a striking similarity. Both contain unusual amino acids such as α,β-unsaturated amino acids, dehydroalanine and

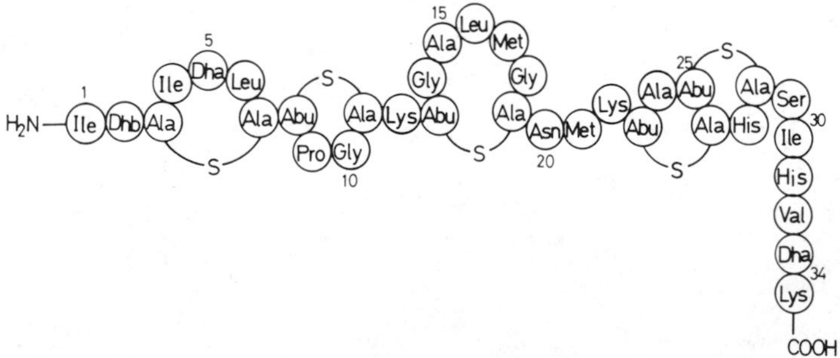

Fig. 3. Structure of nisin. Abu, 2-aminobutyric acid; Dha, dehydroalanine; Dhb, dehydrobutyrine; Ala-S-Ala, lanthionine; Abu-S-Ala, β-methyllanthionine.

dehydrobutyrine, and lanthionine and β-methyllanthionine. The examination of these structures may suggest that they are synthesized by the multienzyme thiotemplate mechanism. However, Hurst (1966) reported that nisin synthesis in growing cells was inhibited by chloramphenicol, puromycin, and actinomycin D, suggesting the participation of a ribosome-mRNA system in its synthesis. Hurst and Paterson (1971) isolated basic peptides from a mutant strain of S. lactis which did not produce nisin and converted them to a product, presumably nisin, by incubating the basic peptide preparations with the wild type cell extract. Ingram (1969, 1970) reported that lanthionine and β-methyllanthionine may be synthesized from cysteine plus serine and threonine, respectively, and that incorporation of cysteine into lanthionine peptides was inhibited by inhibitors of protein synthesis. However, the nature of the precursor or the mechanism of its conversion to nisin has remained unclarified.

We also observed that the biogenesis of subtilin by growing cells was inhibited by chloramphenicol (Kurahashi et al., 1982). Puromycin and actinomycin D were also inhibitory as in the case of nisin. Table 1 shows that [^{14}C]serine incorporation into subtilin by growing cells of Bacillus subtilis ATCC 6633 was inhibited by chloramphenicol and actinomycin D, suggesting that subtilin is synthesized by the protein synthetic mechanism. Subtilin was isolated by butanol extraction of the acidified culture followed by precipitation with acetone.

We purified subtilin by CM-Sephadex column chromatography and high performance liquid chromatography. The purified subtilin was conjugated to keyhole limpet hemocyanine with glutaraldehyde (Reichlin et al., 1968; Baron and Baltimore, 1982) and the complex was used as an antigen to produce antisubtilin antibody in rabbits. The antibody obtained was not able to precipitate subtilin, but neutralized the antibiotic activity of subtilin against Lactobacillus casei, as shown in Figure 5. Antiserum and nonimmune serum (7.5 µl each) were placed in the center wells of an agar plate seeded with L. casei and different concentrations of subtilin were in the

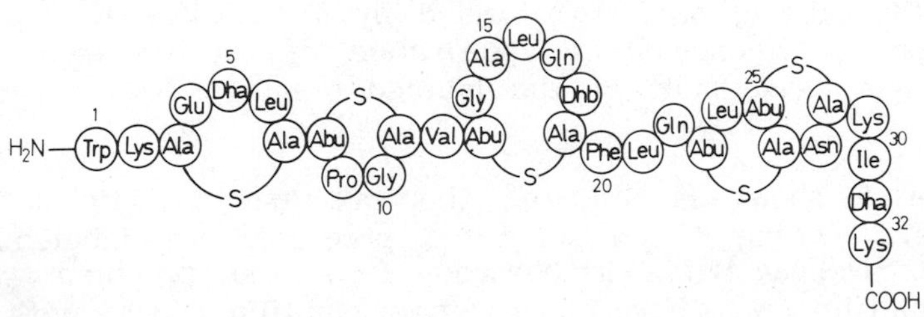

Fig. 4. Structures of subtilin.

Table 1: Effect of Inhibitors on Incorporation of [^{14}C]Serine into Protein and Subtilin by B. Subtilis

Additions	[^{14}C]Serine Incorporated (cpm) 60 min		120 min	
	Subtilin	Protein	Subtilin	Protein
None	5,700	118,000	7,000	121,000
Chloramphenicol 50 μg/ml	240	13,000	270	13,000
Actinomycin D 30 μg/ml	4,200	28,000	4,100	32,000

Incubation for 60 and 120 min at 35° C. L-[U-^{14}C]serine: 0.4 μCi/0.2 μmol/2 ml culture.

surrounding wells. The subtilin-antisubtilin antibody complex can be precipitated with Protein A-sepharose.

Figure 6 shows the time course of incorporation of [^{35}S]cysteine into antisubtilin antibody precipitable materials. Four hundred and fifty μCi of [^{35}S]cysteine was added to a 10-ml culture of B. subtilis. The culture was incubated at 35° C and trichloroacetic acid precipitates were collected at the times indicated and solubilized with sodium dodecyl sulfate (SDS). Immunoprecipitates with antisubtilin antibody and Protein A-sepharose were dissolved in sample buffers and subjected to electrophoresis on a linear gradient of 10 to 20 percent SDS-polyacrylamide gel according to the method of Laemmli (1970). Contrary to our expectation, during a 4 min labeling period no labeled subtilin was formed, but a peptide larger than subtilin was labeled increasingly with time, suggesting the presence of subtilin precursor in the cells.

Figure 7 shows the results of pulse-chase labeling experiments with [^{35}S]cysteine. After 6 min labeling, excess cold cysteine was added. The labeled precursor formed by 6 min decreased with time after chase and concomitant appearance of subtilin was observed. After 24 min chase most of the labeled precursor was converted to subtilin.

Figure 8 shows labeling of the precursor protein with [^{35}S] methionine. During 6 min pulse the precursor was labeled. Upon chase the radioactivity disappeared from the precursor, but no labeled subtilin was formed. Since subtilin does not contain methionine, the results indicate that the portion of the precursor

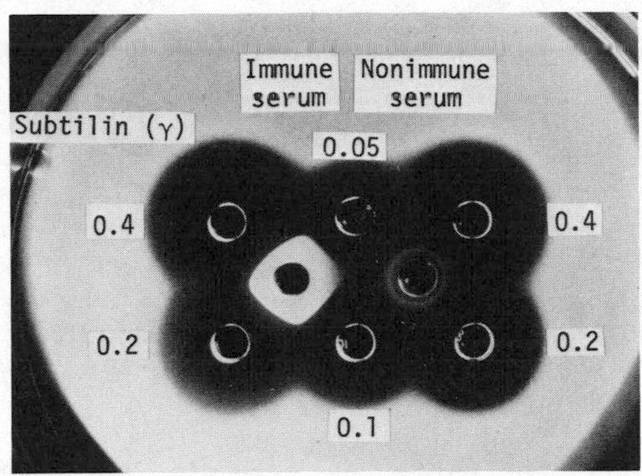

Fig. 5. Inhibition of antibiotic activity by antisubtilin antibody.

peptide which contains methionine was labeled and lost during processing.

Since subtilin was found to be synthesized by processing of precursor peptides, the effect of proteinase inhibitors on subtilin synthesis was studied. Cells were incubated with ethanol that is used as a solvent for phenylmethylsulfonyl fluoride (PMSF), PMSF, ethylene-glycolbis(β-aminoethylether)N,N'-tetraacetic acid (EGTA), and two concentrations of pepstatin in the presence of [^{35}S]cysteine as indicated in Figure 9. After a 10 min incubation, excess unlabeled cysteine was added. The results cannot be interpreted quantitatively, but they show that cells in cultures without inhibitors and with ethanol, EGTA, and 0.7 mM pepstatin synthesized labeled subtilin precursor in a 10 min pulse and converted it to subtilin during 20 min chase. PMSF and 1.7 mM pepstatin were found to be inhibitory for the conversion of the precursor to subtilin as indicated by the much higher radioactivity in the precursor than in subtilin at 30 min incubation.

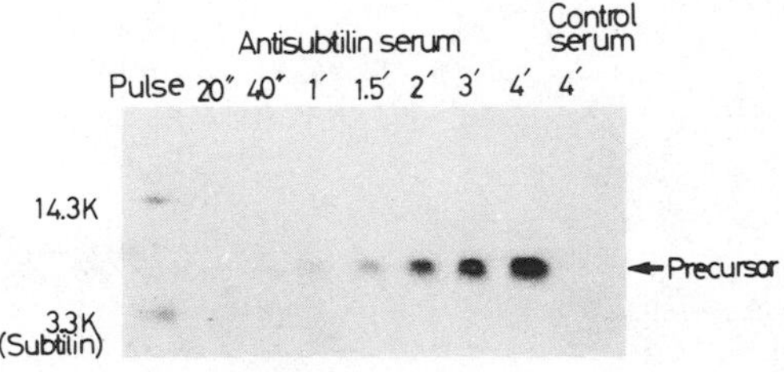

Fig. 6. Labeling of antisubtilin antibody precipitable materials with [^{35}S]cysteine in vivo.

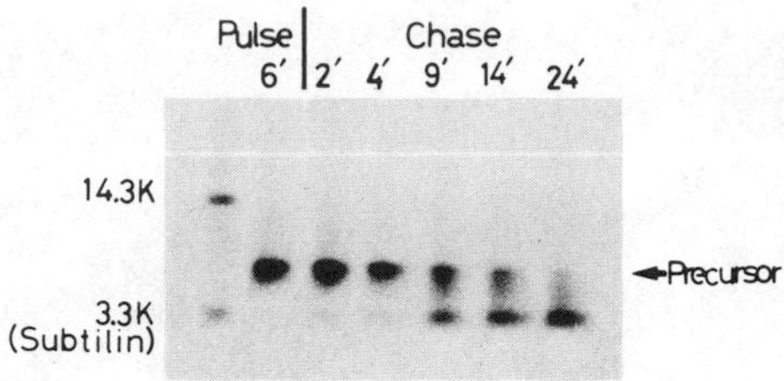

Fig. 7. Conversion of labeled subtilin precursor to subtilin by chase.

In order to confirm the above results, we isolated the subtilin precursor and studied its conversion to subtilin by incubating it with crude extracts of B. subtilis cells. The crude extracts contained an enzyme(s) which converts the precursor to subtilin and the in vitro conversion was inhibited by a mixture of PMSF and pepstatin (Nishio et al., 1983).

The labeled precursor was separated into three radioactive peptides by Sephadex G-50 gel filtration. They have approximate molecular weights of 8,000, 5,000, and 4,000. The structure of subtilin, containing many unusual amino acid residues, and the finding that at least three precursor peptides of subtilin are present in the cells, indicate that posttranslational processing of subtilin precursors to subtilin must be very complicated multistep reactions.

The results presented show that there are two mechanisms for antibiotic peptide synthesis: One is the multienzyme thiotemplate mechanism, and the other is a mechanism involving processing of protein precursors like bioactive peptide syntheses in animals.

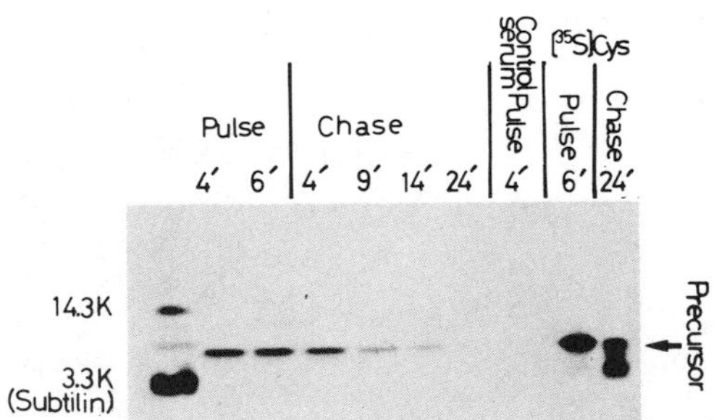

Fig. 8. Pulse-chase labeling of subtilin precursor with [^{35}S]methionine.

REFLECTIONS AND DEDICATION

It is a great pleasure for me to contribute an article to the Kalckar Festschrift in honor of Prof. Herman Kalckar's 75th birthday. This would also be a good celebration of our 30th anniversary of scientific and philosophical friendship as Herman puts it. Merton Utter and myself discovered phosphoenolpyruvate carboxykinase in 1953 which brought about a satisfactory solution to Herman's important observation in 1939 that phosphoenolpyruvate accumulated during the oxidation of malate by kidney preparations (Kalckar, 1939; Utter and Kurahashi, 1954; Krebs, 1954). It was quite a happy coincidence that Herman has become my second wise and inspiring mentor in the United States. I worked with him on galactose metabolism in man and *Escherichia coli* from 1955 through 1958 at the National Institutes of Health. Our work in collaboration with Doctors Joshua and Esther Lederberg revealed the Leloir pathway in *E. coli* and correlated the genetic and enzymatic defects of mutants in galactose metabolism. My association with Herman was not only in the limits of science. I spent many pleasant evening hours at his home discussing music, arts, dramas, and philosophy with him. I shared with Herman devotion to Mozart. He has given me a great impact on extending and deepening my education.

I admire Dr. Kalckar not merely as a great scientist, but also as a man of integrity, culture, philosophy, and above all, a man of romanticism which he inherited from his beloved parents and shared with his brother Fritz whose rich memories he still cherishes.

It gives me the greatest pleasure to extend my warmest congratulations to Herman and Agnete on his 75th birthday, and I wish him good health, productive scientific activity, and letting us benefit from his unique ideas and foresight for many years to come.

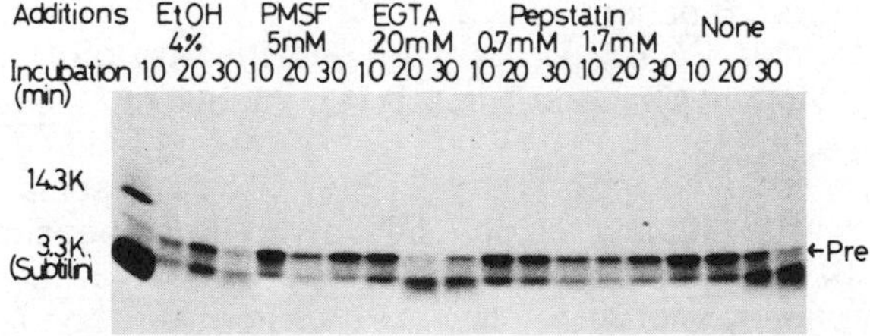

Fig. 9. Effect of proteinase inhibitors on the conversion of subtilin precursor to subtilin in vivo.

REFERENCES

Baron, M.H., and Baltimore, D., 1982, Antibodies against the chemically synthesized genome-linked protein of poliovirus react with naive virus-specific proteins, Cell, 28:395.

Frøyshov, Ø., Zimmer, T.L., and Laland, S.G., 1978, Biosynthesis of microbial peptides by the thiotemplate mechanism, Int. Rev. Biochem., 18:49.

Gevers, W., Kleinkauf, H., and Lipmann, F., 1969, Peptidyl transfers in gramicidin S biosynthesis from enzyme-bound thioester intermediates, Proc. Natl. Acad. Sci. USA, 63:1335.

Gilhuus-Moe, C.C., Kristensen, T., Bredesen, J.E., Zimmer, T.L., and Laland, S.G., 1970, The presence and possible role of phosphopantethenic acid in gramicidine S synthetase, FEBS Lett., 7:287.

Gross, E., Kiltz, H.H., and Nebelin, E., 1973, Subtilin, VI, Die struktur des Subtilins, Z. Physiol. Chem. 354:810.

Hori, K., Kanda, M., Kurotsu, T., Miura, S., Yamada, Y., and Saito, Y., 1981, Absence of pantothenic acid in gramicidin S synthetase 2 obtained from some mutants of Bacillus brevis, J. Biochem., 90:439.

Hori, K., Kurotsu, T., Kanda, M., Miura, S., Yamada, Y., and Saito, Y., 1982, Evidence for single multifunctional polypeptide chain on gramicidin S synthetase 2 obtained from a wild strain and mutant strains of Bacillus brevis, J. Biochem., 91:369.

Hurst, A., 1966, Biosynthesis of the antibiotic nisin by whole Streptococcus lactis organisms, J. Gen. Microbiol., 44:209.

Hurst, A., and Paterson, G.M., 1971, Observation on the conversion of an inactive precursor protein to the antibiotic nisin, Can. J. Microbiol., 17:1379.

Ingram, L., 1969, Synthesis of the antibiotic nisin: formation of lanthionine and β-methyllanthionine, Biochim. Biophys. Acta, 184:216.

Ingram, L., 1970, A ribosomal mechanism for synthesis of peptides related to nisin, Biochim. Biophys. Acta 224:263.

Itoh, H., Yamada, M., Tomino, S., and Kurahashi, K., 1968, The role of two complementary fractions of gramicidin S synthesizing enzyme system, J. Biochem., 64:259.

Kalckar, H.M., 1939, The nature of phosphoric esters formed in kidney extracts, Biochem. J., 33:631.

Katz, E., and Demain, A.L., 1977, The peptide antibiotics of Bacillus: chemistry, biogenesis and possible functions, Bacteriol. Rev. 41:449.

Kleinkauf, H., and Gevers, W., 1969, Nonribosomal polypeptide synthesis: the biosynthesis of a cyclic peptide antibiotic, gramicidin S, Cold Spring Harbor Symp. Quant. Biol., 34:805.

Kleinkauf, H., and Koischwitz, H., 1978, Peptide bond formation in non-ribosomal systems, Prog. Mol. Subcell. Biol., 6:59.

Kleinkauf, H., and Koischwitz, H., 1980, Gramicidin S-synthetase, in: "Multifunctional Proteins," H. Bisswanger and E. Schmincke-Ott, eds., John Wiley and Sons, New York.

Kolschwitz, H., and Kleinkauf, H., 1976, Gramicidin S-sythetase. Electrophoretic characterization of the mutlienzyme, Biochim. Biophys. Acta, 429:1052.

Krebs, H.A., 1954, Considerations concerning the pathways of synthesis in living matter, Bull. Johns Hopkins Hosp., 95:19.

Kurahashi, K., 1981, Biosynthesis of peptide antibiotics, in: "Antibiotics IV. Biosynthesis," J.W. Corcoran, ed., Springer-Verlag, Berlin, Heidelberg, New York.

Kurahashi, K., Komura, S., Akashi, K., Nishio, C., 1982, Biosynthesis of antibiotic peptides polymyxin E and gramicidin A, in: "Peptide antibiotics—biosynthesis and functions," H. Kleinkauf and H.W. Dohren, eds., Walter de Gruyter, Berlin, New York.

Kurahashi, K., Yamada, M., Mori, K., Fujikawa, K., Kambe, M., Imae, Y., Sato, E., Takahashi, H., and Sakamoto, Y., 1969, Biosynthesis of cyclic oligopeptide, Cold Spring Harbor Symp. Quant. Biol., 34:815.

Laemmli, U.K., 1970, Cleavage of structural proteins during the assembly of the head of bacteriophage T4, Nature, 227:680.

Laland, S.G., Frøyshov, Ø., Gilhuus-Moe, C., Zimmer, T.L., 1972, Gramicidin S synthetase, an enzyme with an unusually large number of catalytic functions, Nature New Biol., 239:43.

Laland, S.G., and Zimmer, T.L., 1973, The protein thiotemplate mechanism of synthesis for the peptide antibiotics produced by Bacillus brevis, in: "Essays in Biochemistry," Vol. IX, P.N. Campbell and G.D. Greville, eds., Academic Press, New York.

Lee, S.G., Roskoski, R., Jr., Bauer, K., and Lipmann, F., 1973, Purification of the polyenzymes responsible for tyrocidine synthesis and their dissociation into subunits, Biochemistry, 12:398.

Lipmann, F., 1971, Attempts to map a process evolution of peptide biosynthesis, Science, 173:875.

Lipmann, F., 1973, Nonribosomal polypeptide synthesis on polyenzyme templates, Acc. Chem. Res., 6:361.

Mohr, H., and Kleinkauf, H., 1978, Alamethicin biosynthesis. Acetylation of the amino terminus and attachment of phenylalaninol, Biochim. Biophys. Acta, 526:375.

Nishio, C., Komura, S., and Kurahashi, K., 1983, Peptide antibiotic subtilin is synthesized via precursor proteins, Biochem. Biophys. Res. Common., 116:751.

Reichlin, M., Schnure, J.J., and Vance, V.K., 1968, Induction of antibodies to porcine ACTH in rabbits with nonsteroidogenic polymers of BSA and ACTH, Proc. Soc. Exp. Biol. Med., 128:347.

Roskoski, R., Jr., Ryan, G., Kleinkauf, H., Gevers, W., and Lipmann, F., 1971, Polypeptide biosynthesis from thioesters of amio acids, Arch. Biochem. Biophys., 143:485.

Stoll, E., Frøyshov, Ø., Holm, H., Zimmer, T.L., and Laland, S.G., 1970, On the mechanisms of gramicidin S formation from intermediate peptides, FEBS Lett., 11:348.

Takahashi, H., Sato, E., Kurahashi, K., 1971, Racemization of phenylalnine by adenosine triphosphate-dependent phenylalanine

racemase of Bacillus brevis Nagano, J. Biochem., 69:973.
Tomino, S., Yamada, M., Itoh, H., and Kurahashi, K., 1967, Cell-free synthesis of gramicidin S, Biochemistry, 6:2552.
Utter, M.F., and Kurahashi, K., 1954, Mechanism of action of oxalacetic carboxylase, J. Biol. Chem. 207:821.
Yamada, M., and Kurahasi, K., 1969, Further purification and properties of adenosine triphosphate-dependent phenylalanine racemase of Bacillus brevis Nagano, J. Biochem., 66:529.

TRANSMEMBRANE CHANNELS MADE BY COLICINS*

S. E. Luria

Department of Biology
Massachusetts Institute of Technology
Cambridge, Massachusetts 02139

I believe that Herman Kalckar's interest in membranes, like my own, was first aroused when bacterial mutations were discovered that affected the synthesis of the polysaccharides of the bacterial envelope. His interest was kindled later by his own finding that the same periplasmic binding protein mediated both galactose transport and the chemotactic response to galactose.

My curiosity about membranes also started from a polysaccharide problem, specifically through participation in the work that led Phil Robbins to unraveling the synthesis of the 0-antigens of Salmonella. These, of course, were problems relating to what we now call the outer membrane or cell wall of the gram-negative bacterial cell, not the equivalent of a cytoplasmic membrane. But soon thereafter my interest was stimulated in a new direction by Nomura's beautiful work on the action of a mysterious group of bactericidal proteins, the colicins. These bacterial proteins, coded by plasmid genes, are rather large monomeric molecules that kill the sensitive bacterial cells to which they become absorbed. The mechanism of bacterial killing is specific for each group of colicins, but in all cases it involves interaction with the cytoplasmic membrane of the bacterium.

My curiosity about colicins was directed from the beginning to the question of how these highly water-soluble proteins interacted with the bacterial cytoplasmic membrane. Over the years this study has led me into fields such as oxidative phosphorylation, protonmotive

---

*The author's research was supported by grants from the National Institutes of Health (grant number 5-RO1-AI03038) and the National Science Foundation (grant number PCM-8108866).

force, and the nature of transmembrane channels, at times making me wonder if I was turning into a beginning neurophysiologist. Particularly exciting in recent years has been the possibility that our work and the parallel work in other laboratories may lead to a precise molecular reconstruction of at least one class of transmembrane channels, the one produced by colicins of the E1 group.

Colicin E1, which I shall use here as a prototype, is a 56 Kd protein whose amino acid sequence is completely known. Whether this colicin acts on living bacteria or on artificial membrane, its effects are always due to the action of single molecules, without cooperativity. The gross biochemical effects of this colicin and of others of the same group appear to reflect an overall inhibition of energy metabolism, affecting macromolecular syntheses as well as transport mechanisms and cell mobility. Mutational dissection of energy pathways in Escherichia coli cells revealed that in aerobically growing cells the critical level of inhibition was that of membrane energization. In fact, under appropriate conditions bacteria attacked by the colicins can still carry out macromolecular syntheses, but are unable to perform those functions--active transport and mobility--that utilize as direct energy source the protonmotive force, that is, the energy stored in the proton gradient produced by respiratory proton extrusion.

In the colicin-treated cells respiration continues and the cross-membrane proton gradient persists but its energy becomes unavailable for active transport. The membrane potential rapidly disappears, a disappearance that explains the failure of active transport and which is correlated with loss of potassium and other ions from the cells.

These findings suggest an altered permeability of the cytoplasmic membrane, an interpretation that was soon confirmed by experiments on the action of colicin on artificial membranes, both planar membranes and spherical liposomes. On planar membranes, as shown by S. Schein et al., colicin produced channels--one channel per molecule--that provided specific conductances for ions. On liposomes, the colicin created channels that allowed passage of molecules up to molecular weights of 500 to 600 d. The colicin channels formed on planar membranes are actually gated, that is, they open and close in response to changes in membrane potential. In this the colicin channels resemble the sodium and potassium channels of excitable membranes.

These findings brought the problem close to the molecular level. How does a largely hydrophilic protein interact with a membrane to generate a transmembrane channel? A sequence of 35 hydrophobic amino acids, 14 places removed from the C-terminus of colicins E1, is the reasonable candidate for the site of colicin-membrane interaction. Fragmentation of colicin by enzymatic or chemical means does in fact show that this molecule can be subdivided into three sections, an N-terminal section of unknown function, a central section needed for attachment to specific bacterial receptors, and a C-terminal section that interacts with artificial membranes and produces channels whose properties mimic those of the channels created by intact colicin. The shortest segment so far identified as possessing the channel-forming activity consists of 152 amino acids, including the hydrophobic sequence.

The ultimate goal, to identify the specific amino acid sequence responsible for the channel and uncover the atomic structure of the channel itself, is becoming closer. In our laboratory, the techniques of genetic engineering are currently being applied to obtain "diminutive" variants of colicin E1, that is, deletion mutants that consist of short portions of the C-terminal end of the molecule and yet can generate channels in artificial membrane. These "minicolicins," some of which are currently being purified, will then be candidates for crystallization and x-ray diffraction analysis in the hope of establishing the precise structure of the channel itself.

This may seem to be a modest goal, and a still distant one, for the many years of efforts devoted to the study of a few colicins in a number of laboratories. Yet a success in defining precisely the atomic structure of any protein channel could ultimately throw light on the more general problem of how the conformation of protein molecules relates to the structure of transmembrane channels. The path of ions and other substrates through the channels and the regulation of channel function may throw light on more general questions of the nature of transmembrane transport and on membrane control of cellular metabolism.

REFERENCES

Davidson, V.L., Brunden, K.R., Cramer, W.A., and Cohen, F.S., Studies on the mechanism of action of channel-forming colicins using artificial membranes, J. Membr. Biol. (in press).

Luria, S.E., and Suit, J.L., 1982, Transmembrane channels produced by colicin molecules, in: "Membranes and Transport," vol. 2, A. Martonosi, ed., Plenum Publishing Corporation, New York, pp. 279-284.

SIGNIFICANCE OF LIPOPOLYSACCHARIDE STRUCTURE FOR QUESTIONS OF TAXONOMY AND PHYLOGENETICAL RELATEDNESS OF GRAM-NEGATIVE BACTERIA

Hubert Mayer

Max-Planck-Institute für Immunbiologie
Stübeweg 51
D-7800 Freiburg i. Br.
Federal Republic of Germany

ABSTRACT

Many photosynthetic bactera, especially the sulfur-containing and the sulfur-free purple bacteria (Chromatiaceae and Rhodospirillaceae) possess "unusual" lipid A as part of their O-antigenic lipopolysaccharides. Four different lipid A-types have so far been recognized, which vary from each other by the nature of the backbone sugar, the amide-linked fatty acids and by the absence or presence of phosphate in lipid A. Recent phylogenetical studies based on 16S rRNA and cytochrome c_2 sequences showed that photosynthetic bacteria of the Rhodospirillaceae family comprise different ancient clusters of species which are intermingled with a number of nonphotosynthetic species. A comparison of the 16S rRNA relationship and the respective lipid A-types shows (in most cases) a close correlation, indicating that lipid A structures are valuable for recognizing the phylogenetical relatedness of species.

INTRODUCTION

In 1970, Hiroshi Nikaido published a paper (Nikaido, 1970) entitled "Lipopolysaccharide in the Taxonomy of Enterobacteriaceae," in which he pointed out that lipopolysaccharides (LPS) should be ideal markers for the taxonomy of gram-negative bacteria. They occur in the cell wall of practically all gram-negative bacteria and a number of cyanobacteria, and they possess unusual and characteristic constituents which, at least in combination, are highly indicative for lipopolysaccharides (Rietschel et al., 1982). At that time, however, data were nearly exclusively coming from enterobacterial lipopolysaccharides and the phylogenetical significance of LPS structure was therefore not yet recognizable.

In more recent reviews (Wilkinson, 1977; Rietschel and Lüderitz, 1980) it was demonstrated that some regions of LPS are more variable than others and that taxonomic relationships are best recognized when these conservative regions (inner R core and lipid A) are compared.

It was of importance for the further development of these studies that the phylogenetical analyses based on 16S rRNA (Fox et al., 1980; Stackebandt and Woese, 1981) and on cytochrome c_2 (Ambler et al., 1979; Woese et al., 1980) showed that the classical taxonomy of bacteria in many cases is not reflecting the phylogenetical relationships of species. This became especially evident for the photosynthetic purple bacteria, the Rhodospirillaceae and the Chromatiaceae (Gibson et al, 1979), which comprise very ancient bacterial clusters. Figure 1 (taken from Seewaldt et al., 1982, and Fox et al., 1980) shows a dendrogram based on S_{AB}-values, the association coefficient of 16S rRNA, calculated for each binary pair of strains (Fox et al., 1980). It shows that major bacterial groups comprising Rhodospirillaceae are only related to each other with S_{AB}-values of about 0.35. In contrast, different genera of the Enterobacteriaceae family show S_{AB}-values in the range of $\geq$ 0.7 (compare Fig. 1). The close phylogenetic relationship of all members of this family is also documented by the presence of a family-specific surface antigen, the Enterobacterial Common Antigen (ECA), which is also restricted to this family (Mayer and Schmidt, 1979; Ramia et al., 1982).

The question, whether the phylogenetical scheme depicted in Figure 1 can be used also for explaining differences found in the lipid A composition of species of the Rhodospirillaceae family (Weckesser et al., 1979) was investigated. A number of interesting correlations have been revealed (Weckesser et al., 1979; Mayer et al., in press) and will be discussed below. For this discussion it is necessary to outline the characteristic features of the "usual" Salmonella lipid A-type and the "unusual" lipid A-types found often with photosynthetic prokaryotes.

THE "USUAL" LIPID A

Work by many groups in the world, but especially by the Freiburg group with O. Westphal, O. Lüderitz and E. Rietschel (Westphal et al., 1981; Lüderitz et al., 1971; Rietschel et al., in press) has revealed that almost all LPS's share the same structural architecture in being composed of three distinct subregions: the O-specific chains, the R core, and the lipid A moiety. The O-

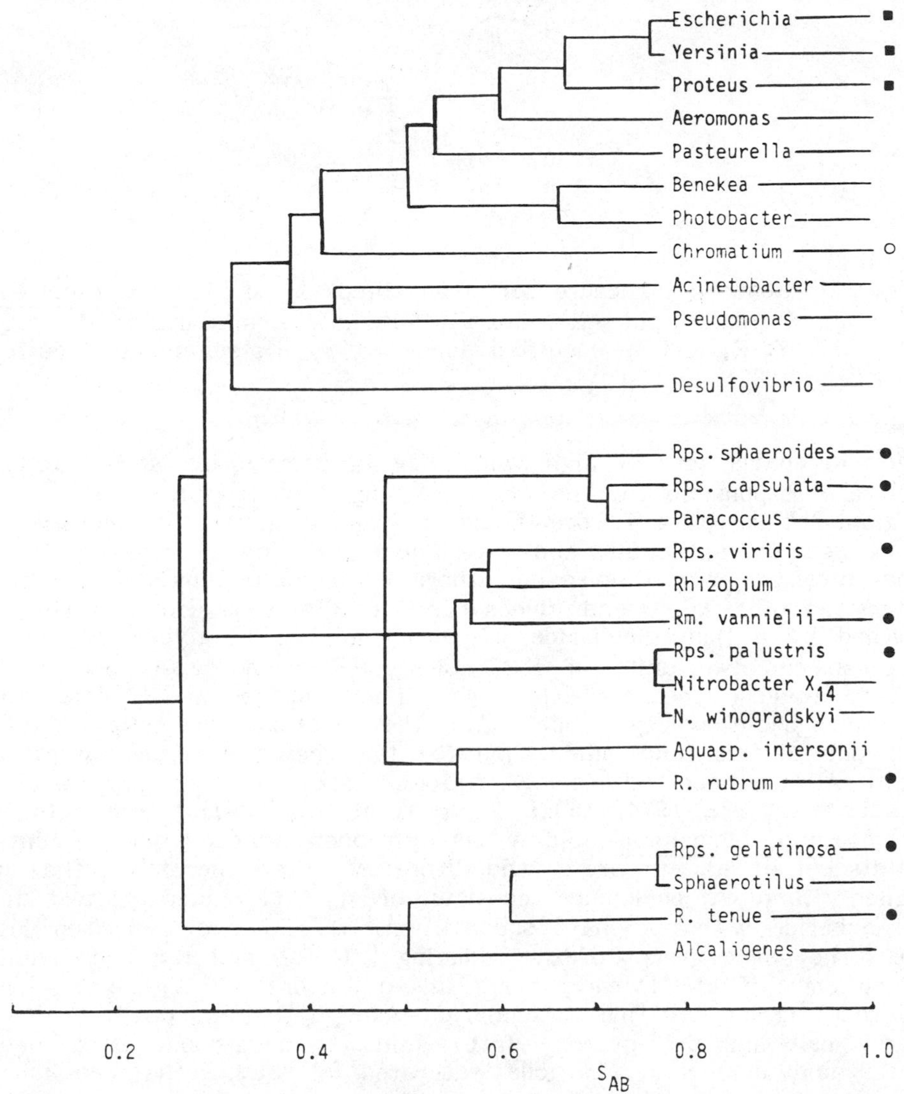

Fig. 1. Dendrogram showing the phylogenetical relationship of gram-negative bacteria based on their 16S rRNA cataloguing. ■, Enterobacteriaceae; ●, Rhodospirillaceae; o, Chromatiaceae (taken from Fox et al., 1980, and Seewaldt et al., 1982).

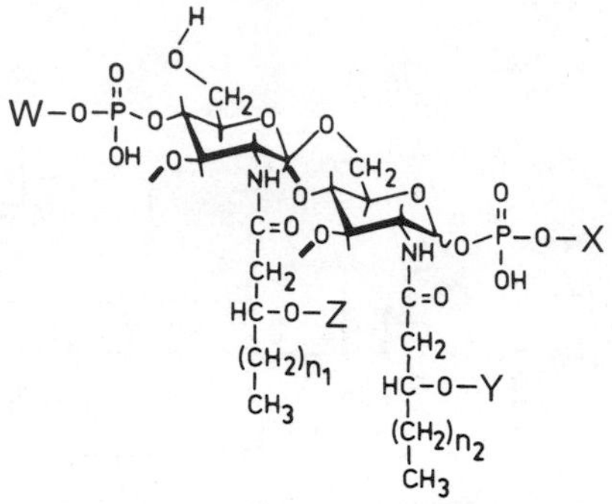

Fig. 2. General structure of free lipid A of Enterobacteriacea ("usual" lipid A). Sites of variations are indicated by X, Y, Z, and W (modified according to Rietschel and Luderitz, 1980).

specific chains show a high variability in composition and structure and are responsible for the many existing O-serotypes. The second region, the so-called R-core, is under separate genetic control and is composed of an outer and an inner part, both differing in a characteristic way from each other. The outer region is mainly composed of hexoses and glucosamine, the inner region, however, is formed by a pentasaccharide of three units of L-glycero-D-manno-heptose and two units of KDO (3-deoxy-D-manno-octulosonic acid). KDO provides the acid-labile ketosidic linkage which links the polysaccharidic moiety to lipid A. The endotoxic properties of LPS are due to this lipid moiety and for this reason structural work on lipid A has been extensively followed for the last two decades (Lüderitz et al., 1971, 1981; Westphal et al., 1981). The detailed structure of Salmonella lipid A has only been reported quite recently (Rietschel et al., in press, and literature cited therein). The so-called "lipid A backbone" consists of a β-1.6-linked glucosamine disaccharide with phosphate substitutions at C-1 and C-4' (see Fig. 2). The amino groups of the reducing (GlcN I) and the nonreducing glucosamine (GlcN II) carry amide-linked 3-hydroxy fatty acids. The hydroxyl groups of this disaccharide—except that in position C-4—carry ester-linked 3-hydroxy fatty acids. Amide- and ester-linked fatty acids may carry additional saturated fatty acids that are linked to the 3-hydroxy groups, thus forming acoyl-acyl residues. The backbone structure can be substituted by additional sugars or amides which are linked to the disaccharide via phosphodiester bridges (see Fig. 2). These substitutions are often incomplete and may depend on conditions of cultivation (media, pH, temperature) (Wollenweber et al., 1982). They are not considered to make lipid A "unusual." Likewise, modifications concerning the chain-length of ester- and amide-linked fatty acids, although being of significance for the chemotaxonomy (Rietschel and Lüderitz, 1980) and being the reason for the considerable heterogeneity found with most lipid A's, are not "unusual" modifications. These sites of "allowed variability" are indicated in Figure 2 by symbols W,X,Y, and Z.

"UNUSUAL" LIPID A'S

Lipid A structures deviating from the "usual" type described above in different ways have been termed "unusual" lipid A's (Mayer et al., in press). This is not implying that "unusual" types are less widely distributed than the "usual" type; the number of investigated species is far too low for such statements.

The following structural variants have been observed so far:

Lipid A can be substituted by additional sugar constituents which are glycosidically linked to the backbone disaccharide at positions C-4 or C-4' (compare Figs. 3 and 5)

The amide-linked fatty acid is not of the usual (R)-3-hydroxy-type but of the rare 3-oxo-type (Fig. 4)

The backbone disaccharide is not phosphorylated (Figs. 5 and 6)

D-glucosamine is replaced by its 3-amino-derivative, 2,3-diamino-2,3-dideoxy-D-glucose, which is now acting as lipid A backbone (Fig. 6)

These variations can occur separately or in combinations. Since most of these variations are leading to a drastic change in endotoxic activities (see Table 1), "unusual" lipid A's are of considerable interest for questions of structure/biological function relationships.

Such lipid A-types have been described from purple bacteria and cyanobacteria (Weckesser et al., 1979), but they do occur also in *Brucella*, *Bacteroidaceae* and in *Haemophilus* and probably in many other families too.

Most structural work has so far been performed with lipid A of *Rhodospirillaceae* (sulfur-free purple bacteria), we will therefore restrict the following discussion to members of this family.

DIFFERENT LIPID A-TYPES IN *RHODOSPIRILLACEAE*

In a systematic study the chemical composition of LPS from a number of different species of the *Rhodospirillaceae* family was investigated. The aim of this study was to compare LPS from a phylogenetically old bacterial family with that of *Enterobacteriaceae*,

Table 1: Characteristic Constituents of Lipid A and the R Core of Rhodospirillaceae Species.

Species	GlcN β 1,6 GlcN	2,3-di-NH_2-Glc	4-NH_2-L-Ara	Neutral Sugars	Amide-Linked Fatty Acids: β-$C_{10}OH$	β-$C_{14}OH$	β-$C_{16}OH$	β-C_{14}-Oxo	KDO	L-glycero-D-manno-heptose	D-glycero-D-manno-heptose	% P in LPS	Serological Reactivity with Salmonella Lipid A	Lethal Toxicity (Mouse)
Rhodospirillum tenue I	+		+	Ara	OR				+	+		0.9-1.1	++	+/++
Rhodospirillum tenue II	+		+	Ara	OR				+	+	+	0.9-1.2	++	+/++
Rhodocyclus purpureus	ND		+	Ara	(OR)				+	+	+	1.0	ND	ND
Rhodopseudomonas gelatinosa	+				OR				+		+	2.0-4.5	+++	+++
Rhodopseudomonas sphaeroides	+					OR'		+	+			2.2-3.1	++	-
Rhodopseudomonas capsulata	(+)					(OR)		+	+			2.5	ND	-
Rhodopseudomonas blastica	ND							+	ND			ND	ND	
Rhodomicrobium vannielii	+			Man			OH		+			0.1	-	+
Rhodopseudomonas acidophila	ND			Man			OH		ND			ND	ND	
Rhodopseudomonas viridis		+				OH			+			0.1	-	-
Rhodopseudomonas sulfoviridis		+				OH			+			0.1	-	-
Rhodopseudomonas palustris		+				(OH)			+	+		(1.5)	-	-

Abbreviations: Glc, glucose; Man, mannose; Ara, D-arabinofuranose; KDO, 3-deoxy-manno-octulosonic acid; OR (OR'), OH-group of β-hydroxy fatty acid is substituted by a saturated (unsaturated) fatty acid; OH, OH-group of 3-OH fatty acid is unsubstituted; ( ), values are prelininary; -, no reactivity; +, ++, ..., extent of serological cross reactivity or lethal toxicity; ND, not determined. The lipid A's of the underlined species (Table) have been investigated in detail (Fig. 3-6).

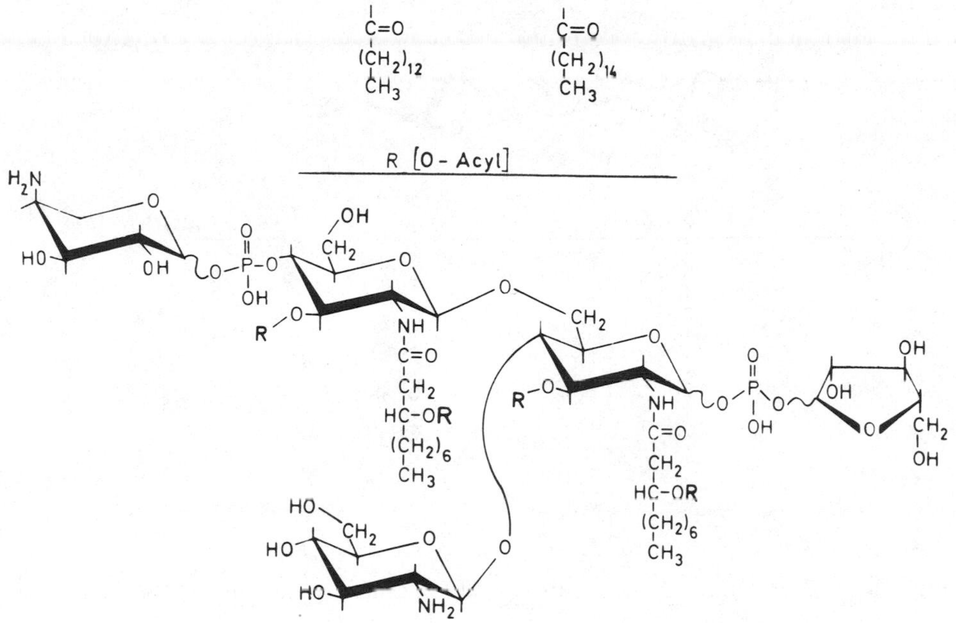

Fig. 3. Free lipid A of Rhodospirillum tenue (Tharanthan et al., 1977).

which as typical endosymbionts are under different evolutionary pressure.

The results of this study (Table 1) allowed a grouping of the species according to the nature of the lipid A backbone sugar and its amide-linked fatty acid constituents (Drews et al., 1978; Mayer et al., in press). Groups I and IV might be subdivided taking into consideration also the presence or absence of heptose(s) in the core region.

The detailed structures of lipid A of one representative of each group (underlined in Table 1) have just been reported (Tharanathan et al., 1978; Salimath et al., 1983; Holst et al., in press) and are depicted in Figures 3-6. Lipid A's of members of groups I-III share the β-1.6-glucosamine disaccharide as lipid A backbone sugar, whereas in group IV, 2,3-diamino-2,3-dideoxy-D-glucose is the backbone sugar (Roppel et al., 1975). Backbone sugars are phosphorylated in groups I and II, whereas phosphate is missing in lipid A of groups III and IV. Members of group II have the very rare 3-oxo-myristic acid (3-0-14:0) as amide-linked fatty acid (Strittmatter al., 1983). It can substitute both amino groups of the glucosamine disaccharide either totally or partially (Salimath et al., 1983) (Fig. 4).

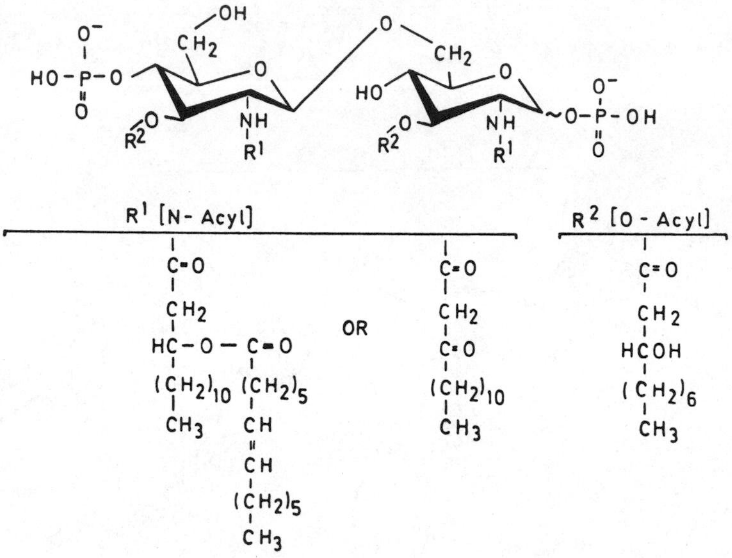

Fig. 4. Free lipid A of Rhodopseudomonas sphaeroides (Salimath et al., 1983).

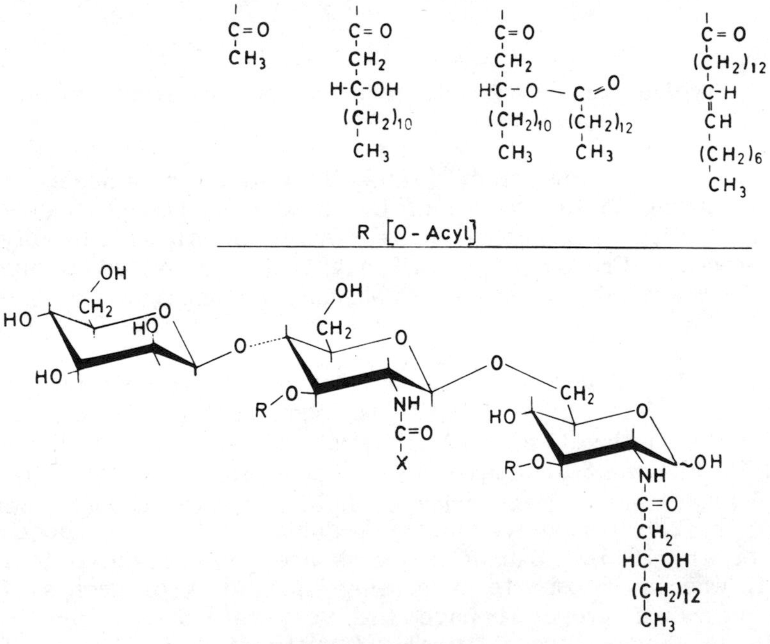

Fig. 5. Free lipid A of Rhodomicrobium vannielii (Holst et al., 1983).

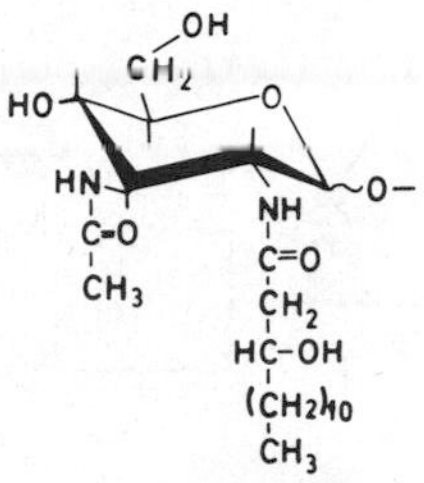

Fig. 6. Free lipid A of Rhodopseudomonas viridis (Roppel et al., 1975, Weckesser et al., 1979).

The structural similarities existing between Salmonella lipid A (Fig. 2) and that of Rhodospirillaceae of groups I and II (Fig. 3 and 4) is also documented by the mutual serological crossreaction of lipid A's with lipid A antisera (Galanos et al., 1977; Strittmatter et al., 1983).

From all species of Rhodospirillaceae only LPS of Rhodopseudomonas gelatinosa shows high lethal toxicity (in the mouse) and pyrogenicity (in the rabbit) (Galanos et al., 1977). Low toxicity (about 1% of that of Salmonella) was found for LPS of Rhodospirillum tenue (group I) and Rhodomicrobium vanniellii (group III) (Holst et al., in press). All other species have nontoxic LPS.

TAXONOMICAL AND PHYLOGENETICAL SIGNIFICANCE OF LIPID A IN THE RHODOSPIRILLACEAE FAMILY

Comparing the different lipid A-types (Table 1) with the 16S rRNA genealogy (Figs. 1 and 7), one recognizes that major parts of this scheme are supported by the lipid A analyses, although some deviations are also evident.

The close phylogenetical relatedness of Rhodospirillum tenue and Rhodocyclus purpureus has been published (Dickerson, 1980; Weckesser et al., 1983) and that of R. sphaeroides and Rhodopseudomonas capsulata is shown in Fig. 1. Furthermore, the common occurrence of the diaminoglucose in Rhodopseudomonas palustris, Rhodopseudomonas viridis and Rhodopseudomonas sulfoviridis supports the genealogical relatedness of these budding bacteria. The glucosamine- and mannose-containing lipid A (group III) of Rhodomicrobium vanniellii, however, does not fit into this cluster, although it is also phosphate-free, like that of group IV. One is tempted to predict a close phylogenetic relationship to exist between R. vanniellii and Rhodopseudomonas acidophila and between Rhodospeudomonas blastica and R. sphaeroides, but the 16S rRNA data are not yet available.

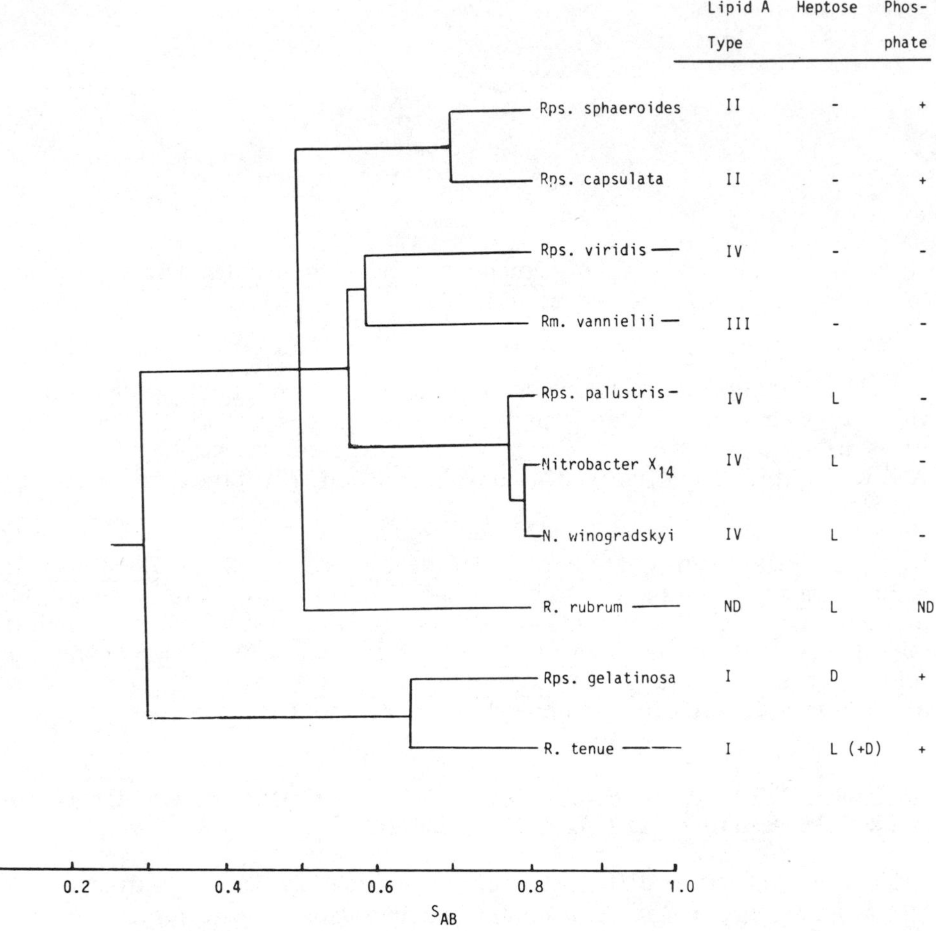

Fig. 7. Section of the dendrogram (Fig. 1) showing only the Rhodospirillaceae species and Nitrobacter. I, II, III, IV, Lipid A type; L, L-glycero-D-manno-heptose; D, D-glycero-D-manno-heptose; +, present; -, absent; ND, not determined (Seewaldt et al., 1982, Mayer et al., 1983a).

Seewaldt et al. (1982) have shown that R. palustris is phylogenetically closely related to a number of nonphotosynthetic bacteria of the Nitrobacter family, namely Nitrobacter winogradskyi and Nitrobacter X_{14}. It was of special interest that they do also possess the 2,3-diamino-2,3-dideoxyglucose-containing lipid A and a L-glycero-D-manno-heptose-containing core as found with R. palustris (Mayer et al., 1983). Furthermore, the same lipid A-type has been established for Pseudomonas diminuta and Pseudomonas vesicularis (Wilkinson and Taylor, 1978) and for a species from a not yet classified group of chloridazon-degrading bacteria (Weisshaar and

Lingens, in press). All these nonphotosynthetic bacteria show high S_{AB}-values with R. palustris (Stackebrandt, personal communication) and, therefore, a close phylogenetical relationship with each other and with R. palustris.

These examples show that lipid A and to some extent also the core constituents (see Fig. 7) have taxonomic relevance and are of importance for phylogenetical studies. It should, however, be pointed out that structurally related lipid A types may occur in phylogenetically distant species. An example is the highly toxic lipid A of Rhodopseudomonas gelatinosa which shows structural similarities to that of Salmonella lipid A.

The correlation existing between the 16S rRNA studies and the lipid A types may be used as a guide in the search for other "unusual" lipid A's. On the other side, the identification of a distinct lipid A-type may be of help for taxonomical classifications (e.g., the chloridazon-degrading species mentioned above). It is quite likely that additional lipid A structures will be found amongst other gram-negative families and prove that lipopolysaccharides are not only ideal markers for taxonomy but also for the phylogenetical relatedness of different species.

REFERENCES

Ambler, R.P., Daniel, M., Hermoso, J., Meyer, T.E., Bartsch, RG., and Kamen, M.D., 1979, Cytochrome c_2 sequence variation among the recognised species of purple nonsulphuric photosynthetic bacteria, Nature, 278:659.

Dickerson, R.E., 1980, Evolution and gene transfer in purple photosynthetic bacteria, Nature (London) 283:210.

Drews, G., Weckesser, J., and Mayer, H., 1978, Cells envelopes, in: "The Photosynthetic Bacteria," R.K. Clayton and W.R. Sistrom (eds.), p. 61, Plenum Publishing Corp., New York.

Fox, G.E., Stackebrandt, E., Hespell, R.B., Gibson, J., Maniloff, J., Dyer, T., Wolfe, R.S., Balch, W., Tanner, R., Magrum, L., Zablen, L.B., Blakemore, R., Gupta, R., Luehrsen, K.R., Bonen, L., Lewis, B.J., Chen, K.N., and Woese, C.R., 1980, The phylogeny of procaryotes, Science, 209:457.

Galanos, C., Roppel, J., Weckesser, J., Rietschel, E.Th., and Mayer, H., 1977, Biological activities of lipopolysaccharides and lipid A from Rhodospirillaceae, Infect. Immun., 16:407.

Gibson, J., Stackebrandt, E., Zablen, L.B., Gupta, R., and Woese, C.R., 1979, A genealogical analyses of the purple photosynthetic bacteria, Curr. Microbiol., 3:59.

Holst, O., Borowiak, D., Weckesser, J., and Mayer, H., in press, Structural studies on the phosphate-free lipid A of Rhodomicrobium vannielii ATCC 17100. Eur. J. Biochem.

Lüderitz, O., Westphal, O., Staub, A.M., and Nikaido, H., 1971,

Isolation and chemical and immunochemical characterization of bacterial lipopolysaccharides, in: "Microbial Toxins: A Comprehensive Treatise," G. Weinbaum, S. Kadis, and S.J. Ajl (eds.), Vol. IV, pp. 369, Academic Press, London and New York.

Lüderitz, O., Freudenberg, M.A., Galanos, C., Lehmann, V., Rietschel, E.Th., and Shaw, D.H., 1981, Lipopolysaccharides of gram-negative bacteria, in: "Microbial Membrane Lipids," S. Razin and S. Rottem (eds.), Current Topics in Membranes and Transport, Academic Press, New York, 17:79.

Mayer, H., and Schmidt, G., 1979, Chemistry and biology of the enterobacterial common antigen (ECA). Curr. Top. Microbiol. Immunol., 85:99.

Mayer, H., Bock, E., and Weckesser, J., 1983, 2,3-Diamino-2,3-dideoxy-glucose containing lipid A in Nitrobacter X_{14}. FEMS Microbiol. L., 17:93.

Mayer, H., Salimath, P.V., Holst, O., and Weckesser, J., in press, Unusual lipid A-types in phototrophic bacteria and related species. Rev. Infect. Dis.

Nikaido, H., 1970, Lipopolysaccharide in the taxonomy of Enterobacteriaceae, Int. J. Sys. Bacteriol., 20:383.

Ramia, S., Neter, E., Brenner, D.J., 1982, Production of enterobacterial common antigen as an aid to classification of newly identified species of the families Enterobacteriaceae and Vibrionaceae, Int. J. Sys. Bacteriol., 32:395.

Rietschel, E.Th., and Lüderitz, O., 1980, Struktur von Lipopolysaccharid und Taxonomie Gram-negativer Bakterien, Forum Mikrobiol., 3:12.

Rietschel, E.Th., Galanos, C., Lüderitz, O., Westphal, O., 1982, Chemical structure, physiological function and biological activity of lipopolysaccharides and their lipid A component, in: "Immunopharmacology and the Regulation of Leukocyte Function," D. Webb (ed.), p. 183, Marcel Dekker, New York.

Rietschel, E.Th., Mayer, H., Wollenweber, H.-W., Zähringer, U., Lüderitz, O., Brade, H., in press, Bacterial lipopolysaccharides and their lipid A component, in: "International Symposium on Chemical and Biomedical Aspects of Bacterial Endotoxins and Related Products," J. Homma and O. Westphal (eds.).

Roppel, J., Mayer, H., and Weckesser, J., 1975, Identification of a 2,3-diamino-2,3-dideoxyhexose in the lipid A component of lipopolysaccharides of Rhodopseudomonas viridis and Rhodopseudomonas palustris, Carbohydr. Res., 40:31.

Salimath, P.V., Weckesser, J., Strittmatter, W., and Mayer, H., 1983, Structural studies on a non-toxic lipid A from Rhodopseudomonas sphaeroides ATCC 17023, Eur. J. Biochem., 136:195.

Seewaldt, E., Schleifer, K.-H., Bock, E., and Stackebrandt, E., 1982, The close phylogenetical relationship of Nitrobacter and Rhodopseudomonas palustris, Arch. Microbiol., 131:287.

Stackebrandt, E., and Woese, C.R., 1981, The evolution of prokaryotes, in: "Molecular and Cellular Aspects of Microbial Evolution," M.J. Carlile, J.F. Collins, and B.E.B. Moseley (eds.), p. 1, Symp.

Soc. Gen. Microbiol., Cambridge University Press, Cambridge.

Strittmatter, W., Weckesser, J., Salimath, P.V., and Galanos, C., 1983, Nontoxic lipopolysaccharide from Rhodopseudomonas sphaeroides ATCC 17023, J. Bacteriol., 155:153.

Tharanathan, R.N., Weckesser, J., and Mayer, H., 1978, Structural studies on the D-arabinose-containing lipid A from Rhodospirillum tenue 2761, Eur. J. Biochem., 84:385.

Weckesser, J., Drews, G., and Mayer, H., 1979, Lipopolysaccharides of photosynthetic prokaryotes, Ann. Rev. Microbiol., 33:215.

Weckesser, J., Mayer, H., Metz, E., and Biebl, H., 1983, Lipopolysaccharide of Rhodocyclus purpureus: Taxonomic implication, Internat. J. Sys. Bacteriol., 33:53.

Weisshaar, R., and Lingens, F., in press, The lipopolysaccharides of a chloridazon-degrading bacterium, Eur. J. Biochem.

Westphal, O., Lüderitz, O., Rietschel, E.Th., and Galanos, C., 1981, Bacterial lipopolysaccharide and its lipid A component: some historical and some current aspects. Biochem. Rev., 9:191.

Wilkinson, S.G., 1977, Composition and structure of bacterial lipopolysaccharides, in: "Surface Carbohydrates of the Prokaryotic Cell," I.W. Sutherland (ed.), p. 97, Academic Press, New York.

Wilkinson, S.G., and Taylor, D.P., 1978, Occurence of 2,3-diamino-2,3-dideoxy-D-glucose in lipid A from lipopolysaccharides of Pseudomonas diminuta, J. Gen. Microbiol., 109:367.

Woese, C.R., Gibson, J., and Fox, G.E., 1980, Do genealogical patterns in purple photosynthetic bacteria reflect interspecific gene transfer? Nature, 283:212.

Wollenweber, H.-W., 1982, Primärstruktur von Lipoid A: Nachweis amidgebundener 3-Acyloxyacyl-Gruppen in Lipopolysacchariden von Salmonella und anderen gram-negativen Bakterien, Thesis, University of Freiburg i. Br., F.R.G.

CLONING AND EXPRESSION OF GENES ENCODING THE nupG NUCLEOSIDE TRANSPORT SYSTEM IN ESCHERICHIA COLI

Agnete Munch-Petersen and Nina Jensen

Enzyme Division
University Institute of Biological Chemistry
Sølvgade, DK-1307
Copenhagen K, Denmark

ABSTRACT

A DNA fragment, directing the synthesis of the nupG nucleoside transport system, has been cloned on a plasmid vector as evidenced by the complementation of chromosomal nupG mutations. The plasmid encoded transport system responded to the cytR and deoR control systems, which are known to act on the nupG transport system. Plasmid-containing minicells were used for specific in vitro synthesis of plasmid encoded proteins. Two proteins, with molecular mass of approximately 10,000 and 80,000, were detected, the occurrence and regulation of which suggested that they are components of the nupG nucleoside transport system.

INTRODUCTION

In 1945 Herman Kalckar demonstrated the phosphorolytic cleavage of nucleosides and the simultaneous formation of a hitherto unknown phosphoric ester, ribose 1-phosphate (Kalckar, 1945). This work was part of his important studies at that time on the function of phosphate in enzymatic synthesis. The readily reversible reaction was catalyzed by enzymes present in liver extracts and the enzymes were termed nucleoside phosphorylases. Based on the specificites of the enzymes, Herman Kalckar developed new and at that time highly sophisticated methods for spectrophotometric determination of minute amounts of different purine compounds in tissue extracts (Kalckar, 1947). These early studies have been the basis for later years extensive work on the characterization and purification of nucleoside catabolizing enzymes from eukaryotic as well as prokaryotic organisms.

Enteric bacteria contain such enzymes and the cells readily utilize exogenous nucleosides as carbon sources. This implies that the cells must also contain transport systems for the nucleosides, but little is known about the actual components of these transport systems.

Escherichia coli contains at least two different active nucleoside transport systems (Komatsu, 1973; Doskocil, 1974; Leung and Visser, 1977), designated the nupC and the nupG systems (Munch-Petersen et al., 1979). The two systems can be separated genetically (B. Mygind, personal communication). They differ in their specificity towards the nucleoside substrates: the nupG system facilitates the transport of all nucleosides, whereas the nupC system mediates transport of nucleosides other than guanosine and deoxyguanosine (Komatsu and Tanaka, 1972; Roy-Burman and Visser, 1972). The nupG system may be the more complex of the two systems; there are indications that overlapping as well as separate components are functioning in the transport of purine and pyrimidine nucleosides (Peterson and Koch, 1966; Peterson et al., 1967; Munch-Petersen and Pihl, 1980). This transport system is known to be regulated by the cytR and deoR genes (Munch-Petersen and Mygind, 1976), which have been previously characterized as the genes controlling the synthesis of the nucleoside

Table 1: Bacterial Strains

Strain No.	Sex	Relevant Genotype	Plasmid	Source or derivation
SØ 421	F^-	nupC,nupG	-	Our collection
MP507	F^-	nupC,nupG,cdd	-	" "
MP509	F^-	nup,nupG,cdd	pMP7	This work
MP585	F^-	nupC,nupG,metB	pMP5	" "
BM401	F^-	nupC,Δ(nupG,speC,glc)	-	B. Mygind, this lab
MP682	F^-	nupC,Δ(nupG,speC,glc)	pMP5	This work
MP683	F^-	nupC,Δ(nupG,speC,glc)	pMP5	" "
MP623	F^-	nupC,nupG,Δdeo,metB	pMP5	" "
MP608	F^-	nupC,nupG,Δdeo,cytR	pMP5	" "
MP628	F^-	nupC,nupG,Δdeo,deoR7	pMP5	" "
BD1854	F^-	minA,minB,his	-	B. Diderichsen, 1980
MP718	F^-	minA,minB,his	pMP5	This work
MP708	F^-	minA,minB,his	pKY2700	" "
MP629	F^-	Δdeo,zjj::Tn10[a]	-	Our collection
SØ 930	HfR	clmA,deoR7,Δdeo11	-	" "
SØ 818	Hfr	cytR,Δdeo11	-	" "

[a]Tn10 inserted in the vicinity of the deo operon.

catabolizing enzymes (Munch-Petersen et al., 1972). The nupC system seems to be regulated by the cytR system alone. In connection with the cytR control, nucleoside transport mediated by either system is subject to catabolite repression through the cyclic AMP-CRP system (Mygind and Munch-Petersen, 1975).

An outer membrane protein, the phage T_6 receptor, encoded by the tsx gene (Hantke, 1976; Krieger-Braun, 1980) has been shown to stimulate the transport of certain nucleosides at low ($< 10^{-6}$ M) concentrations, but at higher substrate concentrations this protein is not required for transport, and membrane vesicles prepared according to Kaback (1971) are able to transport nucleosides, especially pyrimidine nucleosides (Munch-Petersen et al., 1979). Apart from the tsx encoded protein, no single protein connected with nucleoside transport has been isolated or characterized. The present paper reports the cloning on a multicopy plasmid of genes involved in nucleoside transport and related to the nupG transport system. Strains carrying these recombinant plasmids show a highly increased transport of all nucleosides and the capacity for transport responds to the presence of cytR and deoR mutations in the cells. Cytidine and deoxyribose 5-phosphate are the low molecular effectors in these control systems, and analysis of protein synthesis in plasmid-containing minicells in the presence of cytidine or thymidine has served to identify plasmid coded proteins, the synthesis of which is regulated by these two effectors. These proteins are located in the cell envelope and are considered to be part of the nupG transport system.

RESULTS

Isolation of Strains Carrying Recombinant Plasmids with DNA Fragments Complementing Chromosomal nupG Mutations

As a cloning vector the plasmid pKY2592 (Ozaki et al., 1980) was used. It is a ColE1::Tn3 derivative, constructed for direct selection of hybrid clones. This plasmid and plasmid pMP7, isolated from the Clarke-Carbon collection (see Materials and Methods) were digested simultaneously with EcoRI restriction endonuclease. The mixture was ligated 16 hours at 14° C with T4-ligase and used to transform SØ 421, nupC, nupG. Transformants were selected on LB medium, containing ampicillin (25 μg/ml) and replicated onto ampicillin-containing plates with guanosine (1 mg/ml) as sole carbon source. colonies showing rapid growth on guanosine were picked, purified on the same medium and tested for resistance to the nucleoside analog fluorodeoxyuridine (FUdR) (5 μg/ml). Resistance, characteristic for strains unable to transport nucleosides, was lost in those transformants, which were able to grow on guanosine.

Table 2: Plasmids

Plasmid No.	Relevant Genotype	Derivation or Source
PMP7	nupG$^+$,speC$^+$	Clarke and Carbon collection
pMP5	nupG$^+$,speC$^+$,bla$^+$	This paper
pKY2592	ColE1::Tn3[a],ColE1-imm^{-b},bla$^+$	d
pKY2700	ColE1::Tn3[a],Δ tnpA[c],ColE1-imm$^-$, bla$^+$	d

[a]Tn3, transposon conferring ampicillin resistance
[b]ColE1-imm, immunity for colicin E1
[c]tnpA, Tn3 transposase
[d]pKY2592 and pKY2700 are derivatives of pKY2289 (Maeda et al., 1978). pKY2700 differs from pKY2592 in having a deletion of 2.5 KB in the Tn3 transposon.

The phenotype of the plasmid carrying strains indicated that genes encoding the nucleoside transport system were present on the plasmid. One transformant, MP 585, was chosen for further investigation. The location on the E. coli chromosome of nupG has previously been established at 64 min (B. Mygind, personal communication) and the presence of the nupG region in the transformants was verified by transferring the recombinant plasmid pMP5 into a recipient strain, BM 401, which carries a deletion covering the nupG gene and neighboring genes. These include speC and glc, encoding the enzymes ornithine decarboxylase and malate synthetase G, respectively. The two genes cotransduce with nupG (B. Mygind, unpublished data). Two of the transformants as well as the recipient strain were analyzed for the presence of ornithine decarboxylase and for growth on glycolate. (Malate synthetase G is necessary for growth on this compound as sole carbon source (Vanderwinkel and De Vlieghere, 1968). The results are presented in Table 3. Clearly the transformants carry the speC gene but not the glc gene.

Thus the accumulated evidence indicated that the nucleoside transporting ability of the transformant was specified by the presence of a cloned nupG region on plasmid pMP5.

Preliminary results from restriction endonuclease digestion of plasmid pMP5 have indicated that the plasmid contains a chromosomal DNA fragment of 8-9 kilobases. Attempts to subclone the nupG region have been unsuccessful so far.

Phenotypic Properties of Strains Carrying the Recombinant Plasmid pMP5

Transfer of plasmid pMP5, containing the nupG region, into transport negative strains, enabled the cells to grow on a nucleoside as sole carbon source. Simultaneously the cells became sensitive to the toxic nucleoside analog FUdR (5 μg/ml). Transport assays performed on cell suspensions also demonstrated a significant increase in the transport of different nucleosides (Table 4) and the transport was not inhibited by the uridine analog showdomycin (0.1 mM). These observations were in agreement with the presence of a nupG transport system in the cells (Munch-Petersen et al., 1979).

The ability of the plasmid carrying strains to utilize nucleosides as carbon sources did not follow the same pattern as the rate of uptake (Table 4). Thus, deoxycytidine was taken up rapidly by the cells, but this nucleoside was a poor carbon source, unless a cytR or deoR mutation was introduced (not shown). Also, adenosine was effectively transported (as measured in cell suspensions), but growth on agar plates containing adenosine as sole carbon source resulted in only microcolonies being formed (tested in three genetically different backgrounds). In the case of adenosine, introduction of cytR and deoR mutations did not improve growth, indicating in this case that neither the transport rate nor the levels of the nucleoside catabolizing enzymes were limiting factors. The reason for the poor growth on adenosine is not clear.

Introduction of the nupG carrying plasmid gave rise to an increase in the synthesis of nucleoside catabolizing enzymes. A three- to fourfold rise in enzyme levels was observed in the transport negative recipient strain following transformation (shown for cytidine deaminase and thymidine phosphorylase in Table 3). The

Table 3: Complementation of the Chromosomal speC Mutation and the Levels of Nucleoside Catabolizing Enzymes in Transformants Carrying the Recombinant Plasmid.

					Catabolizing Enzymes (units[b] per mg protein)	
Strain No.	Plasmid	Ornithine Decarboxylase Activity[a]	Growth on Glycolate	Growth on Guanosine 1 mg/ml	Cytidine Deaminase	Thymidine Phosphorylase
BM 401	-	88	-	-	105	70
MP 682	pMP5	905	-	++	340	290
MP 683	pMP5	696	-	++	243	380

[a]Assayed as described in Materials and Methods, employing a 3 hr reaction time. Activity is expressed as counts per min per 10^8 cells.
[b]One unit is defined as the amount of enzyme catalyzing the formation of 1 nmol of product per min.

Table 4: Transport and Utilization of Nucleosides by the Plasmid-Carrying Strain MP 585

Nucleoside Substrate	Rate of Transport pmol/10 sec./$4x10^8$ cells		Growth on Nucleosides (1 mg/ml)
Cytidine	213	(< 10)[a]	++
Deoxycytidine	180	(< 10)	(+)
Guanosine	206	(24)	++
Deoxyguanosine	195	(< 10)	nd[b]
Uridine	190	(< 10)	+
Thymidine	90	(< 10)	(+)
Adenosine	131	(18)	(-)
Inosine	165	(28)	++
Deoxyinosine	nd		++

[a]Figures in parentheses indicate rate of transport in the transport negative parent strain.
[b]not determined.

increased enzyme levels may be due to plasmid borne operator sites in the nupG genes. Such sites could be expected to bind an increased number of cytR and deoR repressor molecules thereby derepressing the synthesis of the cytR and deoR controlled enzymes.

Regulation of Nucleoside Transport in Plasmid Carrying Strains

The synthesis of the nucleoside catabolizing enzymes in E. coli is known to be controlled by the cytR and deoR regulatory systems (Munch-Petersen et al., 1972), and it was shown previously that the nupG transport system also responds to mutations in these two regulatory genes (Munch-Petersen and Mygind, 1976). Due to the intimate coupling between transport and catabolism of nucleosides, it was important to avoid interference by the elevated levels of catabolizing enzymes when transport rates were measured in plasmid carrying strains harboring regulatory mutations. Therefore, these measurements were carried out in strains in which a deletion in the deo operon had been introduced (Table 1) in order to reduce catabolism of the transported nucleosides.

In Table 5 the transport rates for deoxycytidine, cytidine, deoxyguanosine and guanosine in the regulatory mutants are given. Whereas a mutation in the cytR gene resulted in an increased transport rate for all four nucleosides, a deoR mutation caused a significant rise in guanosine and deoxyguanosine transport. Transport of pyrimidine nucleosides was almost unchanged in a deoR strain. Other purine and pyrimidine nucleosides follow the same regulatory pattern as guanosine and cytidine respectively (not shown). In glucose-grown cells of strain MP623, transport was reduced 50-70 percent, in agreement with the previously demonstrated catabolite repression of the nucleoside transport exerted through the cytR control system (Mygind and Munch-Petersen, 1975).

Cytidine is a known low molecular effector of the cytR control system, and, as expected, growth of the wild type strain, MP623, in minimal medium containing cytidine gave rise to increased transport (Table 5). However, when a cytR derivative of MP623, MP608, was grown in the presence of cytidine, a dramatic fall in transport capacity was observed (Table 5). This unexpected effect of cytidine required protein synthesis. If chloramphenicol (0.15 mg/ml) was added simultaneously with cytidine, the transport ability was unchanged after 90 min. of incubation (Table 5). Chloramphenicol alone had no effect on the transport. Growth on uridine abolished transport in a similar way. A control experiment with strain SØ 818 (cytR, Δdeo) also showed a decrease in transport of nucleosides after growth on cytidine, but to a lesser degree (not shown). The mechanism behind this effect of cytidine and uridine on nucleoside transport is not clear. It is possible that the presence of nucleosides during growth has a more general effect on cell metabolism. Transport of the amino acid proline, which like nucleoside transport is dependent on the proton motive force of the cells for energy source (Berger and Heppel, 1974), was reduced 15-20 percent after growth on nucleosides and this reduction was also counteracted by simultaneous addition of chloramphenicol (Table 5).

Table 5: Regulation of Nucleoside Transport in Strains Carrying Plasmid pMP5 and Deleted for the deo Operon.

Strain No.	Additional Genotype	Addition to Growth Medium	Rate of Transport pmol/10 sec./$4x10^8$ cells Cytidine	Deoxycytidine	Guanosine	Deoxyguanosine	Proline
MP 623			175	320	125	140	nd
MP 608	cytR		310	475	250	270	66
MP 628	deoR		200	310	225	210	nd
MP 623		cytidine[a]	nd[d]	360	172	nd	nd
MP 623		glucose[b]	nd	80	47	nd	nd
MP 608	cytR	cytidine[a]	nd	36	19	65	54
MP 608	cytR	cytidine[a] +chloramphenicol[c]	nd	404	208	316	70
MP 608	cytR	chloramphenicol[c]	nd	386	203	330	66

Cell were grown in glycerol-salts medium with additions as indicated. After harvest of the cells in the exponential phase, transport was measured as described in Material and Methods.

[a]1 mg/ml.

[b]glucose was substituted for glycerol.

[c]0.15 mg/ml.

[d]not determined.

The plasmid-carrying strains became highly unstable when regulatory mutations were introduced. Plasmid copies were retained, as judged by the continued resistance to ampicillin, and plasmid DNA could be isolated and used for transformation but no longer complemented a nupG mutation. The experiments conducted and described above were all carried out with freshly constructed strains.

Identification of Proteins Related to the nupG Transport System

The DNA fragment, cloned on pMP5, contains approximately 8000 base pairs and encodes several proteins besides those specifying nucleoside transport. The rationale for identifying proteins that might be involved in the nupG transport system, was based on the known regulation of this transport system by the cytR and deoR control systems. Plasmid encoded proteins responding to cytidine and thymidine (the low molecular effectors for the two control systems) by increased synthesis were analyzed for in a minicell protein synthesizing system as follows.

The minicell producing strain BD 1854 (Table 1) was transformed with plasmid pMP5, and freshly prepared minicells from the transformed strain were used for synthesis of proteins in the presence of [^{35}S]-labeled methionine (see Materials and Methods). Protein synthesis was carried out in minimal medium with added cytidine or thymidine. After one hour of incubation, aliquots of the incubation mixtures were fractionated into crude membranes and periplasmic proteins (see Materials and Methods) in order to localize the induced proteins into cellular compartments and thereby specifically detect those present in the membrane fraction.

The autoradiogram of the labeled proteins is shown in Figure 1. In the first three lanes (panel A) are shown the uninduced control sample (lane 1), the cytidine induced sample (lane 2) and the thymidine induced sample (lane 3). Lane 4 represents the vector plasmid without the cloned genes.

In lane 1 seven significant protein bands, which were not detected in lane 4 (the vector plasmid), are numbered. After autoradiography the radioactivity of the bands in lanes 1, 2, and 3 were determined (see Materials and Methods) and normalized in relation to the activity of the corresponding β-lactamase band. Using the normalized values from lane 1 (uninduced minicells) as basal values, the relative increase in synthesis in the presence of cytidine or thymidine was calculated for each of the protein bands.

Table 6 presents mean values from five different experiments. The effect of inducers in the minicell system was rather weak, but the relative intensities of the protein bands were quite reproducible. Interpretation of the results was complicated by the fact that the presence of thymidine repressed the synthesis of some of the plasmid

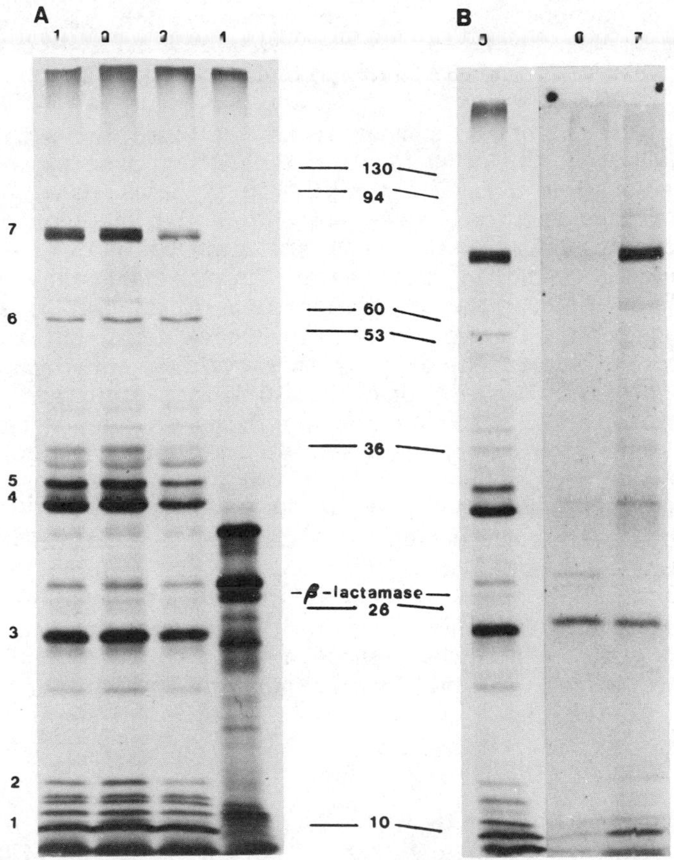

Fig. 1. Autoradiogram of plasmid coded proteins synthesized by minicells. The numbers between panel A and panel B are the molecular weights (x 10^{-3}) of marker proteins with the indicated electrophoretic mobilities. The numbers to the left of panel A indicate proteins, the radioactivity of which was determined as described in the text. **A:** lane 1, pMP5 (nupG$^+$, bla$^+$), uninduced; lane 2, pMP5, induced with cytidine; lane 3, pMP5, induced with thymidine; lane 4, pKY2700 (nupG$^-$, bla$^+$). **B:** lane 5, pMP5, total proteins; lane 6, pMP5, crude membrane fraction; lane 7, pMP5, osmotic shock fluid.

encoded proteins (e.g., proteins No. 4 and 5 in Table 6). However, it appears that a 10 kdal protein (band No. 1) was induced by cytidine and thymidine, while a 57 kdal protein (band No. 6) was induced by thymidine alone. A slight but reproducible induction by cytidine was seen for band No. 7. (This band is in fact a triple band, in which one of the proteins may be the ornithine decarboxylase, encoded by the speC gene, which was shown to be carried by pMP5 [Table 3]. The protein is reported to have a molecular mass of 80,000-82,000 [Applebaum et al., 1977].)

Panel B of Figure 1 compares the total protein sample (lane 5) with proteins from the crude membrane fraction (lane 6) and from the osmotic shock fluid i.e., the periplasmic proteins (lane 7). Several independent fractionations have shown that the 80 kdal proteins are among those appearing in the membrane fraction (lane 6) while the 10 kdal protein is found to some extent in the membrane and in the osmotic shock fluid. The 25 kdal protein (band No. 3) appeared in both the membrane fraction and in the shock fluid, but the synthesis of this protein was affected by neither cytidine nor thymidine (Table 6). Thus at least two proteins, the 10 kdal protein and one of the 80 kdal proteins, respond in their synthesis to the inducers of the deoR and cytR control systems and at the same time appear to be located in the cell envelope. These two proteins are assumed to be part of the nupG transport system. A 13 kdal protein (band No. 2) which responds to both inducers (Table 6) and is faintly detected in the membrane fraction (lane 7, Fig. 1), may be a precursor of the 10 kdal protein. The No. 6 protein is induced by thymidine alone and seems to be a cytoplasmic protein. However, since no thymidine responding genes other than those related to the nupG transport system are known to be located in the nupG region on the

Table 6: Response to the Presence of Inducers During Protein Synthesis in Minicells Carrying Plasmid pMP5

Protein Band[a]	Estimated molecular weight (kdal)	Relative Amount of Radioactivity in Protein Band[b]		
		No Inducer Added	Cytidine Added	Thymidine Added
No. 1	10	3.09	3.77 (1.22)[c]	4.16 (1.35)
No. 2	13	0.48	0.55 (1.14)	0.61 (1.27)
No. 3	25	1.91	1.89 (0.99)	1.85 (0.97)
No. 4	33	2.04	2.01 (0.99)	1.09 (0.53)
No. 5	34	0.96	0.95 (0.99)	0.61 (0.64)
No. 6	57	0.43	0.42 (0.99)	0.66 (1.53)
No. 7	80	1.28	1.48 (1.16)	0.64 (0.50)

[a]Numbers refer to the numbering in Figure 1.
[b]Counting and calculations were carried out as described in the text.
[c]Values in parenthesis indicate the relative increase in synthesis, as compared to uninduced cells.

chromosome, it is possible that this protein is also related to the transport system.

DISCUSSION

Cloning experiments employing the multicopy plasmid vector pKY2592 gave rise to a recombinant plasmid capable of complementing chromosomal nupG mutations. The presence of the nupG locus on the cloned DNA was further verified through the simultaneous complementation of a mutation in a neighboring chromosomal gene, speC, in the transformed recipient strains.

Plasmid pMP5 mediated a highly increased transport of all nucleosides. The cytR and deoR control systems are known to regulate transport as well as the initial catabolism of nucleosides. Accordingly transport in pMP5-carrying strains was further increased by the introduction of a cytR mutation. Introduction of a deoR mutation seemed to affect mainly the transport of purine nucleosides, suggesting that the transport of these nucleosides may involve a component which does not participate in the transport of pyrimidine nucleosides. An additional control of the transport system was indicated by the finding that certain nucleosides, when added to the growth medium, cause a protein synthesis dependent inhibition of the transport (Table 5). The mechanism involved in this type of regulation is unresolved, but the finding adds to the already known complexity of the cellular control of nucleoside utilization.

Protein synthesis in minicells carrying the cloned genes has allowed the identification of two proteins, the properties of which agree with a possible function in the transport of nucleosides: (1) They are synthesized in increased amounts in the presence of cytidine and/or thymidine. (2) They are located in the cell envelope. At present there is no direct proof of a transport function for these proteins, but the response of their synthesis to nucleosides and their cellular location make it highly probable that they are part of, or related to the nupG transport system.

Recently the lacZ gene has been fused to a gene in the nupG region of the E. coli chromosome, resulting in the formation of an altered, now membrane located β-galactosidase (Munch-Petersen, in preparation). The fusion has brought the synthesis of β-galactosidase under cytR and deoR control and has caused a simultaneous loss of nucleoside transport. Two-dimensional gel electrophoresis of the membrane proteins has demonstrated the absence of a 80 kdal protein in the strain containing the protein fusion, as compared to the wild type parent strain. This confirms the notion that a membrane-located protein of this size has an essential role in nucleoside transport.

MATERIALS AND METHODS

Bacterial Strains

The E. coli strains used are listed in Table 1. Unless otherwise specified, the strains were grown in minimal AB medium (Monod et al., 1951) with appropriate requirements and carbon sources as indicated. Growth was measured as an increase in absorbancy at 436 nm.

Materials

Nucleosides, showdomycin and other fine chemicals were purchased from Sigma Chemical Co. (St. Louis). [^{14}C]-labeled nucleosides and amino acids were from the Radiochemical Centre (Amersham, England); [^{35}S]-methionine was from New England Nuclear. Restriction endonucleases and T4 ligase were obtained from Boehringer Mannheim.

Enzyme Assays

The nucleoside catabolizing enzymes were assayed in extracts from glycerol-grown cells as described previously (Hammer-Jespersen et al., 1971). Ornithine decarboxylase activity was assayed according to Tabor et al., 1977, except that addition of toluene to the cell extracts was omitted. The labeled CO_2, formed by the decarboxylation of [^{14}C]-labeled ornithine, was trapped on filter paper as described by these authors, and the positions of radioactive spots were located by autoradiography. For quantitation the spots were cut out and the radioactivity determined by liquid scintillation counting.

Transport Assays

Transport assays were performed as described previously (Mygind and Munch-Petersen, 1975) using suspensions of glycerol-grown cells harvested in the exponential phase. Concentrations of the labeled nucleosides were 0.2-0.5 x 10^{-6} M, and specific activities ranged from 50-90 Ci per mol.

Genetic Techniques

Transductions and conjugations were carried out according to Miller (1972). Mutations in the cytR gene were introduced in the plasmid carrying strains by cotransduction with metB, using a P1 lysate of SØ 818. Introduction of deoR mutations was performed by mating with SØ 930 and selection for resistance to chloramphenicol (7.5 μg/ml). Deletions in the deo operon were introduced by transduction with a P1 lysate from strain MP 622, selecting for resistance to tetracycline (5 μg/ml). In all strains the levels of the nucleoside catabolizing enzymes were assessed by enzymatic assays.

Selection from the Clarke and Carbon Collection of Plasmids Carrying Nucleoside Transport Genes

The more than 2000 clones of the Clarke Carbon collection (Clarke and Carbon, 1976) were mated with a transport negative strain MP 507 nupC nupG cdd. Transconjugants were selected on plates containing guanosine as carbon source (1 mg/ml), methionine, streptomycin (200 μg/ml) and colicin E1, and then purified on the same selective medium. The presence of a plasmid in the recombinants was verified by lysis of the cells, followed by electrophoresis on agarose gels (Eckhardt, 1978). Furthermore, a considerable rise in the ability of the plasmid carrying cells to transport nucleosides could be demonstrated in transport assays using washed cell suspensions. In addition, the transconjugants were able to transfer the acquired phenotype into other transport negative recipients. One of the isolates, MP 509, was used for isolation of the plasmid DNA used in the cloning of nucleoside transport genes.

Preparation of Minicells

The minicell producing strains were grown for 15 hr in 100 ml cultures. To derepress the synthesis of transport proteins, the cells were grown in AB minimal medium with glycerol (0.2%) as carbon source. The minicell fractions were separated from normal cells by two consecutive sucrose gradients as described by Inselburg (1972). Before incubation for protein synthesis the minicells were suspended in AB medium and the cell density adjusted to an A_{436} of approximately 2.0.

Protein synthesis in minicells was carried out as follows: A minicell suspension, usually 1-1.5 ml, was preincubated at 37° C with glycerol (0.2%), biotin (5 μg/ml), thiamine (5 μg/ml), and amino acids (0.01 mM) other than methionine. After 20 min. of equilibration, [^{35}S]-methionine (25 μCi) was added and incubation was continued for 60 min.

Fraction of minicell-synthesized proteins was carried out according to Weiner et al. (1978). After incubation with [^{35}S]-methionine, the minicell suspension (1.5 ml) was divided into three parts. (1) Total Protein: 250 μl were centrifuged, washed with minimal medium and frozen at -70°. (2) Periplasmic Proteins: 750 μl were centrifuged and washed with 2 x 1 ml 10 mM Tris-HCl, 0.5 mM NaCl, pH 7.3 (Weiner and Heppel, 1971). Cells were suspended in 1.0 ml 30 mM Tris-HCl, 0.5 mM EDTA-20% sucrose, pH 7.3, and left 10 min. at room temperature before centrifugation and then rapidly resuspended in 1.0 ml precooled 0.5 mM $MgCl_2$. The suspension was kept on ice for 10 min. and then centrifuged. The supernatant was freeze-dried and stored at -70° C. (3) Crude Membrane Fraction:

500 μl were digested with lysozyme (10 mg/ml) in the presence of 10 mM EDTA for 60 min. at 4° C. A 750 μl volume of Tris-HCl (63.5 mM, pH 6.8) was added and the suspension sonicated, followed by centrifugation for 15 min. at 3000 g. The supernatant was then centrifugation for 12 hr at 110,000 g. The resulting membrane pellet was stored at -70° C.

Isolation of plasmid DNA was carried out according to the procedures of Clewell (1972) and Jørgensen et al. (1977).

Polyacrylamide gel electrophoresis was performed on sodium dodecyl sulfate polyacrylamide slab gels (12.5%) according to Laemmli (1970) and Ames (1974). Before electrophoresis, the minicell fractions were suspended in 100 μl sample buffer and boiled for 2 min. Samples (15-20 μl) were applied to the gel. The protein bands were located by autoradiography and by staining with Coomassie brilliant blue.

ACKNOWLEDGEMENTS

This work was supported by a grant from the Danish National Science Research Foundation (Grant No. 11-2526). One of us (A.M.-P.) was a fellow of the Japanese Society for Promotion of Science in 1980 and gratefully acknowledge the hospitality and useful advice from professor Kiyoshi Kurahashi, Institute for Protein Research, Osaka University, and Professor Yasuyuki Takagi, Department of Biochemistry, Kyushu University, where the initial part of the work was carried out.

REFERENCES

Ames, G., 1974, J. Biol. Chem., 249:634.

Applebaum, D.M., Dunlap, J.C., and Morris, D.R., 1977, Biochemistry, 16:1580.

Berger, E.A., and Heppel, L.A., 1974, J. Biol. Chem., 249:7747.

Bolivar, F., Rodriquez, R.L., Greene, P.J., Betlach, M.C., Heyneker, H.L., nad Boyer, H.W., 1977, Gene, 2:95.

Clarke, L., and Carbon, J., 1976, Cell, 9:91.

Clewell, D., 1972, J. Bacteriol., 110:667.

Diderichsen, B., 1980, Dissertation, University of Copenhagen.

Eckhardt, T., 1978, Plasmid, 1:584.

Hafner, E.W., Tabor, E.W., and Tabor, H., 1977, J. Bacteriol., 132:832.

Hammer-Jespersen, K., Munch-Petersen, A., Nygaard, P., and Schwartz, M., 1971, Eur. J. Biochem., 19:533.

Hammer-Jespersen, K., and Munch-Petersen, A., 1975, Molec. Gen. Genet., 137:327.

Hantke, K., 1976, FEBS Lett., 70:109.

Inselburg, J., 1970, J. Bacteriol., 102:642.
Jørgensen, P., collins, J., and Valentin-Hansen, P., 1977, Molec. Gen. Genet., 155:93.
Kaback, H.R., 1971, Methods of Enzymology, 22:99.
Kalckar, H.M., 1945, Fed. Proc., 4:248.
Kalckar, H.M., 1947, J. Biol. Chem., 167:429.
Komatsu, Y., and Tanaka, K., 1972, Biochim. Biophys. Acta, 288:390.
Krieger-Brauer, H.J., and Braun, V., 1980, Arch. Microbiol., 124:233.
Laemmli, v.K., 1970, Nature, 227:680.
Maeda, S., shimada, K., and Takagi, Y., 1978, Gene, 3:1.
Miller, J.H., 1972, Experiments in Molecular Genetics, Cold Spring Harbor Laboratory Press, New York.
Monod, J., Cohen-Bazire, G., and Cohn, M., 1951, Biochim. Biophys. Acta, 7:585.
Munch-petersen, A., Nygaard, P., Hammer-Jespersen, K., and Fiil, N., 1972, Eur. J. Biochem., 27:208.
Munch-Petersen, A., and Mygind, B., 1976, J. Cell. Physiol., 89:551.
Munch-Petersen, A., Mygind, B., Nicolaisen, A., and Pihl, N.J., 1979, J. Biol. Chem., 254:3730.
Munch-Petersen, A., nad Pihl, N.J., 1980, Proc. Natl. Acad. Sci. USA, 77:2519.
Myind, B., and Munch-Petersen, A., 1975, Eur. J. Biochem., 59:365.
Ozaki, L.S., Maeda, S., Shimada, K., and Takagi, Y., 1980, Gene, 8:301.
Tabor, H., Tabor, C.W., and Hafner, E.W., 1976, J. Bacteriol., 128:485.
Petersen, R.N., and Koch, A.L., 1966, Biochim. Biophys. Acta, 126:129.
Petersen, R.N., Boniface, J., and Koch, A.L., 1967, Biochim. Biophys. Acta, 135:771.
Vanderwinkel, E., and De Vlieghere, M., 1968, Eur. J. Biochem., 5:81.
Weiner, J.H., and Heppel, L., 1971, J. Biol. Chem., 246:6933.
Weiner, J.H., Lohmeier, E., and Schryvers, A., 1978, Can. J. Biochem., 56:611.

FROM HERMAN KALCKAR'S GALACTOSE RECAPTURE TO THE GALACTOSE CHEMORECEPTOR

Winfried Boos

Department of Biology
University of Konstanz
Federal Republic of Germany

I joined Herman Kalckar's group at the Massachusetts General Hospital in January, 1968, and left it in May, 1974, at the time when the famous Biochemical Research Laboratory was shut down for good.

During this time with Herman Kalckar I learned English, I learned Kalckarian, I published my first papers on my own responsibility, I made my first big scientific mistakes, but most of all, I had the privilege to learn an attitude towards science which nowadays has become rather rare, but nevertheless is still necessary. Kalckar's idea of science is not the production of endless data, for data are for him only painful necessity. Even more painful for him is the necessity of doing control experiments. Once, when I insisted that I carry out a certain control, he put his arm around me and said patiently "But Winfried, don't you understand, I am not interested in what is not." Kalckar's idea of science is the creation of free floating ideas and playing with them like a piece of music, where certain themes come and go, reappear in different variations, but in the end create a symphony. Of course, this type of free floating ideas is not done to find a solution, but is done for the joy attained by playing with ideas. Kalckar always regarded himself as a Dionysian rather than an Apollonian type of scientist.

The very few times when I made a real discovery, instead of just producing data I came to it by awakening from a loose daydream, Kalckarian style. Of course, Kalckar's scientific daydreaming has its serious drawbacks. It is not always as pleasurable for his listeners as it is for him. You have to understand Kalckarian in order to derive the conclusion. He never really tells you the conclusion, he always stopped short of it, probably because he feels it insults your intelligence to draw the conclusion for you, or to deprive you from the pleasure of drawing it yourself. And in fact, sometimes I have the suspicion that for him it is enough to have conceived the idea about the solution of a problem—to prove it is not necessary. It is not even unpleasant for him when an idea of his turns out to be wrong. The beauty is in the idea, not in its proof.

The sad part of the story is that he often has the solution to a problem in his mind, but he does not do much about it until somebody else proves it. It is even sader when he does something about it, but nobody will listen or understand. Since I was witness to one of these sad stories and since it was I who did not understand, or understood only too late, I feel obliged to tell you on this occasion that it was really Herman Kalckar who first perceived the idea that a transport related binding protein is a chemoreceptor in *E. coli*, even though the published record may suggest otherwise.

As you all know, Herman Kalckar is a galactose fan. In 1968 his graduate student, Henry Wu, was working on a particular galactose transport system in *E. coli*. There are several galactose transport systems in *E. coli*, but this one was special since it was operating at very low substrate concentrations (even below 1 µM). It was not essential for growth on galactose, but mutants lacking this system were unable to retain galactose (Fig. 1) (Wu et al., 1969). Therefore, Kalckar used to call it the recapture system. Galactose leaking out the cell would be recaptured and transported back into the cell. Recapture was an active process, not the locking up of doors.

I had joined Kalckar's group because of Henry Wu's work on this recapture system (Wu, 1967). During the time I was working on my Ph.D in Freiburg I had found that the reason why a certain sugar, glycerolgalactoside, was a very good inducer for the *lac* system, even in lactose transport negative strains, was that this galactoside was transported by Kalckar's recapture mechanism (Boos and Wallenfels, 1968). When I came to Kalckar's laboratory, I wanted to do some biochemistry and to isolate the protein that was responsible for the recapture mechanism. At that time, Anraku, a graduate student in Leon Heppel's laboratory, had isolated from the periplasm of *E. coli* cells a protein that could bind galactose with high affinity and he proposed that this protein was involved in galactose transport (Anraku, 1968).

My first achievement in Kalckar's laboratory was to demonstrate that this galactose-binding protein was in fact part of Kalckar's recapture mechanism. I demonstrated that the galactose-binding protein exhibited the same substrate specificity as the recapture system, notably it bound glycerolgalactose. I also could show that some mutants that were defective in recapture did not synthesize

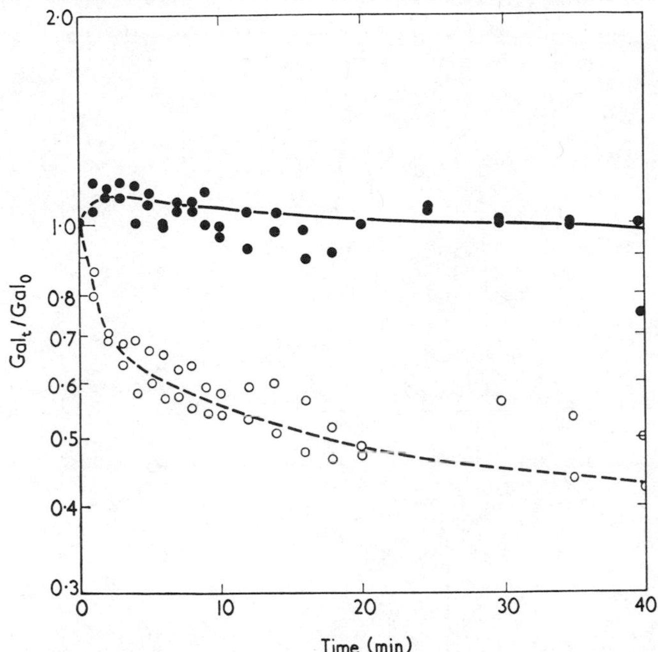

Fig. 1. Loss of [^{14}C]galactose from strain W3092i (recapture) as compared with strain W3092c (recapture$^+$). Bacterial cells grown in minimal A succinate were harvested at late logarithmic phase of growth. Cells had been loaded with [^{14}]galactose (4×10^{-4} M for strain W3092i and 8×10^{-5} M for strain W3092c) at 37° C for 20 min in minimal medium A containing 20 µg chloramphenicol/ml. After centrifugation the cells were resuspended in fresh medium and the dilute suspension (100 µg dry wt of cells/ml) was incubated at 37° C. At various intervals of time, 2-ml portions of cell suspension were filtered on Millipore filters and the radioactivity remaining inside the cells was measured. The ordinate (in logarithmic scale) represents the fraction of intracellular galactose retained at time t (in min) in relation to the initial galactose concentration at t = 0 (●—●, strain W3092c $P^+{}_{mg}K^-$; o—o, strain W3092i $P^-{}_{mg}K^-$) (from Wu et al, 1969).

active galactose-binding protein. As an easy binding test I developed an acrylamide gel technique that allowed the direct measurement of binding activity (Fig. 2) (Boos, 1969). Kalckar was fascinated by this technique, by the high affinity of the binding protein, and by the idea of this protein sitting in the periplasm waiting for, catching and recapturing galactose.

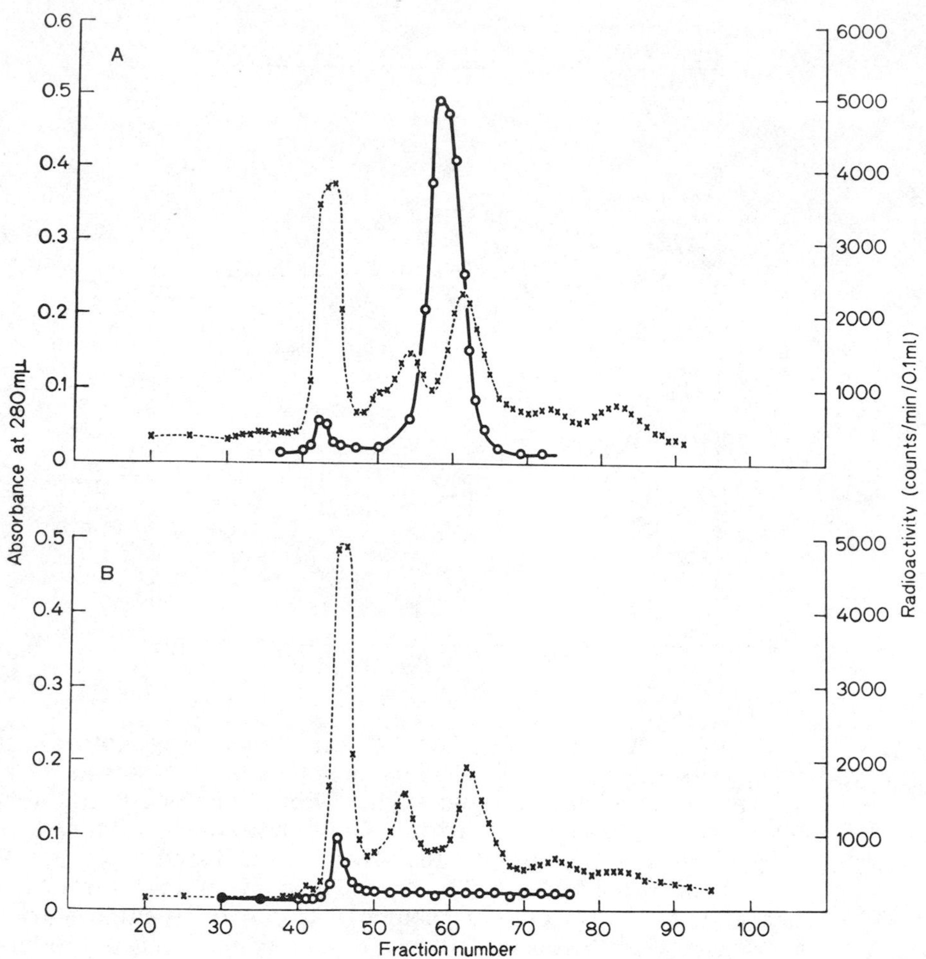

Fig. 2. Polyacrylamide electrophoresis of the shock protein preparation of W3092i and W4345 in the presence of labeled galactose. Both strains are recapture negative. The elution rate was 0.3 ml/min; 5.0 ml were fractionated. The gels contained 1.2 μM [^{3}H]galactose (2.5 x 10^6 counts/min). A: preparation from strain W3092i. B: preparation from strain W4345 (x—x, absorance at 280 nM; o—o, radioactivity of 0.1 ml aliquots) (From Boos, 1969).

Just shortly before that time, Julius Adler gave a talk at Harvard about chemotaxis in E. coli. Both Kalckar and I attended. Julius Adler showed how E. coli cells were attracted to amino acids and sugars, among them galactose. He proposed the existence of specific chemoreceptors on the cell envelope and suggested that it was the recognition of the sugar by the chemoreceptor that elicited the chemotactic response (Adler, 1969).

Kalckar was very excited after Adler's lecture and he talked about neurobiology, E. coli and galactose recapture (maybe Mozart was mixed in too). It was clear to him that galactose recapture and chemotaxis towards galactose were related processes. Kalckar was not bothered by the fact that Adler had looked at mutants defective in galactose transport or metabolism and that these mutants still exhibited galactose chemotaxis. At the end of 1968, Kalckar convinced Adler that galactose recapture was something more special than simple galactose transport, and Adler agreed to test two strains for galactose chemotaxis, one containing the recapture activity, the other not. Kalckar was so eager for the results that he even sent a telegram to ask for the outcome. They were negative, both strains still were fully active for galactose chemotaxis. I suppose it was these data that made Julius no longer think about galactose recapture.

However, Kalckar did not give up. As soon as I had demonstrated with my gel technique that there were two types of recapture negative strains, one containing the galactose-binding protein, the other not, the solution was very clear to him. The binding protein, and only the binding protein, must be the chemoreceptor (Kalckar, 1971). Thus, only part of the recapture machinery constituted the chemoreceptor, the sensory input. Kalckar kept calling Julius these days, and I had to send him my mutants. But the combined effect of Julius' poor understanding of Kalckarian and his apparent distrust in galactose recapture let him be a polite man who listened but did nothing for a long time. In his desperation, Kalckar asked me to do chemotaxis assays. At that time I was heavily involved in biochemistry of binding protein, in particular its conformational change when in contact with substrates (Boos et al., 1972) and I did not care much about chemotaxis. Thus I too was polite, I listened and did nothing about his nagging thoughts. Finally, Kalckar gave up and the weekly phone calls to Madison ceased.

It was not until 1971 that Hazelbauer and Adler finally arrived on their own at the same conclusion as Kalckar had done two years earlier. In their famous paper published in Nature (Hazelbauer and Adler, 1971), they demonstrated that the galactose-binding protein functions as the chemoreceptor in galactose chemotaxis. In their acknowledgement they thank Herman Kalckar for having led them in the right direction by sending a mutant lacking the galactose-binding protein and suggesting to test it for galactose chemotaxis. In retrospect I always feel somewhat sad. The paper of Hazelbauer and Adler was the starting point for the very exciting story of bacterial chemotaxis. Several groups joined in: It was discovered that bacterial flagella rotate like propellers, and that chemotaxis was mediated by regulating the frequency of changes in the direction of rotation (Berg, 1974). The basic mechanism of signal adaptation could be traced to a methylation-demethylation reaction of the signal transducers, the intimate and membrane-bound partners of the chemoreceptor (Springer et al., 1979). The powerful methods of genetics and recombinant DNA technique contributed as well. By now, all chemotaxis genes are probably known (Parkinson and Houts, 1982) and their products have been seen on gels (Silverman and Simon, 1977). They have been cloned and some have been sequenced (Boyd et al., 1983). Finally, the galactose chemoreceptor has experienced noble treatment. It has been sequenced (Mahoney et al., 1981) and its crystal structure has been determined (Quiocho and Pflugrath, 1980).

I feel sad, because with all this excitement Herman Kalckar's name is no longer heard, yet I believe it was his dreaming that initiated it all.

REFERENCES

Adler, J., 1969, Chemoreceptor in bacteria, Science, 166:1588.

Anraku, Y., 1968, Transport of sugars and amino acids in bacteria. III. Studies on the resaturation of active transport, J. Biol. Chem., 243:3128.

Berg, H.C., 1974, Dynamic properties of bacterial flagellar motors, Nature, 249:77.

Boos, W., 1969, The galactose binding protein and its relationship to the β-methylagalactoside permease from Escherichia coli, Eur. J. Biochem., 10:66.

Boos, W., and Wallenfels, K., 1968, Untersuchungen zur Induktion der Lac Enzyme; 2. Die Permeation von Galaktosylglyzerin in Escherichia coli, Eur. J. Biochem., 3:360.

Boos, W., Gordon, A.S., Hall, R.E., and Price, H.D., 1972, Transport properties of the galactose-binding protein of Escherichia coli: substrate induced conformational change, J. Biol. Chem., 247:917.

Boyd, A., Kendall, K., and Simon, M.I., 1983, Structure of the serine chemoreceptor in Escherichia coli, Nature, 301:623.

Hazelbauer, G.L., and Adler, J., 1971, Role of the galactose-binding protein in chemotaxis toward galactose, Nature New Biol., 230:101.

Kalckar, H.M., 1971, The periplasmic galactose-binding protein of E. coli, Science, 174:565.

Mahoney, W.C., Hogg, R.W., and Hermodson, M.A., 1981, The amino acid sequence of the D-galactose-binding protein from Escherichia coli B/r, J. Biol. Chem., 256:4350.

Parkinson, J.S., and Houts, S.E., 1982, Isolation and behavior of Escherichia coli deletion mutants lacking chemotaxis functions, J. Bacteriol., 151:106.

Quiocho, F., and Pflugrath, J.W., 1980, The structure of D-galactose-binding protein at 4.1 A resolution looks like L-arabinose-binding protein, J. Biol. Chem., 255:6559.

Silverman, M., and Simon, M., 1977, Identification of polypeptides necessary for chemotaxis in Escherichia coli, J. Bacteriol., 130:1317.

Springer, M.S., Goy, M.F., and Adler, J., 1979, Protein methylation in behavioral control mechanisms and in signal transduction, Nature, 280:279.

Wu, H.C.P., 1967, Role of the galactose transport system in the establishment of endogenous induction of the galactose operon in Escherichia coli, J. Mol. Biol., 24:213.

Wu, H.C.P., Boos, W., and Kalckar, H.M., 1969, Role of the galactose transport system in the retention of intracellular galactose in Escherichia coli, J. Mol. Biol., 41:109.

GLYCEROL UTILIZATION BY FACILITATED DIFFUSION COUPLED TO PHOSPHORYLATION IN BACTERIA

E.C.C. Lin

Harvard Medical School
Department of Microbiology and Molecular Genetics
25 Shattuck Street
Boston, Massachusetts 02115

ABSTRACT

Escherichia coli and probably most other bacteria entrap glycerol by the tandem action of a cytoplasmic membrane protein which catalyzes facilitated diffusion and an ATP-dependent kinase subject to feedback inhibition. The facilitator protein behaves as though it provides an aqueous channel with an effective pore diameter of about 0.4 nm. The kinase is also not highly specific, since, in addition to glycerol, the enzyme can phosphorylate dihydroxyacetone and L-glyceraldehye. Under physiological conditions, however, the two proteins appear to function as a complex, and together they impose a more stringent substrate specificity.

Whereas wild-type *E. coli* can grow at a maximal rate on glycerol at concentrations well below 0.05 mM, mutants lacking the facilitator require at least 5 mM of the compound to achieve full growth rate. Utilization of glycerol as sole carbon and energy source by wild-type cells is rate-limited by the action of fructose-1,6-bisphosphate as a noncompetitive inhibitor. Utilization of glycerol in the presence of glucose is prevented at least in part by increased concentration of dephosphorylated factor III^{Glc} of the phosphoenolpyruvate phosphotransferase system. This effect seems to be exerted either on the kinase alone or on both the kinase and the facilitator.

Glycerol is bactericidal to mutants synthesizing high levels of a kinase which is insensitive to feedback control by fructose-1,6-bisphosphate. The cells are killed by the copious production of methylglyoxal from the elevated pool of dihydroxyacetone phosphate.

In contrast, glycerol is bacteriostatic to mutants blocked in the dehydrogenation of sn-glycerol 3-phosphate.

It is suggested that the evolution of a concentrative mechanism for the uptake of glycerol in bacterial and other kinds of cells is prevented by the high intrinsic permeability of biological membranes to the compound, a property which would make active transport a Sisyphean process.

INTRODUCTION

The phosphorylation of glycerol played a part in the discovery of oxidative phosphorylation by Kalckar (1937) working with extracts of cat kidney cortex. Esterification of glycerol was proportional to the oxygen consumption stimulated by dicarboxylic acids (Kalckar, 1938). The ester was identified (Kalckar, 1939) as the levorotatory α-glycerophosphate or sn-glycerol 3-phosphate (G3P) according to the nomenclature of Hirschmann (1960), whose structure had previously been determined (Karrer and Benz, 1926). Glycerol kinase, the enzyme responsible for catalyzing the esterification, was later purified from the rat liver and was shown to utilize ATP as the phosphoryl donor (Bublitz and Kennedy, 1954). Since then, this enzyme activity was found in a wide variety of sources (Thorner and Paulus, 1973b).

The phosphorylation of glycerol at the expense of ATP is employed by cells ranging from hepatocytes to bacteria as a way of trapping the carbohydrate from the surrounding medium (Li and Lin, 1975; Lin, 1976, 1977; Li and Lin, 1983). What is unusual about the utilization of this carbohydrate in both eukaryotic and prokaryotic cells is the mode of its transfer across the cytoplasmic membrane: instead of being pumped into the cell against a concentration gradient, the substrate diffuses across the cytoplasmic membrane with the help of an energy-independent carrier.

PERMEATION OF GLYCEROL

Simple and Facilitated Diffusion

Phospholipid bilayers are intrinsically permeable to small non-electrolytes, such as glycerol, by a process of simple diffusion. The rate-limiting step seems to be the penetration of the paraffin core by glycerol without the company of water. The barrier increases with the length of the fatty acid side chains and the degree of their saturation, and the rate of penetration is highly temperature-dependent (de Gier et al., 1968, 1971). The term "facilitated diffusion" was coined to describe a carrier mediated process in which net flow of the solute is down a concentration gradient, as in the case of simple diffusion (Danielli, 1954). Knowledge of this facilitated process came from observations that the red cells of many birds, rodents, and primates are much more permeable to glycerol (two orders of magnitude) than expected on the basis of simple diffusion (Jacobs, 1931, 1950). The existence of special carriers for the substrate was evidenced by noncompetitive inhibition of permeability to glycerol by traces of copper ions (Jacobs and Corson, 1934; Jacobs and Stewart, 1946) and competitive inhibition by ethylene glycol (Jacobs, 1954). In contrast to simple diffusion, facilitated diffusion is not highly temperature-dependent (de Gier et al., 1966).

High intrinsic permeability of bacterial cell membranes to glycerol was initially inferred by the poor ability of this compound to sustain osmotic pressure across the cell membrane in experiments on plasmolysis caused by hypertonic solutions (Fischer, 1903; Avi-Dor et al., 1956; Mager et al., 1956; Mitchell and Moyle, 1956; Bovel et al., 1963) and on hypotonic osmotic shock (Britten and McClure, 1962). Later it was shown by an optical method based on transient plasmolysis caused by a hypertonic solution of glycerol (increase in the turbidity of a cell suspension) that there are two ways by which this compound can enter the cell: a nonspecific route and a specific one. The latter can be induced in wild-type Escherichia coli grown on glycerol, and mutants in which the dissimilation system for glycerol is derepressed express this special permeation pathway constitutively (Sanno et al., 1968). More rapid measurements by stopped-flow spectrophotometry showed that whereas in noninduced cells the glycerol permeability is less than a tenth that of water, in induced cells the glycerol permeability approaches that of water. Furthermore, the rate of nonspecific diffusion decreases as the membrane lipids are converted to the gel state by decreasing temperature, in contrast to the rate of specific diffusion which is relatively insensitive (Alemohammad and Knowles, 1974; Eze and McElhaney, 1981). These data together with the inability of mutants lacking glycerol kinase to accumulate the substrate against a concentration gradient (see below) indicate that, like the erythrocytes of certain animal species, cells of E. coli can also synthesize a membrane carrier capable of mediating facilitated diffusion.

Physiological Characterization of Glycerol Facilitator

The lack of a concentrative uptake mechanism in E. coli was anticipated by genetic studies (Hayashi and Lin, 1965). More than 40 independent mutants of E. coli lacking glycerol kinase, but induced in the glycerol system, failed to accumulate radioactive material when incubated with 10 μM ^{14}C-labeled substrate at 30° C. In contrast, when a mutant blocked in aerobic G3P dehydrogenase was used instead, rapid linear accumulation of radioactivity occurred. Essentially all of the labeled material accumulated within 5 minutes was recovered as G3P, with no detectable free glycerol (less than 1

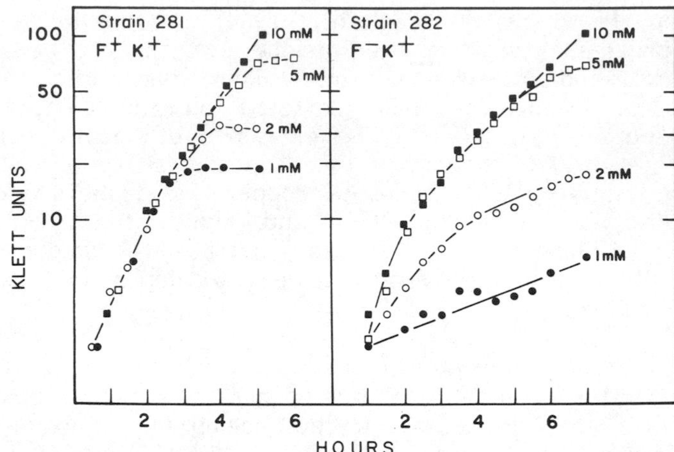

Fig. 1. Growth of E. coli strains with the facilitator (F^+K^+) or without the facilitator F^-K^+) on glycerol as the sole source of carbon and energy at different concentrations. Cells growing exponentially (to about 100 Klett units) in 10 mM glycerol were collected by centrifugation, resuspended in minimal medium, and diluted into minimal medium containing glycerol at the indicated concentrations. Growth curves were obtained from 10 ml of culture (incubated in a 300-ml flask fitted with a side arm and shaken at 37° C) by measurement of turbidity in a Klett colorimeter with a #42 filter (From Richey and Lin, 1972).

percent of the total radioactivity).

When the growth rate of wild-type cells was measured as a function of the concentration of glycerol as the sole source of carbon and energy, half-maximal growth rate was observed at about 1 μM, which was close to the apparent K_m of the kinase measured in the cell extract. Thus, even a nonconcentrative mechanism seems to be able to supply the cell with the carbon and energy source at a nonlimiting rate (Hayashi and Lin, 1965), a result consistent with the calculated capacity of the facilitator (see below).

The phenotypic difference between cells with or without the glycerol facilitator protein was revealed by the discovery that an isolate of Shigella flexneri (strain M4243), which lacks the facilitator but possesses a glycerol kinase with a substrate K_m close to that of the E. coli enzyme, requires 5-10 mM glycerol in the medium to grow rapidly on the compound. To ascertain that this growth trait reflected the lack of the facilitator, rather than some unknown factors, phage P1 was raised on this organism and used to transduce an E. coli mutant lacking both glycerol facilitator and glycerol kinase (encoded by neighboring genes). Transductants were selected on agar containing 20 mM glycerol. When an isolate which acquired the kinase was tested in liquid culture, full growth rate was observed only when the substrate concentration exceeded 5 mM (Fig. 1), a feature characteristic of the S. flexneri donor. Vice versa, when a strain of S. flexneri (strain 24570) lacking both the facilitator protein and glycerol kinase was transduced with phage P1 raised on wild-type E. coli, growth on glycerol occurred at maximal rate at 1 mM or below (Richey and Lin, 1972). These experiments show that despite the high intrinsic permeability of the cell membrane to glycerol, the rate of diffusion is inadequate for effective exploitation of the substrate.

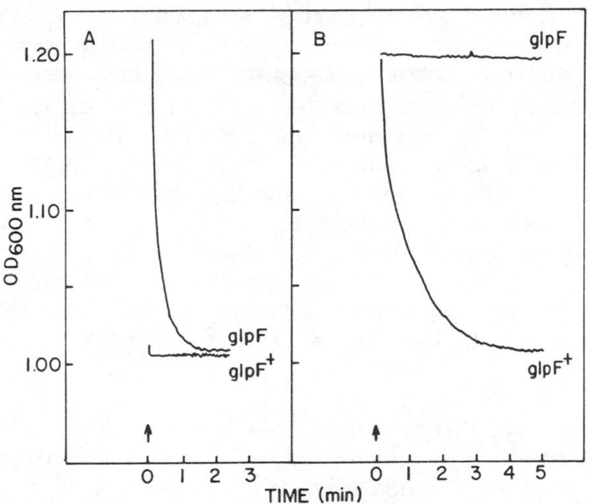

Fig. 2. Optical changes associated with shrinkage and swelling of E. coli. Suspended wild-tupe (glpF$^+$) or mutant cells lacking glycerol facilitator (glpF) grown in a medium with G3P as inducer were exposed to 250 mM solute (hypertonic). The rapid plasmolysis due to osmotic water movement out of the cytoplasm resulted in an increase in optical density (too rapid to be recorded in the spectrophotometer used). Entry of the solute (and reentry of water) led to reswelling of the original cytoplasmic volume which resulted in a decrease in optical density. Since solute entry is rate limiting in reswelling, the decrease in optical density was taken as a measure of substrate permeability. (A) Glycerol was used as the substrate. (B) Xylitol was used as the substrate. A reading of 1.2 O.D. indicates full shrinkage of the cell. In induced glpF$^+$ cells the half time of equilibration for glycerol was too short to be recorded; for xylitol the half time was 0.7 min (Heller et al, 1980).

Glycerol Facilitator Protein as a Channel

By comparing E. coli strains with or without glycerol facilitator, using the optical osmotic assay shown in Figure 2, it was found that the membrane protein accelerated the entry of not only glycerol, but also a number of small molecules including DL-glyceraldehyde, glycine, urea, erythritol, D-arabitol, L-arabitol, ribitol, xylitol, D-galactitol, D-mannitol, and D-sorbitol (thiourea and imidazole entered too rapidly to be measured even in noninduced cells). Not detectably transported by the facilitator were sn-glycerol 1-phosphate, G3P, glycylglycine, inositol, erythrose, D-arabinose, L-arabinose, D-ribose, D-xylose, D-galactose, D-mannose, and D-sorbose. Since cells lacking the facilitator were practically impermeable to xylitol, and those with the protein gave a convenient half time of equilibration of 0.7 min at 25° C, [^{14}C]xylitol was used as a substrate analog for exploring some of the properties of this transport mechanism. First, it was confirmed that uphill transport of the permeant did not occur. Second, the facilitator catalyzed exit of the compound as well as its entry. Third, the entry of [^{14}C]xylitol at 5 mM was not significantly inhibited by unlabeled glycerol up to 500 mM; and the entry of [^{14}C]glycerol at 20 μM was not strongly inhibited by unlabeled xylitol up to 20 mM (maximal inhibition observed was 40 percent). Fourth, the uptake rate of xylitol was only slightly reduced by lowering the temperature from 25 to 2° C (Fig. 3). In contrast, the basal entry rate of the compound into cells lacking the facilitator was depressed more than 10-fold (to a nondetectable level) by the same temperature change (Heller et al., 1980).

Substrate permeability depending more on molecular size than on chemical structure, nonsaturability by the substrate and lack of competitive inhibition (in contrast to the carrier of red blood cells), and insensitivity of substrate entry rate to temperature, taken together, suggest that the bacterial facilitator protein provides an aqueous pore (estimated diameter of 0.4 nm) for substrate diffusion. Although nonspecific channels are well known in the outer membrane of gram-negative bacteria (Boehler-Kohler et al., 1979; Luckey and Nikaido, 1980), the glycerol facilitator is the first example of a channel across the cytoplasmic membrane. On the assumption of equal synthesis of the facilitator and the kinase (the two proteins are encoded by the same operon), the turnover number of the facilitator with glycerol as the substrate was extrapolated to be about 200 molecules per second at a concentration gradient of 0.5 mM. (The turnover number of β-galactoside permease was estimated to be 200 per sec at 25° C by the same kind of calculation.) At that flow rate, about 10^6 molecules per second should enter a cell, or about 0.5 μmol of glycerol will be admitted per min per mg of cell protein. A glycerol consumption rate of about 0.4 μmol per min per mg of protein is expected to support growth with a 90-minute doubling

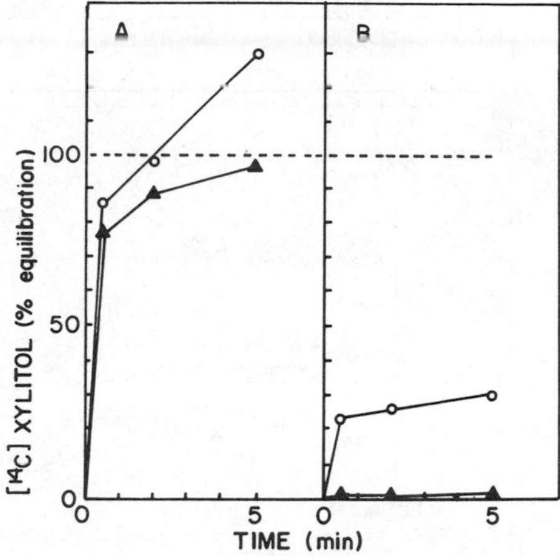

Fig. 3. Temperature dependence of xylitol entry. Uptake of 100 μM [^{14}C]xylitol was measured in cells possessing (A) and lacking (B) the facilitator at 25° C (o) and 2° C (Δ). Equilibration of 100% indicates equal concentrations in the intracellular and extracellular compartments (Heller et al, 1980).

time. The transport rate extrapolated from the osmotic measurement therefore falls within physiological range (Heller et al., 1980).

THE ATP-DEPENDENT GLYCEROL KINASE

Although effective scavenging of glycerol requires the facilitator, the trapping process depends on glycerol kinase which utilizes ATP as the driving force. The E. coli enzyme was purified to homogeneity (crystallized) and extensively characterized. It catalyzes also the phosphorylation of dihydroxyacetone (DHA) and L-glyceraldehyde (Hayashi and Lin, 1967). The enzyme activity is noncompetitively inhibited by the remote product fructose-1,6-bisphosphate (FDP) and mutants were obtained that produce enzymes which are insensitive to this inhibition (Zwaig and Lin, 1966; Berman and Lin, 1971). The apparent K_m for glycerol is 10 μM and the K_i for FDP is about 0.5 mM. The enzyme consists of four identical subunits of about 55,000 to 57,000 daltons. Apparently each subunit has a binding site for glycerol and an allosteric site for FDP (Thorner and Paulus, 1971; Thorner and Paulus, 1973a; de Riel and Paulus, 1978a, 1978b, 1978c).

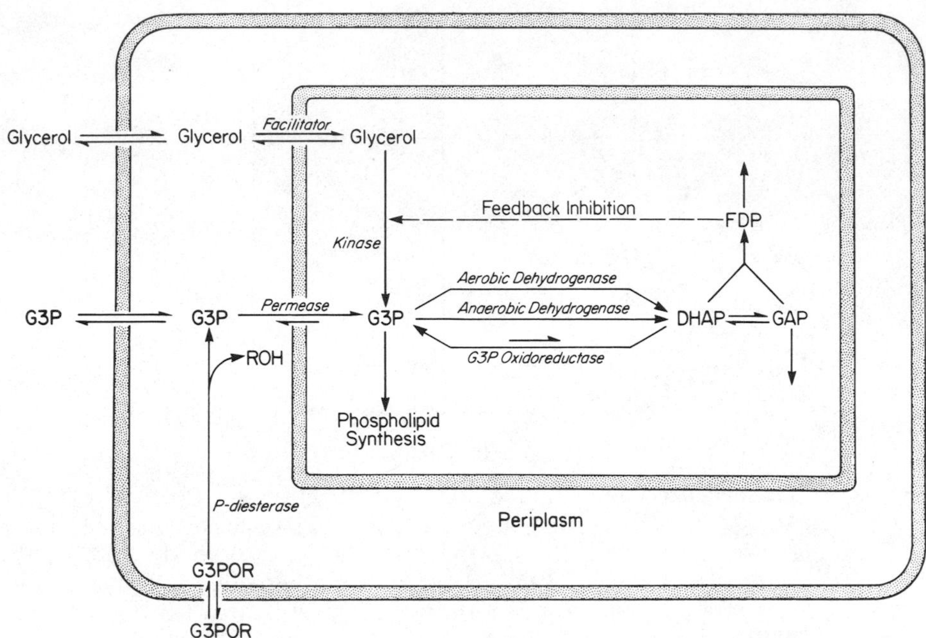

Fig. 4. Scheme for the utilization of glycerol, G3P, and glycerophosphodiesters by E. coli. The outer membrane allows free passage of the small substrates through porin channels. Abbreviations: G3P, sn-glycerol 3-phosphate; G3POR, glycerophosphodiester; DHAP, dihydroxyacetone phosphate; GAP, D-glyceraldehyde 3-phosphate; and FDP, fructose-1,6-bisphosphate.

THE GENETIC SYSTEM

In E. coli the gene encoding the glycerol facilitator (glpF) and that encoding the kinase (glpK) belong to the same operon (Cozzarelli and Lin, 1966; Berman-Kurtz and Lin, 1971). The kinase gene is promoter proximal (Conrad et al., 1984). It is interesting that of the two independent isolates of S. flexneri studied, one lacks only glpF and the other lacks both glpF and glpK. Possibly the ancestral S. flexneri possessed a complete glpKF operon and that regressive evolution resulting in genetic loss had taken place (Richey and Lin, 1972). A similar hypothesis was proposed for the lactose system of Shigella dysenteriae which lost the lacY gene encoding β-galactoside permease and apparently accumulated missense mutations in the lacZ gene encoding a partially active β-galactosidase. However, the function of the lacI gene, encoding the repressor, remains intact, presumably because loss of the repressor function would result in gratuitous protein synthesis (Luria, 1965).

The operon glpKF in E. coli is a member of the glp regulon (Cozzarelli et al., 1968) which specifies a set of reactions shown in Figure 4. It may be seen that in addition to the direct utilization of glycerol via the facilitator and the kinase, the cell can also exploit external G3P (Hayashi et al., 1964) and phosphodiesters of glycerol which are released as G3P by a periplasmic hydrolase (Boos et al., 1977; Larson et al., 1983). G3P is transported against concentration gradient into the cytoplasm by an oligomeric integral membrane protein (Larson et al., 1982; Ludtke et al., 1982). Inside the cell G3P is converted to DHAP by either of two inner membrane associated flavoprotein dehydrogenases: the aerobic enzyme using molecular oxygen as terminal hydrogen acceptor, or the anaerobic enzyme using nitrate or fumarate as the acceptor (Lin, 1976).

So far eight proteins encoded by the glp system have been identified in E. coli. They are encoded by five operons in three widely separated clusters as shown in Figure 5. Located in clockwise order are the glpTQ operon encoding respectively the G3P transport protein and the periplasmic glycerophosphodiesterase (W. Boos, personal

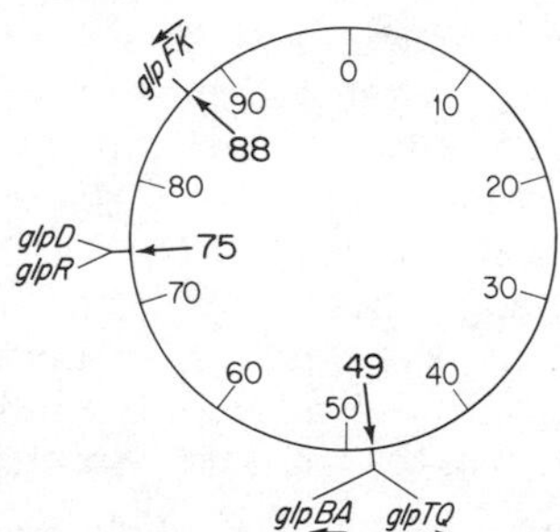

Fig. 5. Genetic map of the glp regulon of E. coli. The glpTQ operon encodes respectively the G3P transport protein in the cytoplasmic membrane and the periplasmic glycerophosphodiester phosphodiesterase. The glpAB operon encodes respectively anaerobic G3P dehydrogenase and its membrane anchor protein. These two operons are adjacent to each other but are oriented in opposite directions. Another pair of operons, glpR and glpD, are also adjacent. They encode respectively the repressor and aerobic G3P dehydrogenase. The glpKF operon encodes respectively glycerol kinase and glycerol facilitator. Arrows indicate polarity of transcription (Lin, 19767; Larson et al, 1982; Ludtke et al, 1982; Conrad et al., 1984; Kuritzkes et al., 1984; W. Boos, personal communication).

communication); the glpAB operon encoding respectively anaerobic G3P dehydrogenase and its membrane anchor protein (Kuritzkes et al., 1984); the glpR operon encoding the repressor which responds to G3P as the true inducer (Hayashi and Lin, 1965); the glpD operon encoding aerobic G3P dehydrogenase, and the glpKF operon encoding respectively glycerol kinase and glycerol facilitator. Although all four operons specifying the structural genes are under the control of the same repressor and are subject to catabolite repression, the sensitivity to each of these two controls vary from operon to operon. In addition, the expression of the glpD and glpAB operons are subject to respiratory control whose nature is not yet understood at the mechanistic level (Freedberg and Lin, 1973; Kuritzkes et al., 1984). A curious observation that accumulation of galactose 1-phosphate inhibits the induction of the glp system (Sundararajan, 1963) also deserves further study.

REGULATION OF THE RATE OF GLYCEROL CONSUMPTION

Concerted Control

Without an active transport mechanism to amplify the intracellular concentration of glycerol when its external supply is scarce, the cell compensates for the shortage by stepping up the synthesis of glycerol kinase, thereby increasing the extracting power. This can be accomplished by the gradual release from self-catabolite repression (Koch et al., 1964). It is significant that of the four operons encoding the catalytic proteins of the glp system, the glpKF operon encoding the proteins responsible for glycerol input is the most sensitive to catabolite repression. The effect of increased gene expression is probably augmented by the diminution of feedback inhibition of the kinase, as the metabolic pool of FDP subsides with carbon source limitation. (Under affluent conditions, the intracellular concentration of FDP fluctuates in the vicinity of 3 mM [Lin, 1976]; with a K_i of 0.5 mM, the kinase should be largely inhibited.) Maintaining a critical rate of phosphorylation might also be important for keeping the G3P level sufficiently high for strong induction. On the other hand, should external glycerol suddenly become abundant, over-phosphorylation can be averted by the rising tide of FDP, inhibiting the kinase noncompetitively. In time this constraint will be reinforced by enzyme dilution caused by enhanced catabolite repression. Regulation of gene expression and enzyme catalysis thus probably act in concert to poise a prudent balance of potential substrate extracting power against redundant protein synthesis and the risk of metabolic flooding.

The cell is so well guarded against excessive phosphorylation that both the controls of gene expression and enzyme action must be lifted for lethal metabolism of glycerol to occur. When a mutant lacking the repressor and feedback inhibitability of glycerol kinase (glpR$^-$,glpK^i) was made to produce elevated levels of the desensitized enzyme by prior growth on carbon sources exerting low catabolite repression (succinate or casein hydrolysate), exposure to glycerol resulted in rapid fatality.

Lethal Metabolism and Methylglyoxal. The lethal effect of glycerol metabolism is not attributable to excessive accumulation of G3P, which can only be bacteriostatic when it occurs (Cozzarelli et al., 1965). This conclusion was substantiated by imposing a further genetic lesion at the step of G3P dehydrogenation (glpR$^-$, glpK^i, glpD$^-$). Cells of this triple mutant pregrown in a medium of low catabolite repression were merely growth arrested when exposed to glycerol (Zwaig et al., 1970). Thus the flooding of the DHAP pool is a necessary condition for killing the cells. Cells undergoing metabolic suicide in this way excrete a bactericidal substance, methylglyoxal (Zwaig and Diegez, 1970; Freedberg et al., 1971; Krymkiewicz et al., 1971). DHAP is converted to methylglyoxal by a pathway believed to play ordinarily a physiological role in replenishing intracellular inorganic phosphate from organic intermediates during phosphate starvation. This idea is supported by multi-order feedback inhibition of methylglyoxal synthase by orthophosphate. The inhibition is overcome in a cooperative manner by increasing the concentration of DHAP, the substrate (K_m of 0.5 mM) (Cooper and Anderson, 1970; Hopper and Cooper, 1971).

There are two interesting aspects of the lethal production of methylglyoxal by mutants making a decontrolled kinase. First, lethal synthesis takes place only aerobically; anaerobically, the consequence is no more serious than growth inhibition. (Spontaneous conversion of DHA to methylglyoxal seems to occur more rapidly aerobically than anaerobically [Riddle and Lorenz, 1968; R.Z. Jin, personal communication]. Perhaps spontaneous formation of methylglyoxal from DHAP also occurs more rapidly aerobically, thus contributing to the enzymatically catalyzed reaction.) Second, the phenomenon is manifested by male (Hfr and F$^+$) but not female cells (Zwaig and Dieguez, 1970).

Wild-type E. coli, as well as the mutant strain synthesizing constitutively the decontrolled glycerol kinase, yields spontaneous mutants (at a frequency of about 10^{-7}) that are resistant to 1 mM methylglyoxal. The apparent one-step mutation not only confers resistance to the highly reactive chemical, but actually permits growth in a medium containing 1 mM methylglyoxal as sole carbon and energy source. Resistance to methylglyoxal also renders the mutant producing constitutively the decontrolled glycerol kinase immune to glycerol exposure during growth on succinate. Crude extracts of resistant strains catalyzed the disappearance of methylglyoxal (with lactate as final product) at four to eight times the normal rate. This increase in activity reflected higher constitutive levels of glyoxalase I (lactoyl-glutathione lyase); the constitutive level of glyoxalase II (hydroxyacylglutathione hydrolase) was unchanged (Freedberg et al., 1971). Thus a detoxifying or salvaging system has been mutated to a catabolic pathway, another illustration of the adaptive potential of the bacterial genome.

Increased Growth Rate on Glycerol. When the mutant producing constitutively the decontrolled glycerol kinase was first grown in a glucose medium (to reduce the level of the kinase by strong catabolite repression) and then transferred to a glycerol medium, no bactericidal effect was seen. Indeed, growth resumed at a rate (doubling time of 100 min) that was more rapid than that of wild-type cells (doubling time of 130 min). Evidently catabolite repression alone is sufficient to prevent metabolic flooding. (Feedback inhibition alone is also sufficient, since cells producing the wild-type enzyme constitutively are immune to the lethal metabolism when transferred from a casein hydrolysate to a glycerol medium.) The higher growth rate of cells with the mutant enzyme shows that in wild-type cells the pace of glycerol utilization (under abundant conditions) is limited by the internal FDP level (Zwaig et al., 1970). However, it should be borne in mind that the inability of wild-type cells to use glycerol at a maximal rate does not imply an imperfection in the evolution of E. coli, since natural environments cannot be expected to supply a single carbon and energy source in abundance.

Abolition of Glucose-Glycerol Diauxie. When wild-type cells are presented with a mixture of glucose and glycerol, classic diauxic growth occurs with a lag period of about 1 hr for adaptation to glycerol. Either demolishment of the specific repressor ($glpR^-$) or the kinetic decontrol of glycerol kinase ($glpK^i$) shortens the transition period without preventing the biphasic growth. Diauxie is no longer apparent if both repressor and feedback controls are lifted by mutations. However, such a cell probably still preferentially utilized glucose (see the Interaction with PTS below).

Glycerol, in contrast to glucose or mannitol, exerts little catabolite repression on other inducible systems, but if glycerol kinase is released from feedback inhibition, the power of glycerol to exert catabolite repression approaches that of glucose, as indicated by the equal effect of glycerol and glucose in antagonizing the induction of the lactose operon by isopropyl-β-thiogalactoside at a high concentration (to overcome inducer exclusion effect). Whether the decontrolled kinase is produced inducibly or constitutively is immaterial in this regard (Zwaig et al., 1970). This observation provides further evidence that catabolite repression does not act exclusively through the PTS system (see below).

Interaction with the Phosphoenolpyruvate Phosphotransferase System (PTS). Glucose prevents the utilization of a number of carbohydrates such as lactose, galactose, and glycerol by impeding substrate entry and by catabolite repression, and the PTS plays a complex role in assuring this preference (Saier and Feucht, 1975; Postma and Roseman, 1976; Meadow et al., 1982). One might suppose that in the case of glycerol, feedback inhibition by the FDP pool arising from glucose would be sufficient to cut off glycerol consumption and along with this, the formation of the inducer G3P. However, it was pointed out that FDP at 10 mM and above inhibited glycerol kinase (E. coli strain ML) to a maximum of about 90 percent, whereas glycerol utilization by the cell was completely inhibited in the presence of glucose. Therefore, glucose might also be affecting the permeation step (Edgar et al., 1972). This conjecture seems to be vindicated by the discovery of the dual role of factor III^{Glc} protein of the PTS system as a catalytic component of vectorial phosphorylation of glucose and as an effector molecule which interacts allotropically with permeases of the non-PTS sugars. Factor III^{Glc} can exist in two states: nonphosphorylated and phosphorylated. The latter serves as a phosphoryl donor in the glucose uptake process. Depletion of phosphorylated factor III^{Glc} during this process increases the concentration of its nonphosphorylated form which inhibits the activity of other kinds of membrane transport proteins, such as β-galactoside permease (Osumi and Saier, 1982). This substrate (or inducer) exclusion effect is abolished in mutants lacking the factor III^{Glc} protein (Saier and Roseman, 1976).

Methyl α-glucoside (analog of glucose) inhibition of glycerol uptake by a mutant with a feedback insensitive kinase led to the suggestion that factor III^{Glc} interacts also with glycerol facilitator (Saier et al., 1978). However, the recent finding that factor III^{Glc} in the dephosphorylated form inhibited glycerol kinase activity in vitro, raises the question of whether the inhibition by methyl-α-glucoside really acts at the level of the facilitator. The enzyme which is desensitized to FDP might retain its sensitivity to factor III^{Glc}. Since the inhibitory effect of methyl α-glucoside was studied by assaying the accumulation of radioactivity in cells incubated with labeled glycerol, it is not possible to distinguish inhibition of the facilitator from that of the kinase (Postma, P.W., Epstein, W., Schuitema, A.R.J., and Nelson, S.O., personal communication).

In addition to the inducer exclusion effect, mutations abolishing the function of enzyme I or HPr (the histidine-containing protein) of the PTS interfere with induction of catabolic pathways for a number of carbohydrates, including that for glycerol. The growth impairments apparently result from deficiency of cyclic AMP (Perlman and Pastan, 1969; Postma and Roseman, 1976). The inability of an enzyme I mutant to grow on glycerol can be remedied in four ways: (1) by provision of 5 mM cyclic AMP in the growth medium; (2) by making the glp system constitutive; (3) by freeing glycerol kinase from feedback control and thus allowing the basal enzyme to be fully engaged in the formation of the inducer, G3P; and (4) by an up-promoter mutation in the glpKF operon which also increases the basal glycerol kinase activity (Berman et al., 1970; Berman and Lin, 1971; Berman-Kurtz et al., 1971). In this framework it might be noted that a promoter-like mutation in the glpKF operon can also enable an adenyl cyclase mutant to grow on glycerol, although the level of expression of glycerol kinase remains subnormal in the absence of exogenous cyclic AMP (Fraser and Yamazaki, 1980).

INTERACTION OF THE FACILITATOR AND THE KINASE

Since glycerol facilitator displays little substrate specificity towards uncharged compounds with low molecular weight and glycerol kinase has a sufficiently broad specificity to catalyze the phosphorylation of DHA, a mutant of Klebsiella pneumoniae, strain ECL54, expressing the glp system constitutively (and therefore synthesizing glycerol facilitator and glycerol kinase constitutively), was tested for its growth response with DHA as the sole source of carbon energy. It should be added that K. pneumoniae has a second system, the dha regulon, which includes a gene encoding an ATP-dependent kinase whose physiological function is for DHA utilization and which has no activity on glycerol. However, all the genes of the dha system are inactivated in this mutant. Therefore, growth of this mutant on DHA is possible only if the compound enters through glycerol facilitator and is phosphorylated by glycerol kinase. Indeed, this was found to occur (Fig. 6B). However, an unexpectedly high concentration of DHA was required to achieve moderate growth rates. On the basis of the K_m of the kinase for DHA--and assuming a rapid rate of substrate entry--half maximal growth rate should be achieved in the vicinity of 1 mM (the K_m of the kinase). Instead, the growth rate was dependent on substrate concentration up to 50 mM (doubling time of 80 min). In contrast, when glycerol was provided as the carbon and energy source, growth occurred at the same exponential rate (doubling time of 50 min) with substrate concentration ranging from 2 to 50 mM (Fig. 6C).

To exclude the possibility that DHA was spuriously converted by the mutant to glycerol, which was then acted on by glycerol kinase, a strain lacking aerobic G3P dehydrogenase was derived from strain ECL54. This derivative, strain ECL61, could no longer grow on glycerol (or G3P), but the ability to grow on DHA remained. The dependence of DHA phosphorylation on glycerol kinase was further substantiated by the instant growth arrest of strain ECL61 by introducing 0.5 mM glycerol (preferential phosphorylation of glycerol led to toxic accumulation of G3P because of the absence of its dehydrogenase) into a culture containing 50 mM DHA as carbon and energy source (Fig. 6D). In another control experiment, strain ECL12 was used. In this mutant the glp regulon is completely inactivated (super-repressed by a $glpR^n$ mutation) and all the genes in the dha operon are inactivated except the operon encoding the

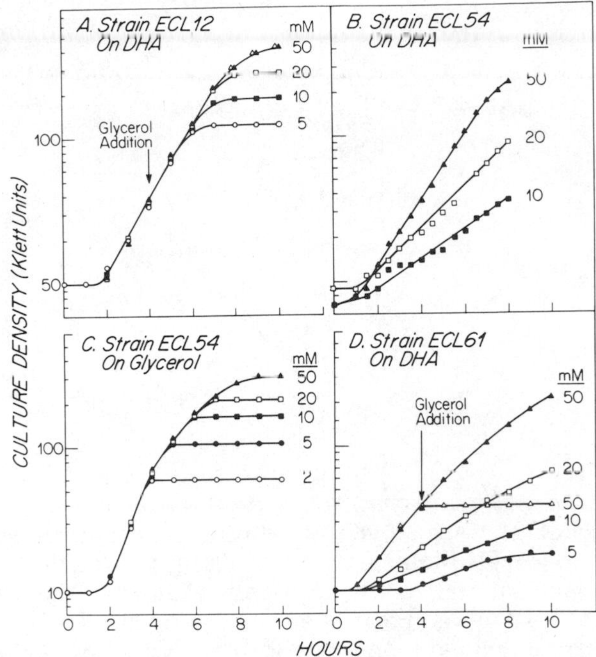

Fig. 6. Aerobic growth properties of cells employing glycerol kinase or DHA kinase. (A) Growth of strain ECL12 (DHA kinase$^+$, glycerol kinase$^-$) on DHA. At the time indicated by the arrow, 5 mM glycerol was added to one of the two cultures initially containing 50 mM DHA. (B) Growth of strain ECL54 (glycerol kinase$^+$, aerobic G3P dehydrogenase$^+$, DHA kinase$^-$) on DHA. (C) Growth of strain ECL54 on glycerol. (D) Growth of strain ECL61 (glycerol kinase$^+$, aerobic G3P dehydrogenase$^-$, DHA kinase$^-$) on DHA. At the time indicated by the arrow, 0.5 mM glycerol was added to one of the two cultures initially containing 50 mM DHA (Jin et al, 1982).

ATP-dependent DHA kinase and presumably also a corresponding facilitator. The rate of exponential growth of strain ECL12 on DHA remained constant even below 5 mM. Since DHA kinase has no affinity of glycerol, the addition of glycerol did not prevent growth on DHA (Fig. 6A).

The above results suggest that in vivo glycerol kinase acts in close physical association with glycerol facilitator (Fig. 7) and that the protein complex exhibits more stringent substate discrimination than each component acting by itself (Jin et al., 1982). This model of glycerol kinase, functioning as a peripheral protein of the inner membrane of the cell, fits well with the ability of factor III^{Glc} to

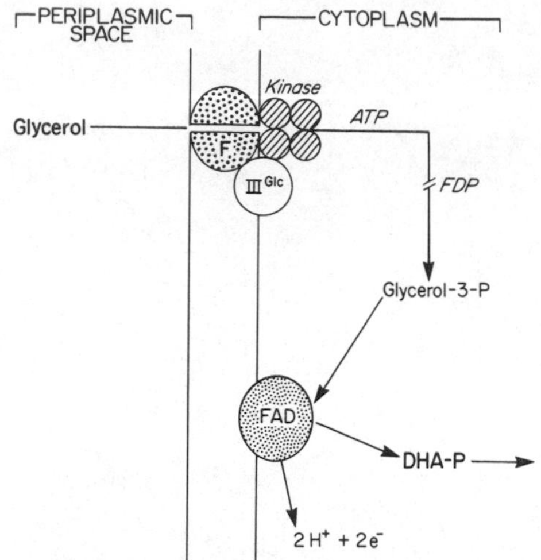

Fig. 7. A model showing glycerol kinase functioning as a complex with glycerol facilitator to explain the stronger discrimination against DHA by the cell than by the free enzyme. Factor III^{Glc} is positioned between the facilitator and the kinase to indicate the possibility that this regulator protein might interact with both the glycerol proteins.

inhibit cellular glycerol uptake and glycerol phosphorylation in vitro. Physical interaction between facilitator molecules and kinase molecules seems also to occur in the uptake of hexoses by yeast (Bisson and Fraenkel, 1983a, 1983b). Therefore the phenomenon might be quite general.

ON THE STRATEGY OF GLYCEROL TRANSPORT

Two aspects of cellular transport of glycerol call for attention. First, energy coupled concentrative mechanisms for the compound have not been observed in any kind of cells. Second, in bacteria PTS-dependent vectorial phosphorylation of glycerol has so far not been discovered (Richey and Lin, 1973). The first feature is readily understood in the framework of natural selection. The intrinsic permeability of biological membranes, which makes it difficult to retain glycerol, would render active transport of the compound a futile process. Internal trapping by phosphorylation at the expense of ATP, on the other hand, is an economical mechanism. With a single high energy phosphate, capture and activation of the substrate for further metabolism are accomplished in one coup. Hence the facilitator-kinase coupled system is functionally analogous to the PTS system, with ATP serving as the drive instead of PEP. With respect to the absence of vectorial phosphorylation of glycerol by the PTS in bacteria, there is no obvious teleonomic explanation. In all the bacterial species studied, an ATP dependent glycerol kinase was found. This includes Bacillus subtilis (Lindgren and Rutberg, 1976), Pseudomonas aeruginose (Siegel and Phibbs, 1979; McCowen et al., 1981), Rhodopseudomonas capsulata (Lueking et al., 1973, 1975; Pike, 1982), and Rhodopseudomonas spheroides (Pike and Sojka, 1975). An interesting report that a binding protein is associated with glycerol uptake in P. aeruginosa hints at a novel mechanism for the transport of this compound (Tsay et al., 1971). A broader study of uptake mechanisms for glycerol uptake by bacteria might reveal their course of evolution, and cast some light on whether chance or necessity played the dominant role in the shaping of this process.

ACKNOWLEDGEMENT

I thank Sarah Monosson for editorial assistance. This work was supported by grant PCM79-24046 from the National Science Foundation and by Public Health Service grant 5 ROI GM11983 from the National Institute of General Medical Sciences.

REFERENCES

Alemohammad, M.M., and Knowles, C.J., 1974, Osmotically induced volume and turbidity changes of Escherichia coli due to salts, sucrose and glycerol, with particular reference to rapid permeation of glycerol into the cell, J. Gen. Microbiol., 82:125.

Avi-Dor, Y., Kuczynski, M., Schatzberg, G., and Mager, I., 1956, Turbidity changes in bacterial suspensions: kinetics and relation to metabolic state, J. Gen. Microbiol., 14:76.

Berman, M., and Lin, E.C.C., 1971, Glycerol-specific revertants of a phosphoenolpyruvate phosphotransferase mutant: suppression by the desensitization of glycerol kinase to feedback inhibition, J. Bacteriol., 105:113.

Berman, M., Zwaig, N., and Lin, E.C.C., 1970, Suppression of a pleiotropic mutant affecting glycerol dissimilation, Biochem. Biophys. Res. Commun., 38:272.

Berman-Kurtz, M., Lin, E.C.C., and Richey, D.P., 1971, Promoter-like mutant with increased expression of the glycerol kinase operon of Escherichia coli, J. Bacteriol., 106:724.

Bisson, L.F., and Fraenkel, D.G., 1983a, Involvement of kinases in glucose and fructose uptake by Saccharomyces cerevisiae, Proc. Natl. Acad. Sci. USA, 80:1730.

Bisson, L.F., and Fraenkel, D.G., 1983b, Transport of 6-deoxyglucose in Saccharomyces cerevisiae, J. Bacteriol., 155:995.

Boehler-Kohler, B.A., Boos, W., Dieterle, R., and Benz, R., 1979, Receptor for bacteriophage lambda of Escherichia coli forms larger pores in block lipid membranes than the matrix protein (porin), J. Bacteriol., 138:33.

Boos, W., Hartig-Beecken, I., and Altendorf, K., 1977, Purification and properties of a periplasmic protein related to sn-glycerol-3-phosphate transport in Escherichia coli, Eur. J. Biochem., 72:571.

Bovell, C.R., Packer, L., and Helgerson, R., 1963, Permeability of Escherichia coli to organic compounds and inorganic salts measured by light-scattering, Biochim. Biophys. Acta, 75:257.

Britten, R.J., and McClure, F.T., 1962, The amino acid pool in Escherichia coli, Bacteriol. Rev., 26:292.

Bublitz, C., and Kennedy, E.P., 1954, Synthesis of phosphatides in isolated mitochondria. III. Enzymatic phosphorylation of glycerol, J. Biol. Chem., 211:951.

Conrad, C.A., Stearns, G.W., Prater, W.E., Rheiner, J.A., and Johnson, J.R., 1984, Characterizatin of a glpK tranducing phage. Mol. Gen. Genet., 193:376.

Cooper, R.A., and Anderson, J A., 1970, The formation and catabolism of methylglyoxal during glycolysis in Escherichia coli, FEBS Lett., 11:273.

Cozzarelli, N.R., Freedberg, W.B., and Lin, E.C.C., 1968, Genetic control of the L-α-glycerophosphate system in Escherichia coli, J. Mol. Biol., 31:371.

Cozzarelli, N.R., Koch, J.P., Hayashi, S., and Lin, E.C.C., 1965, Growth stasis by accumulated L-α-glycerophosphate in Escherichia coli, J. Bacteriol., 90:1325.

Cozzarelli, N.R., and Lin, E.C.C., 1966, Chromosomal location of the structural gene for glycerol kinase in Escherichia coli, J. Bacteriol., 91:1763.

Danielli, J.F., 1954, The present position in the field of facilitated diffusion and selective active transport, p. 1, in: "Recent Developments in Cell Physiology," J.A. Kitching, ed., Academic Press, New York.

De Gier, J., Mandersloot, J.G., Hupkes, J.V., McElhaney, R.N., and Van Beek, W.P., 1971, On the mechanism of non-electrolyte permeation through lipid bilayers and through biomembranes, Biochim. Biophys. Acta, 233:610.

De Gier, J., Mandersloot, J.G., and Van Deenen, L.L.M., 1968, Lipid composition and permeability of liposomes, Biochim. Biophys. Acta, 150:666.

De Gier, J., Van Deenen, L.L.M., and Van Senden, K.G., 1966, Glycerol permeability of erythrocytes, Experimentia, 22:20.

de Riel, J.K., and Paulus, H., 1978a, Subunit dissociation in the allosteric regulation of glycerol kinase from Escherichia coli. I. Kinetic evidence, Biochemistry, 17:5134.

de Riel, J.K., and Paulus, H., 1978b, Subunit dissociation in the allosteric regulation of glycerol kinase from Escherichia coli. 2. Physical evidence, Biochemistry, 17:5141.

de Riel, J.K., and Paulus, H., 1978c, Subunit dissociation in the allosteric regulation of glycerol kinase from Escherichia coli. 3. Role in desensitization, Biochemistry, 17:5146.

Edgar, W., Forrest, I.S., Holms, W.H., and Jasani, B., 1972, The control of glycerol utilization by glucose metabolism, Biochem. J., 127:59.

Eze, M.O., and McElhaney, R.N., 1981, The effect of alterations in the fluidity and phase state of the membrane lipids on the passive permeation and facilitated diffusion of glycerol in Escherichia coli, J. Gen. Microbiol., 124:299.

Fischer, A., 1903, "Vorlesungen über Bakterien," Gustav Fischer, Jena.

Fraser, A.D.E., and Yamazaki, H., 1980, Characterization of an Escherichia coli mutant which utilizes glycerol in the absence of cyclic adenosine monophosphate, Can. J. Microbiol., 26:393.

Freedberg, W.B., Kistler, W.S., and Lin, E.C.C., 1971, Lethal synthesis of methylglyoxal by Escherichia coli during unregulated glycerol metabolism. J. Bacteriol., 108:137.

Freedberg, W.B., and Lin, E.C.C., 1973, Three kinds of controls affecting the expresiosn of the glp regulon in Escherichia coli, J. Bacteriol., 115:816.

Hayashi, S.I., Koch, J.P., and Lin, E.C.C., 1964, Active transport of L-α-glycerophosphate in Escherichia coli, J. Biol. Chem., 239:3098.

Hayashi, S.-I., and Lin, E.C.C., 1965, Product induction of glycerol kinase in Escherichia coli, J. Mol. Biol., 14:515.

Hayashi, S., and Lin, E.C.C., 1965, Capture of glycerol by cells of Escherichia coli, Biochim. Biophys. Acta, 94:479.

Hayashi, S., and Lin, E.C.C., 1967, Purification and properties of glycerol kinase from Escherichia coli, J. Biol. Chem., 242:1030.

Heller, K.B., Lin, E.C.C., and Wilson, T.H., 1980, Substrate specificity and transport properties of the glycerol facilitator of Escherichia coli, J. Bacteriol., 144:274.

Hirschmann, H., 1960, The nature of the substrate asymmetry in stereoselective reactions, J. Biol. Chem., 235:2762.

Hopper, D.J., and Cooper, R.A., 1971, The regulation of Escherichia coli methylglyoxal synthase: a new control site in glycolysis? FEBS Lett., 13:213.

Jacobs, M.H., 1931, The permeability of the erythrocyte, Ergebn. Biol., 7:1.

Jacobs, M.H., 1950, Surface properties of the erythrocyte, Ann. N.Y. Acad. Sci., 50:824.

Jacobs, M.H., 1954, A case of apparent physiological competition between ethylene glycol and glycerol, Biol. Bull., 107:314.

Jacobs, M.H., and Corson, S.A., 1934, The influence of minute traces of copper on certain hemolytic processes, Biol. Bull., 67:325.

Jacobs, M.H., and Stewart, D.R., 1946, Observations on an oligodynamic action of copper on human erythrocytes, Am. J. Med. Sci., 211:246.

Jin, R.Z., Forage, R.G., and Lin, E.C.C., 1982, Glycerol kinase as a substitute for dihydroxyacetone kinase in a mutant of Klebsiella pneumoniae, J. Bacteriol., 152:1303.

Kalckar, H., 1937, Phosphorylation in kidney tissue, Enzymologia, 2:47.

Kalckar, H., 1938, Formation of a new phosphate ester in kidney extracts, Nature, 142:871.

Kalckar, H., 1939, The nature of phosphoric esters formed in kidney extracts, Biochem. J., 33:631.

Karrer, P., and Benz, P., 1926, Die Spaltung der Glycerin-α-phosphorsaure in optisch aktive Formen, Helvet. Chim. Acta, 9:23.

Koch, J.P., Hayashi, S.-I., and Lin, E.C.C., 1964, The control of dissimilation of glycerol and L-α-glycerophosphate in Escherichia coli, J. Biol. Chem., 239:3106.

Krymkiewicz, N., Dieguez, E., Rekarte, U.D., and Zwaig, N., 1971, Properties and mode of action of a bactericidal compound (= methylglyoxal) produced by a mutant of Escherichia coli, J. Bacteriol., 108:1338.

Kuritzkes, D.R., Zhang, X.-Y., and Lin, E.C.C., 1984, Use of Φ(glp-lac) in studies of respiratory control of the Escherichia coli anaerobic sn-glycerol-3-phosphate dehydrogenase genes (glpAB), J. Bacteriol., 157:591.

Larson, T.J., Ehrmann, M., and Boos, W., 1983, Periplasmic glycerophosphodiester phosphodiestrase of Escherichia coli, a new enzyme of the glp regulon, J. Biol. Chem., 258:5428.

Larson, T.J., Schumacher, G., and Boos, W., 1982, Identification of the glpT-encoded sn-glycerol-3-phosphate permease of Escherichia coli, an oligomeric integral membrane protein, J. Bacteriol., 152:1008.

Li, C.-C., and Lin, E.C.C., 1975, Uptake of glycerol by tumor cells and its control by glucose, Biochem. Biophys. Res. Commun., 67:677.

Li, C.-C., and Lin, E.C.C., 1983, Glycerol transport and phosphorylation by rat hepatocytes, J. Cell. Physiol., 117:230.

Lin, E.C.C., 1976, Glycerol dissimilation and its regulation in bacteria. Ann. Rev. Microbiol., 30:535.

Lin, E.C.C., 1977, Glycerol utilization and its regulation in mammals, Ann. Rev. Biochem., 46:765.

Lindgren, V., and Rutberg, L., 1976, Genetic control of the glp system in Bacillus subtilis, J. Bacteriol., 127:1047.

Luckey, M., and Nikaido, H., 1980, Specificity of diffusion channels produced by lambda phage receptor protein of Escherichia coli, Proc. Natl. Acad. Sci. USA, 77:167.

Lueking, D., Tokuhisa, D., and Sojka, G., 1973, Glycerol assimilation by a mutant of Rhodopseudomonas capsulata, J. Bacteriol., 115:897.

Lueking, D., Pike, L., and Sojka, G., 1975, Glycerol utilization by a mutant of Rhodopseudomonas capsulata, J. Bacteriol., 125:750.

Ludtke, D., Larson, T.J., Beck, C., and Boos, W., 1982, Only one gene is required for the glpT-dependent transport of sn-glycerol-3-phosphate in Escherichia coli, Mol. Gen. Genet., 186:540.

Luria, S.E., 1965, On the evolution of the lactose utilization gene system in enteric bacteria, p. 357, in: "Evolving Genes and Proteins," H.J. Vogel, ed., Academic Press, New York.

Mager, J., Kuczynski, M., Schatsberg, G., and Avi-Dor, Y., 1956, Turbidity changes in bacterial suspensions in relation to osmotic pressure, J. Gen. Microbiol., 14:69.

McCowen, S.M., Phibbs, P.V., Jr., and Feary, T.W., 1981, Glycerol catabolism in wild-type and mutant strains of Pseudomonas aeruginosa, Cur. Microbiol., 5:191.

Meadow, N.D., Saffen, D.W., Dottin, R.P., and Roseman, S., 1982, Molecular cloning the the crr gene and evidence that it is the structural gene for III^{Glc}, a phosphocarrier protein of the bacterial phosphotransferase system, Proc. Natl. Acad. Sci. USA, 79:2528.

Mitchell, P., and Moyle, J., 1956, Osmotic function and structure in bacteria, p. 150, in: "Bacterial Anatomy," Sixth Symp. Soc. Gen. Microbiol., Cambridge University Press, London.

Osumi, T., and Saier, M.H., Jr., 1982, Regulation of lactose permease activity by the phosphoenolpyruvate:sugar phosphotransferase system: evidence for direct binding of the glucose-specific enzyme III to the lactose permease, Proc. Natl. Acad. Sci. USA, 79:1457.

Perlman, R.L., and Pastan, I., 1969, Pleiotropic deficiency of carbohydrate utilization in an adenyl cyclase deficient mutant of Escherichia coli, Biochem. Biophys. Res. Commun., 37:151.

Pike, L., 1982, Glycerol-utilizing mutants of Rhodopseudomonas capsulata, J. Bacteriol., 151:500.

Pike, L., and Sojka, G.A., 1975, Glycerol dissimilation in Rhodopseudomonas sphaeroides, J. Bacteriol., 124:1101.

Postma, P.W., and Roseman, S., 1976, The bacterial phosphoenolpyruvate:sugar phosphotransferase system, Biochim. Biophys. Acta, 457:213.

Richey, D.P., and Lin, E.C.C., 1972, Importance of facilitated diffusion for effective utilization of glycerol by Escherichia coli, J. Bacteriol, 112:784.

Richey, D.P., and Lin, E.C.C., 1973, Phosphorylation of glylcerol in Staphylococcus aureus, J. Bacteriol., 114:880.

Riddle, V., and Lorenz, F.W., 1968, Nonenzymic, polyvalent anion-catalyzed formation of methylglyoxal as an explanation of its presence in physiological systems, J. Biol. Chem., 243:2718.

Saier, M.H., Jr., and Feucht, B.U., 1975, Coordinate regulation of adenylate cyclase and carbohydrate permeases by the phosphoenolpyruvate:sugar phosphotransferase system in Salmonella typhimurium, J. Biol. Chem., 250:7078.

Saier, M.H., Jr., and Roseman, S., 1976, Sugar transport. The crr mutation: its effect on repression of enzyme synthesis, J. Biol. Chem., 251:6598.

Saier, M.H., Jr., Straud, H., Massman, L.S., Judice, J.J., Newman, M.J., and Feucht, B.U., 1978, Permease-specific mutations in Salmonella typhimurium and Escherichia coli that release the glycerol, maltose, melibiose, and lactose transport systems from regulation by the phosphoenolpyruvate:sugar phosphotransferase system in Salmonella typhimurium, J. Bacteriol., 133:1358.

Sanno, Y., Wilson, T.H., and Lin, E.C.C., 1968, Control of permeation to glycerol in cells of Escherichia coli, Biochem. Biophys. Res. Commun., 32:344.

Siegel, L.S., and Phibbs, P.V., Jr., 1979, Glycerol and L-α-glycerol-3-phosphate uptake by Pseudomonas aeruginosa, Cur. Microbiol., 2:251.

Sundararajan, T.A., 1963, Interference with glycerokinase induction in mutants of E. coli accumulating Gal-1-P, Proc. Natl. Acad. Sci. USA, 50:463.

Thorner, J.W., and Paulus, H., 1971, Composition and subunit structure of glycerol kinase from Escherichia coli, J. Biol. Chem., 246:3885.

Thorner, J.W., and Paulus, H., 1973a, Catalytic and allosteric properties of glycerol kinase from Escherichia coli, J. Biol. Chem., 248:3922.

Thorner, J.W., and Paulus, H., 1973b, Glycerol and glycerate kinases, p. 487, in: "The Enzymes," vol. 8, P.D. Boyer, ed., Academic Press, New York.

Tsay, S.-S., Brown, K.K., and Gaudy, E.T., 1971, Transport of glycerol by Pseudomonas aeruginosa, J. Bacteriol., 108:82.

Zwaig, N., and Dieguez, E., 1970, Bactericidal product obtained from a mutant of Escherichia coli. J. Bacteriol., 102:753.

Zwaig, N., and Lin, E.C.C., 1966, Feedback inhibition of glycerol kinase, a catabolic enzyme in Escherichia coli, Science, 153:755.

MECHANISMS OF CELLULAR PROTEIN LOCALIZATION

S.A. Benson, E. Bremer, S. Garrett, D.R. Kiino,
J.W. Shultz, T.J. Silhavy, E.J. Sodergren,
R.K. Taylor, and N.J. Trun

Laboratory of Genetics and Recombinant DNA
LBI-Basic Research Program
NCI-Frederick Cancer Research Facility
P.O. Box B
Frederick, Maryland 21701

With the exception of the limited biosynthetic activity exhibited by mitochondria and chloroplasts, the translation of mRNA occurs in the cytoplasm. Nevertheless, all cells synthesize proteins that are exported to various noncytoplasmic locations. In addition, many cells are capable of true protein secretion. These processes of protein localization are selective and efficient in that proteins are strictly compartmentalized to a particular cellular location. During the past decade, considerable effort has been directed towards elucidating the molecular mechanisms by which cells accomplish these processes. One important outcome of this work was the realization that all cells employ similar mechanisms of protein localization. This realization has fostered a solid interaction among biologists working with numerous different organisms.

Essentially everything known about the process of protein localization stems from the work of Palade (1975). In pioneering studies, Palade and his coworkers traced the intracellular routing of several of noncytoplasmic proteins in mammalian cells. They demonstrated, for example, that proteins destined to be secreted are synthesized initially by ribosomes tightly bound to the rough endoplasmic reticulum. Such proteins are found in the lumen of this organelle immediately after synthesis is completed. Subsequently, the protein is routed to the Golgi apparatus before secretion.

In molecular terms, perhaps the most difficult step to envision in the process of protein secretion is the transfer of a large, water-soluble protein through a hydrophobic membrane. Palade's work suggested that this step occurs at the level of the rough endoplasmic reticulum membrane. Furthermore, it became apparent that this process occurs in a manner tightly coupled to, if not inseparable from, translation. The model shown in Figure 1 represents the current thinking on the mechanisms of coupling of translation and export. This model is drawn to depict the various steps in eucaryotic cells (Blobel and Dobberstein, 1975; Walter and Blobel, 1981). A similar series of events occurs in bacteria except that the membrane to which the ribosomes bind is the cytoplasmic membrane and transfer is directed toward the periplasm (Silhavy et al., 1983).

GOALS

We wish to understand, in molecular terms, how proteins are directed to various cellular locations. Since the process of protein localization appears to have been conserved in the evolution of cells of all species, we are able to use Escherichia coli as a model system for the study of this mechanism. The availability of genetic methods, such as gene fusion, and the ease with which recombinant DNA technology can be applied in E. coli make this problem particularly amenable to study in this organism. Our approach has been (1) to define, at the level of the DNA sequence, the signals within a gene that direct the gene product to its sequence, the signals within a gene that direct the gene product to its final location, (2) to identify and characterize unlinked genetic loci that are involved in the process of cellular protein localization, and (3) to determine whether multiple export mechanisms are employed.

Our studies have focused on proteins destined for the outer membrane, in particular the export of the major outer membrane protein LamB. This protein is a component of the maltose transport system, and as such, its synthesis is induced by the presence of maltose in the growth medium. The protein also serves as the receptor for certain bacteriophages. Mutants lacking this protein are unable to grow on maltodextrins and are resistant to the bacteriophage λ .

INTRAGENIC EXPORT SIGNALS

Perhaps the most significant contribution of gene fusion technology to studies of protein localization stems from the unusual phenotypes often exhibited by fusion strains. These phenotypes are the consequence of the cell's attempt to export a hybrid protein containing sequences of β-galactosidase, and they can be exploited to isolate export-defective mutants. More importantly, because of the properties of lacZ fusions, a mechanism to distinguish mutations that block export from those that prevent synthesis is provided.

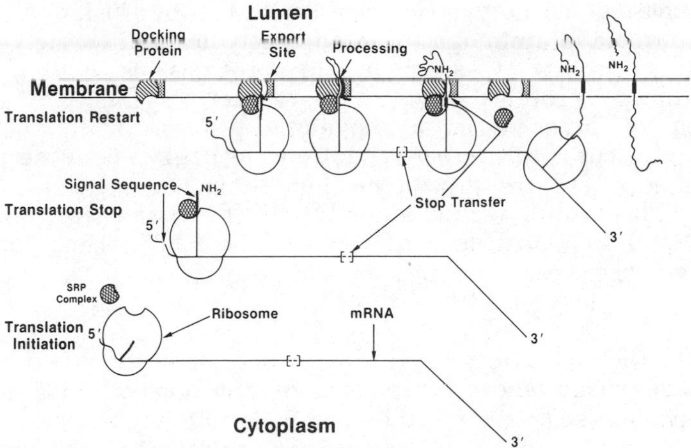

Fig. 1. Schematic Illustration of Cotranslational Export. The export process begins at the bottom of the illustration: the ribosome assembles and initiates translation of the mRNA at the 5' end. The signal sequence (represented by the wavy line in the mRNA or by small open circles in the protein) emerges from the ribosome and is recognized by the signal recognition protein (SRP), which interacts with the ribosome and the nascent polypeptide chain, stopping translation (middle illustration). This translational block is relieved when the complex (ribosome, SRP, and nascent polypeptide chain) interacts with the membrane-associated docking protein at the export site (top illustration). The signal sequence is composed of two segments, an NH_2-terminal charged segment and a hydrophobic segment. During the initial stages of polysome binding to the export sites, the positively charged segment interacts with either the inner leaflet of the membrane bilayer or a component of the export site. The hydrophobic segment then loops into the bilayer and a functional export site is formed. As translation proceeds, the nascent chain is transferred vectorially across the membrane bilayer. Proteolytic processing of the signal sequence from the polypeptide chain is achieved by a peptidase activity located at the outer face of the membrane. Such processing may occur before synthesis of the protein is complete. The model shows the existence of a second information signal (stop transfer) located within the protein. As this signal emerges from the ribosome, it results in a dissociation of the ribosome from the membrane and release of the SRP complex. Subsequent translation of the mRNA completes the COOH-terminal end of the protein in the cytoplasm, leaving the protein embedded in a transmembrane fashion with the NH_2 terminus facing the lumen and the COOH terminus facing the cytoplasm.

One unusual phenotype is characteristic of fusion strains that produce a hybrid protein that is incorrectly or inefficiently exported. High-level synthesis of such hybrid proteins is lethal, probably because these proteins jam the export machinery preventing localization of other essential envelope proteins. In the case of lamB-lacZ fusions, this overproduction lethality is observed when maltose is added to the growth medium to induce high-level synthesis of the hybrid protein. Accordingly, such strains are sensitive to maltose (Mal^s). Mutations that relieve the Mal^s phenotype, but do not prevent synthesis of the hybrid protein (Lac^+), are export-defective.

All of the mutations that were obtained using the selection procedure described above alter one of the amino acids present in the hydrophobic segment of the LamB signal sequence. All block export at an early step, causing the accumulation of the precursor in the cytoplasm. These mutations provided the first conclusive evidence that the signal sequence is required for the initiation of the export process (Emr and Silhavy, 1982).

Although these previous studies were very successful, they yielded only a single class of signal sequence mutations, i.e., all of the mutations block the same step in the export process. Since the signal sequence probably performs more than one function (Fig. 1), multiple classes of mutations should exist. New selection procedures must be devised to provide insight into the functions performed by the various components of the LamB signal sequence.

INTRAGENIC INFORMATION OUTSIDE THE SIGNAL SEQUENCE SPECIFYING EXPORT AND MEMBRANE INSERTION

The signal sequence functions in the initiation of protein export. Although this sequence is essential, studies with lamB-lacZ fusions, for example, suggest that the signal sequence is not sufficient to cause export from the cytoplasm; other information located within lamB is also required. Moreover, there must be information within lamB and other genes that code for membrane proteins that causes membrane insertion (Fig. 1) and sorting signals that specify localization to the correct cellular membrane. Since the signal sequence is removed during the export process, such information must lie elsewhere. Our studies indicate that most of this export information must lie between amino acids 1 and 70 of mature LamB (Benson and Silhavy, 1983). To define these export and sorting signals more precisely, we have again exploited an unusual phenotype exhibited by a lamB-lacZ fusion strain.

The largest lamB-lacZ fusion encodes a chimeric protein containing a complete LamB signal sequence, 241 amino acids of mature LamB and a functional β-galactosidase moiety. This fusion is localized with > 85 percent efficiency to the outer membrane. Strains in which the lac genes are deleted and which contain this fusion exhibit two useful and distinct phenotypes. Such strains are Lac$^-$ and show a modest, but characteristic, growth sensitivity in the presence of maltose. The Lac$^-$ phenotype appears to result from the inability of the β-galactosidase moiety to function in the environment of the outer membrane. Mutations that prevent localization to the outer membrane confer a Lac$^+$ phenotype, as described above. The growth sensitivity in the presence of maltose appears to result from jamming of the export machinery. Mutations that block export at an early step prevent jamming, thus relieving maltose sensitivity.

By selecting for a Lac$^+$ phenotype, we were able to isolate numerous deletions within the lamB-lacZ fused gene. In this manner, we have been able to collect a large number of lamB-lacZ fusions that differ only in the amount of lamB sequences present in the hybrid gene (Fig. 2). By scoring the Mal^S phenotype of the various fusion strains and by determining the cellular location of the hybrid protein produced, we have been able to define several important export signals in the lamB gene (S. Benson, E. Bremer, and T. Silhavy, unpublished results).

The new lamB-lacZ fusions isolated in this manner allow us to more precisely define the minimum amount of lamB required to confer a Mal^S phenotype. Fusions that specify a hybrid protein containing the complete signal sequence plus as many as 27 amino acids of LamB do not confer sensitivity to maltose. In contrast, fusions that specify a hybrid protein containing the signal sequence plus as few as 39 amino acids of LamB confer a high degree of maltose sensitivity (Fig. 3). Since the Mal^S phenotype is correlated with export, we suggest that at least a portion of an essential export signal is contained in a region of the lamB gene corresponding to amino acids 27 to 39 of the mature protein. If this region is not present, export does not occur and the protein remains in the cytoplasm.

In order to direct β-galactosidase to the outer membrane, a lamB-lacZ fusion must contain more information than that which corresponds to the signal sequence and 39 amino acids of mature protein. The smallest fusion we have isolated that directs appreciable amounts of product to the outer membrane specifies a hybrid protein containing the signal sequence plus 49 amino acids of LamB. This result suggests that at least a portion of the information that specifies an outer membrane location must be present between amino acids 39 and 49 of the mature protein. If these amino acids are absent, export from the cytoplasm begins but very little hybrid protein reaches the outer membrane.

Neither the results described here nor sequence analysis of signal-containing regions provides any clues as to the nature or role of the intragenic information required for export from the cytoplasm

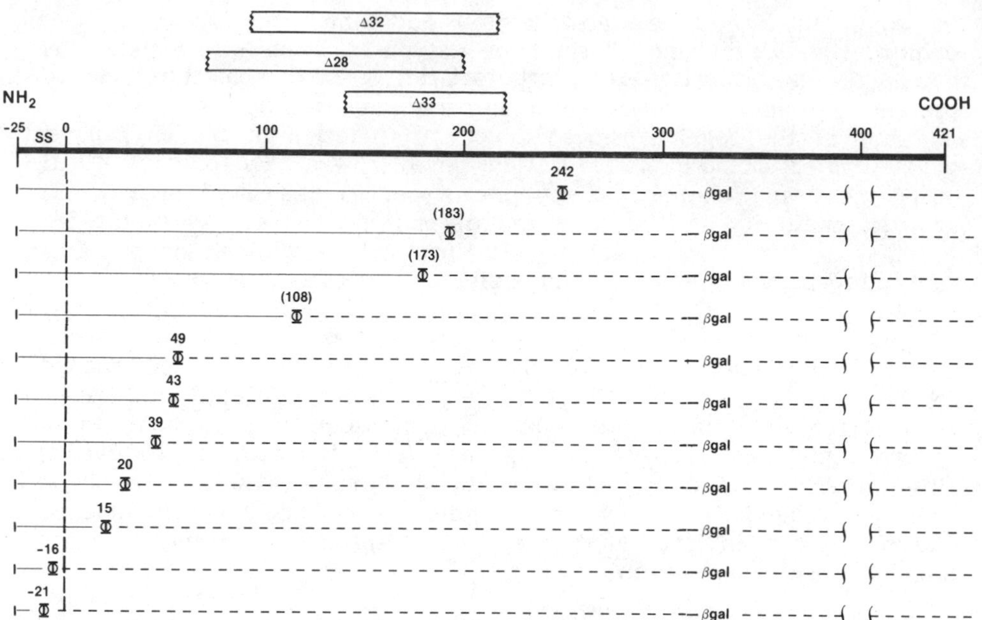

Fig. 2. Location of Internal Deletion and Fusion Joint Within LamB. The blocks above the heavy line, which represents the LamB protein, show areas of the protein that can be deleted and do not alter location of LamB itself, nor the largest LamB-LacZ fusion. The various lamB-lacZ fusions are shown below the heavy line. The number above the fusion joint symbol (Φ) indicates the amount of mature LamB present on the fusion. The values in parenthesis were determined by restriction analysis and thus are accurate only within ± 10%. All other fusion joints were determined by DNA sequence or protein sequence. The size of the β-galactosidase (β-gal) moiety is approximately the same in all cases.

or sorting to the outer membrane. Although the signal sequence and the sorting signal (amino acids 39 to 49) are likely to function at the amino acid level, other information, such as the region corresponding to amino acids 27 to 39, may be read from the mRNA. Moreover, the regions identified may function in an indirect manner by allowing the formation of specific, recognizable conformational domains. It can always be argued that by fusing β-galactosidase to various segments of LamB we have perturbed the system to the point where results are not physiological. Given past successes with gene fusion techniques, we think this unlikely. However, conclusive proof must await specific modification of the regions identified in an otherwise wild-type lamB gene.

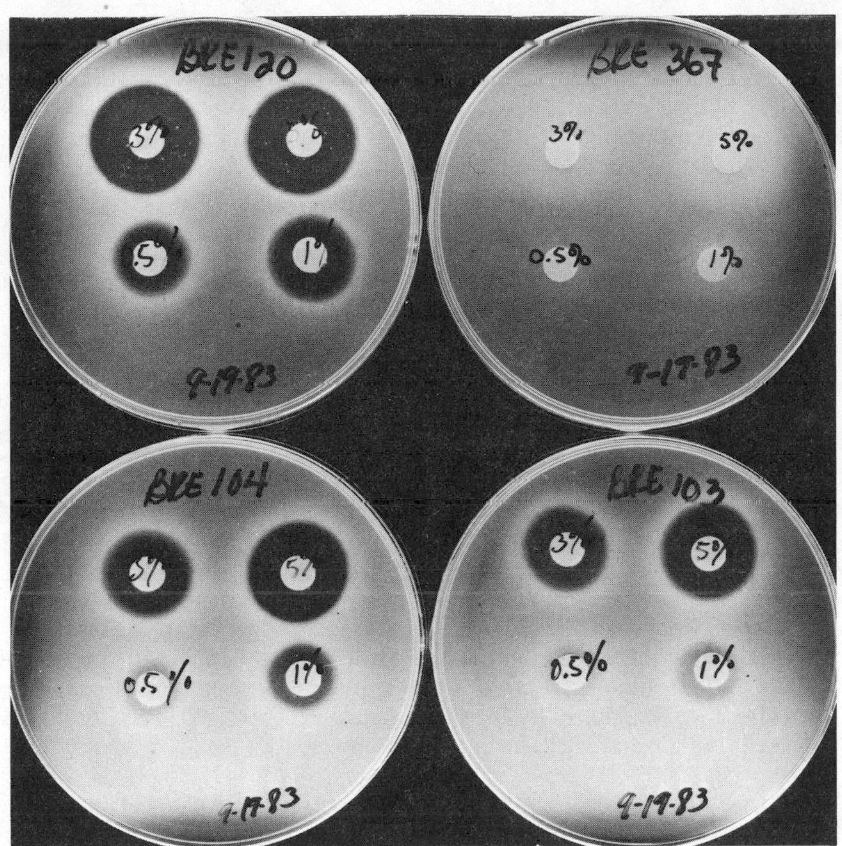

Fig. 3. Plate Assay for Maltose Sensitivity. The various lamB-lacZ fusion strains BRE 103, 104, 330, 367 were grown to mid-log growth in L broth at 37 °C. Cell aliquots (0.2 ml) were mixed with 3 ml of minimal F-top agar and overlayed on a 0.2% glycerol M63 plate. Filter discs (7 mm) of Whatman number 1 filter paper were placed on the agar overlay and 10 µl aliquots of 5%, 3%, 1%, and 0.5% maltose were placed on individual disc. The plates were then incubated overnight approximately 16 to 20 hours at 30 °C. Zones of inhibition appear as area around the discs where growth is prohibited.

CELLULAR COMPONENTS OF THE PROTEIN EXPORT MACHINERY

An important goal of studies on protein export is the identification and characterization of cellular components of the protein export machinery. One approach to this problem is to isolate mutants that exhibit a pleiotropic export-defective phenotype. Another approach is to devise a selection of mutants in which an internalized protein is exported. The mutants in which the precursor of the lamB gene product is found in the cytoplasm provide a genetic selection for the export of an internalized protein. These mutant strains do not localize LamB to the outer membrane because the export machinery cannot recognize the mutationally altered signal sequence. Therefore, one should be able to alter the export machinery by mutation to restore recognition of the signal sequence. Such mutations would define components of the export machinery. The availability of mutants in which the export process is altered should allow the identification of important gene products, and this, in turn, should provide a means to analyze the export pathway biochemically.

In mutant strains that produce a LamB protein with a defective signal sequence, the protein is found in soluble form in the cytoplasm with the altered signal sequence still attached. Consequently, such strains exhibit a typical LamB$^-$ phenotype, e.g., inability to utilize maltodextrins as a carbon source and resistance to bacteriophage λ. Using these strains, second-site pseudorevertants were isolated by selecting for the ability to grow on maltodextrins. The mutation responsible for pseudoreversion suppresses the defect caused by the altered signal sequence, restores export, and in most cases restores normal processing of the LamB protein. Three different second-site suppressor mutations have been identified. These suppressors, termed prl (protein localization) A, B, C, should define cellular components that interact with the signal sequence during the export process.

prlA. One of the suppressors, prlA, has been characterized in some detail. It causes phenotypic suppression of all export-defective signal sequence mutations in lamB, malE, and phoA. In some cases, prlA restores export to levels that are > 85 percent of that seen in wild-type strains. Despite this powerful suppression, prlA causes no growth defects nor does it alter normal protein export. Since the prlA suppressors restore export sequences, the cellular component altered by the prlA mutation probably interacts with this sequence during the export process. Genetic mapping shows that the gene altered by the suppressor mutation is a component of the spc operon (Emr et al., 1981). This operon, which maps at 72 minutes on the E. coli chromosome, has been shown to specify ten different ribosomal proteins. Using a combination of genetic, biochemical, and recombinant DNA techniques, we have shown that the prlA mutations define a heretofore unidentified gene that lies at the extreme promoter-distal end of the spc operon (Shultz et al., 1982).

Recent DNA sequence analysis by Cerretti and coworkers demonstrates that the prlA gene specifies an extremely basic protein containing 443 amino acids (Cerretti et al., 1983). Our results show that the prlA^4 allele, a particularly strong suppressor, is a point mutation that changes a isoleucine residue near the COOH-terminus to an asparagine (J. Shultz, unpublished results).

The function of the prlA protein remains to be elucidated. Although the prlA gene does not appear to code for a known ribosomal protein, its location in an operon that specifies ten other ribosomal proteins suggests that PrlA may be associated with the machinery of protein synthesis. The genetic data suggests that PrlA may be involved in signal sequence recognition. Accordingly, PrlA may function in the coupling of translation and export. Using a PrlA-LacZ hybrid protein as an antigen, we have attempted to raise an antibody that cross-reacts with the wild-type PrlA protein. With this reagent, we have, in collaboration with D. Oliver (State University of New York, Stoneybrook) and H. Liepke (Yale University), obtained preliminary data that PrlA may be a component of the procaryotic equivalent of SRP (Fig. 1). If this can be proven, it would provide the necessary biochemical evidence to support our predictions, which are based on genetic studies.

prlB. The prlB suppressor is unusual in several respects. First of all, it suppresses only LamB signal sequence mutations. Second, the suppressor restores export of the mutant LamB proteins but does not restore processing. Only a single prlB allele has been isolated. The mutation is a small deletion (approximately 250 bp) in the gene coding for the periplasmic ribose-binding protein that lies at 84 minutes in the E. coli chromosome (S. Emr and J. Garvin, unpublished results). We do not yet understand why a mutation in the gene coding for a protein involved in ribose transport can restore export of a LamB protein with a defective signal sequence. Moreover, since export in the presence of the prlB suppressor occurs without processing, the mutant LamB protein may not be localized by the normal route. For these reasons, we suspect that prlB is not a component of the export machinery and that it suppresses the lamB signal sequence mutations by a bypass mechanism.

prlC. Two alleles of prlC have been isolated. These suppressors resemble prlA because they suppress signal sequence mutations in both malE and lamB and restore both export and processing. However, because suppression is very weak, the prlC mutations have been difficult to characterize. Recently we determined that prlC maps at 68 minutes on the E. coli chromosome and that it is cotransducible at frequencies of 1-2 percent with argG. This map location provides no clues as to the nature of the cellular component altered by the suppressor mutation. Using a "brute-force" screening procedure, we have identified several clones from a library prepared from a prlC strain that suppresses lamB signal sequence mutations. Currently, we are in the process of characterizing these clones.

prlE*. The largest lamB-lacZ fusion specifies a hybrid protein that is exported to the outer membrane with high efficiency. Since the hybrid protein has little or no β-galactosidase activity when present in this cellular location, export-defective mutations can be obtained by selecting cells with the Lac$^+$ phenotype. As described above, most of these mutations are deletions in the lamB portion of the hybrid gene. However, we have obtained several mutations from this selection that are not linked genetically to the fusion. All of these mutations map between 9 and 10 minutes on the E. coli chromosome, tightly linked to the gene tsx. Moreover, all of these mutations have pleiotropic effects, including a pronounced growth defect at low (30° C) but not high (37° C) growth temperatures.

If these mutations prevent export of the LamB-LacZ hybrid protein as predicted, then they must be quite unusual. All of the known export-defective mutations block export at an early step causing the accumulation of precursor in the cytoplasm. With the maltose-sensitive lamB-lacZ fusions, the export block is evidenced by loss of the characteristic overproduction lethality. The mutations that are linked to tsx do not relieve overproduction lethality; indeed, they enhance it. Thus, if these mutations block export, they must do so at a step later in the* temporal export pathway. This possibility is exciting because it would provide, for the first time, a method for examining these steps. Currently, we are examining these mutations biochemically to find out whether they cause an export defect and, if this can be demonstrated, to gain some insight into the nature of the block.

prlF. The discovery of SRP and the demonstration of its role in protein export (Fig. 1) revealed a serious shortcoming in the selection procedures used previously for the isolation of export-defective mutants. With the objective of reducing the ever-present background mutations that block protein synthesis, these selections were designed to identify the rarer mutations that blocked protein export but did not affect protein synthesis. However, it has been shown that when certain steps in the export process are blocked, synthesis of exported proteins is decreased markedly (Oliver and Beckwith, 1982; Hall et al., 1983). We concluded, therefore, that we may have been eliminating an important class of mutation during our selections because we demanded that the β-galactosidase activity of the hybrid protein be expressed at high levels.

Consequently we have altered our selections to include maltose-resistant mutants that exhibit decreased β-galactosidase activities. The procedure that we employed is based on the observation that mutations that hinder translation, unlike mutations that terminate translation (nonsense), are only marginally polar. With gene fusions and the sophisticated lac genetic techniques, one can exploit this observation simply. Growth on lactose requires both the lacZ and the lacY (lactose permease) gene products. In the fusion strains we

*prlD was isolated by P. Bassford (University of North Carolina) as a suppressor of malE signal sequence mutations. The suppressor maps between 2 and 3 minutes in the E. coli chromosome.

use and in the lactose operon itself, lacY is located downstream from lacZ. In a fusion strain, any mutation that prevents transcription of the hybrid gene or any polar mutation that prevents synthesis of the hybrid protein will also prevent the expression of lacY. Since melibiose is a substrate of the lactose permease and since the normal melibiose transport system is not expressed at growth temperatures above 37° C, lacY function can be simply monitored independently from lacZ by scoring for growth on melibiose at 37° C. Thus by starting with a maltose-sensitive lamB-lacZ fusion strain (Mal^s, Lac^+, Mel^+ at 37° C), we were able to greatly enrich the cell population for the desired mutant by selecting for Mal^r, Lac^-, Mel^+ at 37° C. We grew the fusion strain for an extended period in medium containing maltose and then plated for survivors on minimal medium containing melibiose as the sole carbon source and a sensitive indicator for β-galactosidase, thus selecting for Mal^r, Mel^+ mutants. We examined those that had simultaneously become $LacZ^-$ as well (D. Kiino and T. Silhavy, unpublished results).

Twenty mutants were isolated initially and two were found that exhibited the expected phenotypes and that had a mutation that was unlinked genetically to the lamB-lacZ gene fusion. One of the mutations, prlF, maps at 70 minutes in the E. coli chromosome near argG. (It is now, however, linked to prlC, see above.) The other has not yet been mapped. Both mutations confer similar phenotypes except that prlF, like prlE, causes a pronounced growth defect at low temperature (30° C).

Although we have no direct biochemical proof that the prlF mutation affects a component of the cellular export machinery, we think it likely. As described above, maltose sensitivity has been a useful and sensitive indicator of hybrid protein export. High-level synthesis of exported LacZ hybrids blocks the localization of other proteins at some undefined step and this leads ultimately to cell death. Presumably this block occurs because sequences within the normally cytoplasmic protein LacZ are tolerated poorly by the cellular components that mediate the localization process. The prlF1 mutation relieves the blockade and restores normal protein export without preventing export of the hybrid itself.

The prlF1 mutation dramatically decreases the β-galactosidase activity that is associated with lamB-lacZ fusions that specify exported hybrid proteins, and the LacZ phenotype provides a means to study the prlF1 mutation independent of maltose sensitivity. If, as we propose, the mutation alters a cellular component involved in protein export, then mutations that prevent export should alter this phenotype. This is clearly the case. The prlF1 mutation has no effect on lamB-lacZ fusions that carry a mutation that alters or removes the signal sequence. Since signal sequence mutations block an early step in protein localization, it would appear that prlF acts subsequently to the initiation of the export process.

At present we do not know the nature of the prlF gene product nor do we understand its relationship to maltose sensitivity or protein export. It is possible that the prlF1 mutation causes increased production of some component of the export pathway that is rate-limiting during high-level expression of exported hybrid proteins. Alternatively, it may slow the kinetics of protein localization so that the hybrid can be exported from the cytoplasm with greater efficiency. Since prlF1 confers a growth defect that is independent of the presence of a gene fusion, it must affect some normal cellular function and not only hybrid protein export. We have not, however, been able to detect any noticeable effect on normal protein export. Further insights into the function of the prlF gene product must await the isolation of defined null mutations.

MULTIPLE EXPORT PATHWAYS

There is substantial evidence both in eucaryotic and procaryotic systems to indicate the existence of multiple export pathways. The most striking examples of this are the import proteins into mitochondria in eucaryotic cells and the export of certain proteins to the inner membrane of E. coli. In both of these cases, export appears to occur after synthesis of the protein is complete, i.e., postranslational export. This raises the possibility that there are multiple pathways of protein export to the outer membrane. Indeed, many outer membrane proteins (e.g., OmpF) first appear in discrete patches over the entire cell surface, whereas others (e.g., LamB) first appear in the septal region of dividing cells. Moreover, certain export-defective mutations isolated elsewhere appear to block strongly the export of some (e.g., OmpF), but not all (e.g., LamB), outer membrane proteins. For these reasons we have begun to study the export of another major outer membrane protein, OmpF.

Early in the course of these studies, an interesting ompF-lacZ gene fusion was constructed. This fusion specifies a hybrid protein that contains the complete OmpF signal sequence (22 amino acids) plus 12 amino acids of mature OmpF fused to a large functional COOH-terminal fragment of β-galactosidase. Unlike lamB-lacF fusions of similar size, the hybrid protein specified by this fusion is exported to the outer membrane. The efficiency of export, however, is very low (15-20 percent). In wild-type strains the fusion confers no unusual phenotypes.

Recently we discovered that although λ transducing phages that carry the ompF-lacZ gene fusion grow normally on wild-type strains, they appear not to grow on strains that carry on $ompR_2$ mutation. OmpR is a regulatory gene that specifies a positive activator required for ompF expression. The $ompR_2$ mutations cause constitutive high-level expression of ompF (for a recent review, see Hall and Silhavy, 1981). At first glance, this phenomenon is reminiscent of the overproduction lethality observed with lamB-lacZ fusions that specify an exported hybrid protein, i.e., high-level expression of the OmpF-LacZ hybrid protein is apparently lethal and dead cells cannot support phage growth. On closer examination, however, the two phenomena are quite dissimilar.

One-step growth experiments reveal that the ompF-lacZ fusion transducing phages grow normally in $ompR_2$ strains. The defect in phage production is due to a defect in cell lysis, not to the ability of infected cells to support phage growth. If cells are lysed artificially, a normal burst of phages can be detected. Although we do not yet understand fully the molecular mechanism responsible for this phenomenon, it is possible that the overproduction of the OmpF-LacZ hybrid protein that occurs in $ompR_2$ strains prevents export of the lysis proteins specified by phage λ. Preventing export of these proteins prevents lysis, and, unless cells are broken by some other method, no phage production can be detected. This is in marked contrast to results obtained with λ transducing phages carrying lamB-lacZ fusions. These phages grow normally in all strains, even those that constitutively express lamB. This may reflect a difference in the export pathways for LamB and OmpF.

If the explanation offered above is correct, then we should be able to exploit this phenotype to isolate mutations in the ompF portion of the hybrid gene that prevent the synthesis or that prevent the export of the OmpF-LacZ hybrid protein. This could be done simply by looking for mutations on the λ transducing phage that permit phage production, i.e., formation of a plaque on lawns of an $ompR_2$ strain. Mutations that prevent synthesis of the hybrid protein should result in a transducing phage that grows productively on the indicator strain but which produces little or no β-galactosidase activity. Mutations that prevent export of the hybrid protein should also allow productive growth of the phages on the indicator strain, however, these phages should produce high levels of β-galactosidase activity. By using a lac deletion strain that contains on $ompR_2$ mutation and medium that contains an indicator for β-galactosidase activity, one should be able to distinguish these two mutant classes.

Many mutant transducing phages have been isolated and characterized. As expected, these mutations fall into two general classes: those that prevent or greatly decrease expression of the hybrid protein and those that do not. Among the first class we found nonsense and frameshift mutations that lie in the ompF portion of the hybrid gene and cis-acting promoter or regulatory site mutations that decrease ompF expression. An ompF frameshift mutation is described in the following section. The promoter or regulatory mutations will be described elsewhere (Taylor, 1983). So far, at least, all of the mutations that allow productive phage growth, but do not decrease expression of β-galactosidase activity, appear to lie outside of the bacterial DNA carried by the fusion transducing phage. This finding may indicate a fallacy in our logic or may indicate that a mutation in λ DNA can suppress the detrimental effects of the OmpF-LacZ hybrid protein. Based on results described in the following section, we favor the latter explanation.

REVERTANTS OF AN ompF FRAMESHIFT MUTATION THAT ALTER THE SIGNAL SEQUENCE

Mutations obtained with the selection procedure described above have been analyzed at the level of the DNA sequence. Most of these are promoter or regulatory site mutations. One, however, turned out to be a frameshift mutation at a position corresponding to amino acid 14 of the OmpF signal sequence. We reasoned that, by selecting revertants of this mutation that restored expression of the ompF-lacZ hybrid protein, we were selecting for mutations only on the basis that they restore a correct reading frame, we expected that at least some would destroy the export information contained within the OmpF signal sequence.

Thirteen such revertants have been isolated and characterized. Most of these revertants are indistinguishable from wild type in that the transducing phage grows normally on wild-type strains but fails to grow productively on $ompR_2$ strains. Five mutant phages show an altered phenotype: four grow as well on the wild-type strain as on $ompR_2$ strains and one grows normally on the wild-type strain but forms plaques only poorly on $ompR_2$ strains.

Although we do not yet have DNA sequence data or any biochemical results to demonstrate that these revertants have altered the OmpF signal sequence, we think it likely. First, the correct reading frame to the ompF portion of the hybrid gene cannot be restored without altering the signal sequence unless the mutation resulted in true reversion. Since the phenotype is different than that seen with the original fusion transducing phage, we can discard this possibility. Second, two of the revertants that restore expression of β-galactosidase activity and yet still allow productive infection of $ompR_2$ strains are affected by prlA. These mutations are suppressed by prlA, i.e., the transducing phage will grow productively on $ompR_2$ strains but not on ompR2 prlA strains. Since prlA is a specific suppressor of signal sequence mutations, we conclude that these revertants have altered the OmpF signal sequence. By analogy, we suspect that the other revertants also have altered this sequence.

Since prlA suppresses the ompF signal sequence mutations as well as lamB signal sequence mutations, we conclude that at least this step in the export pathway must be the same for both proteins. If the export pathways for these two proteins are different, then they must diverge at a later step. Currently, these mutations are being analyzed to determine the precise nature of the mutation and the export defect.

CONCLUSIONS/PROJECTIONS

Results described here provide an example of the enormous progress that has been made in the last decade towards understanding the molecular mechanisms of protein localization. We have learned that the genetic information specifying correct cellular localization is contained within the structural gene. In many cases, this export information is composed of small discrete sequences arranged in a defined order in the primary structure of the protein. Each of these export signals mediates a particular step in the localization process. Alteration or removal of these signals by mutation causes the protein to be incorrectly localized to a different cellular compartment.

Specific selection procedures have been described that have permitted the identification of several genetic loci whose products are components of the cellular protein export machinery. With existing technology, this knowledge can be exploited to identify the gene products and raise specific antisera. Such reagents will allow protein isolation and biochemical characterization. Information gained in such studies will form the basis for reconstituting the export machinery in vitro from purified components.

Despite these advances, much remains to be done. Although we have identified intragenic export signals, our knowledge is descriptive. The mechanisms by which the various molecular components of these export signals function remain to be elucidated. With respect to the cellular export machinery, we have only scratched the surface. In all probability, most of the essential components have yet to be identified. Solutions to these problems will require more extensive mutant analysis and increasingly sophisticated in vitro systems.

ACKNOWLEDGEMENTS

Research sponsored by the National Cancer Institute, DHHS, under Contract No. NO1-CO-23909 with Litton Bionetics, Inc. The contents of this publication do not necessarily reflect the views or policies of the Department of Health and Human Services, nor does mention of trade name, commerical products, or organizations imply endorsement by the U.S. Government.

S. Benson is sponsored by the National Cancer Institute Postdoctoral Fellowship, E. Bremer by the Deutsche Forschungsgemeinschaft (West Germany); S. Garrett by the Natural Sciences and Engineering Research Council of Canada Postgraduate Fellowship, D. Kiino by the Damyon Runyon-Walter Winchell Cancer Fund, and J. Shultz by the Jane Coffin Childs Memorial Fund for Medical Research.

REFERENCES

Benson, S., and Silhavy, T., 1983, Information within the mature LamB protein necessary for localization to the outer membrane of Escherichia coli K-12, Cell, 32:1325.

Blobel, G., and Dobberstein, D., 1975, Transfer of proteins across membranes. II. Reconstitution of functional rough microsomes from heterologous components, J. Cell Biol., 67:852.

Cerretti, A., Dean, D., Davis, G.R., Bedwell, D.M., and Nomura, M., 1983, The SPC ribosomal protein operon of Escherichia coli: Sequence and cotranscription of the ribosomal protein genes and a protein export gene, Nucl. Acids Res., 11:2599.

Emr, S.D., and Silhavy, T.J., 1982, The molecular components of the signal sequence that function in the initiation of protein export, J. Cell Biol., 95:689.

Emr, S.D., Hanley-Way, S., and Silhavy, T.J., 1981, Suppressor mutations that restore export of a protein with a defective signal sequence, Cell, 23:79.

Hall, M.N., and Silhavy, T.J., 1981, Genetic analysis of the major outer membrane proteins of Escherichia coli, Ann. Rev. Genet., 15:91.

Hall, M.N., Gabay, J., and Schwartz, M., 1983, Evidence for a coupling of synthesis and export of an outer membrane protein in Escherichia coli, Cell, 30:311.

Oliver, D.B., and Beckwith, J., 1983, Regulation of a membrane component required for protein secretion in Escherichia coli, Cell, 30:311.

Palade, G.E., 1975, Intracellular aspects of the process of protein synthesis, Science, 189:347.

Shultz, J., Silhavy, T.J., Berman, M.L., Fiil, N., and Emr, S.D., 1982, A previously unidentified gene in the spc operon of Escherichia coli K-12 specifies a component of the protein export machinery. Cell, 31:227.

Silhavy, T.J., Benson, S.A., and Emr, S.D., 1983, Mechanisms of protein localization. Mibrobiol. Rev., 47:313.

Taylor, R.K., 1983, "Transcription Regulation of the Synthesis of the Major Outer Membrane Porin Proteins of E. coli K12," Ph.D. dissertation, University of Maryland Baltimore County.

Walter, P., and Blobel, G., 1981, Translocation of proteins across the endoplasmic reticulum. III. Signal recognition protein (SRP) causes signal sequence-dependent and site-specific arrest of chain elongation that is released by microsomal membranes, J. Cell Biol., 91:557.

THE CELL MEMBRANE OF EUKARYOTES

EKTOBIOLOGY, GROWTH CONTROL AND CANCER

Arthur B. Pardee

Dana-Farber Cancer Institute
44 Binney Street
Boston, Massachusetts 02115

ABSTRACT

An enormous and varied amount of experimental information has shown over the years that cells' surfaces are altered both by conditions of growth and by neoplastic transformation. Clearly, every aspect of ektobiology can change in either of such circumstances. The regulation of mammalian cell growth by external agents such as the substratum on which cells are growing, neighboring cells, nutrients, and particularly growth factors has been well established. Transport systems for nutrients as well as receptors for growth factors have been identified, and in a few cases they have been isolated. The numbers and properties of these receptors also can change under different conditions of growth and following transformation. The question remains, though, do such changes cause alterations in growth or are they simply secondary adaptations of tumorigenic primary events?

The experiments briefly summarized here lead us to the conclusion that membrane receptors for nutrients and for factors are certainly important and necessary for growth, but that their quantities are probably not limiting fibroblast growth. Thus, contrary to the original hypothesis, growth control and its variation following transformation probably is not mediated by these ektobiological properties. The secrets of growth regulation, indeed influenced by external factors that are utilized via ekto systems, must reside within the cell. It is in this domain that our more recent efforts have been focused.

Aided by a grant GM/CA24571 from the National Institutes of Health. We thank Marjorie Rider for preparing the manuscript.

Herman Kalckar and I became acquainted just 20 years ago, in 1963. We participated in a cancer meeting at Lima, at which I first heard Herman use the term "Ektobiology" (Kalckar, 1964). He suggested that in tumor cells the formation of galactose might be inhibited sufficiently so that altered galactosyl compounds on the surface could have a striking effect in permitting neoplastic growth. In the adjacent article I proposed that membrane transport properties of cells must be important for growth regulation, and that alterations of the cancer cell membrane could be important in this regulation (Pardee, 1964). We thus had an immediate common scientific interest in the surfaces of cells, Herman's being oriented structurally and mine towards functions of the membrane in terms of receptors for growth regulatory molecules present in the environment.

We went together to Cuzco on a side trip. Among many events that we experienced during that trip was one that was more an endo- rather than an ekto- affair. At a folk festival in the local high school, the climax was shooting off of numerous fireworks within the building. This was very exciting, particularly when it seemed quite possible that the wooden building itself might burn down.

Enormous and varied experimental information has shown over the intervening years that cells' surfaces are altered by both conditions of growth and by neoplastic transformation (Pardee, 1975; Nicolson, 1982). Clearly, every aspect of ektobiology can change in either of such circumstances. The regulation of mammalian cell growth by external agents such as the substratum on which cells are growing, neighboring cells, nutrients, and particularly growth factors has been well established. Transport systems for nutrients as well as receptors for growth factors have been identified, and in a few cases they have been isolated. The numbers and properties of these receptors also can change under different conditions of growth and following transformation. The question remains, though, do such changes cause alterations in growth or are they simply secondary adaptations of tumorigenic primary events?

If a function is to be considered as growth regulatory it should show certain characteristics. One criterion is that the function must change appropriately with the growth state. For example, nutrient transport should be lower in resting than in growing cells if transport controls growth. A second criterion is that this connection should be relaxed in cells with defective growth control, i.e., in tumorigenic cells.

Boris Ephrussi (1972) divided cell functions into housekeeping and luxury. The former are functions necessary for maintenance and growth of cells—such as processes of energy production and macromolecular synthesis. The luxury functions are a class found in only certain cells. They include hormones and other specialized products, and also receptors, such as, for example, acetylcholine receptors of nerve cells. Another distinction is between functional and regulatory cell constitutents. This categorization evidently can overlap the other. That is, some housekeeping functions could be regulatory, for example, the level of some key enzyme or transport system could limit growth; or such a function might simply be necessary but not rate-limiting under physiological conditions. A decrease of any necessary function will of course limit growth, for example if the aerobic energy producing pathway is poisoned by cyanide. But this can hardly be considered as evidence of relevance to physiological regulation.

Since these initial proposals that characteristics of cells' surfaces might regulate growth and that surface derangements could be important in cancer, we have tested our hypothesis. Do ekto systems for nutrient transport or growth factor reception play a primary controlling role, limiting functioning of the external agents that interact with them? Or are these ekto components simply present in amounts that vary according to needs of the cell, but do not themselves limit the growth rate? In the remainder of this essay I will summarize results bearing on this question, first those related to nutrient transport systems and second to growth factor receptors.

NUTRIENT TRANSPORT AND GROWTH CONTROL

Alterations in transport activities of animal cells' plasma membranes may, it has been suggested, play an important role in the regulation of cell growth and multiplication, and may be altered in tumorigenic cells so as to relax growth control (Foster and Pardee, 1969; Holley, 1972). Consistent with this possibility, we found (Fig. 1) transport rates of several metabolizable and nonmetabolizable amino acids to be increased when cells were taken from confluent quiescence to growth, or after they were transformed by a DNA virus (Foster and Pardee, 1969; Cunningham and Pardee, 1969). Also, a marked stimulation of both transport and also incorporation of phosphate into membrane phospholipids was observed following stimulation of quiescent cells with serum (Cunningham, 1972). This stimulated phosphorylation recalls stimulations of membrane protein phosphorylation dependent on various growth factors including epidermal growth factor (EGF) (Carpenter and Cohen, 1979). These changes in transport activity occur quickly, within minutes after adding serum to quiescent cells. Changes were seen for transport of some but not for all amino acids, indicating that a general membrane alteration had not modified permeability and transport. Rather, changes were found only for special metabolites.

Numerous subsequent studies have not provided definitive evidence supporting a controlling role of nutrient transport for growth, and some have supplied negative evidence regarding specific

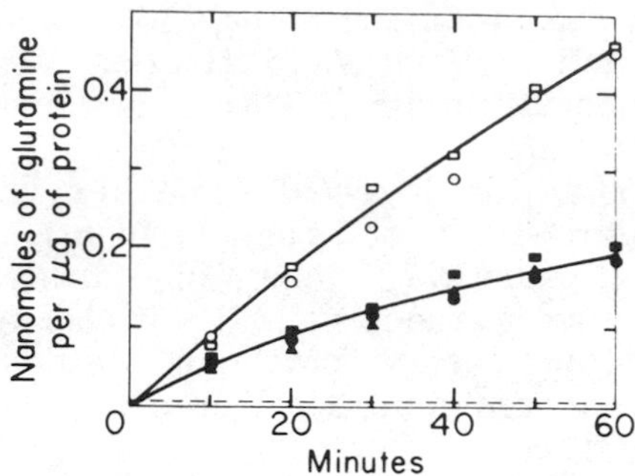

Fig. 1. Uptake of L-glutamine by 3T3 cells (closed symbols) and Py3T3 cells (open symbols) at different densities on glass coverslips: ●, 2.2 x 10^4 cells (nonconfluent) and 18 μg of protein per coverslip; ▲, 3.8 x 10^4 cells (nonconfluent) and 26 μg of protein; ■, 9.2 x 10^4 cells (confluent) and 47 μg of protein; o, 2.1 x 10^4 cells (nonconfluent) and 20 μg of protein; □, 7.0 x 10^4 cells (confluent) and 44 μg of protein. The concentration of glutamine in the incubation medium was 2.0 mM. The broken horizontal line represents the level expected from passive diffusion (Foster and Pardee, 1969).

nutrients (see Pardee, 1982). There are, however, some noteworthy relationships. Cells arrested by nutrient limitations can be stimulated to grow by adding only growth factors (Kamely and Rudland, 1976). Also, lower concentrations of serum are required for growth after the concentrations of four common ions and pyruvate are optimized (McKeehan and McKeehan, 1980).

Transport changes were seen as results of both transformation and state of growth, and it is important to distinguish these. Changes in transport activities in nontransformed cells decreased dramatically well before their growth was arrested, whereas transformed cells at the same density showed no such change (Fig. 2). Differences in transport rates for glucose between untransformed and transformed cells that have been observed by several workers could be based on such differences in growth rates, although the two kinds of cells might have been at the same density. Thus, one can be misled into believing that a transport difference between transformed and untransformed cells exists even if measurements are made before the untransformed cells reach confluence, whereas in actuality the difference could be related to slowing of growth.

In order to decide whether a transport decrease is truly growth-limiting, or is merely a secondary consequence of slower metabolism as the cells' growth is otherwise limited, one needs at least to know the transport rate required to just maintain growth. Nutrient

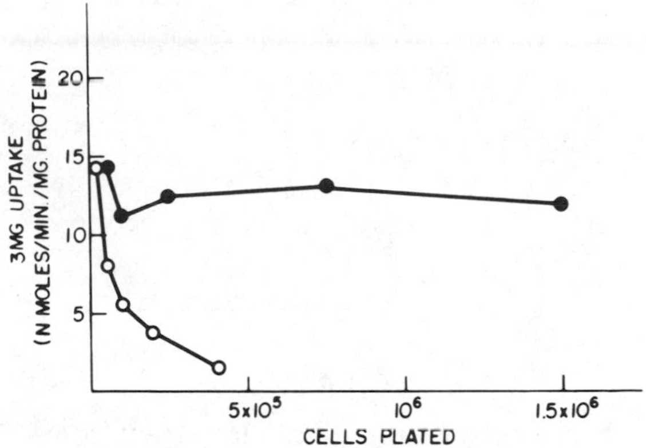

Fig. 2. 3MG transport as a function of cell density. Cells were plated over a range of densities and assayed for 3MG uptake on the following day. 3T3 o---o; SV101 ●---● (Dubrow, et al., 1978).

concentrations have to be decreased as much as 100-fold to arrest growth (Kamely and Rudland, 1976); transport rates must be reduced corresponding to these low concentrations. But as 3T3 cells approach confluence and quiescence, their rates of transport of amino acids and of sugars drop only a few fold, and approach values that can be calculated as possibly being somewhere near rate-limiting for rapid growth (Pardee, 1982). Several workers have observed that pools of amino acids actually rise when cells become confluent; therefore rate of utilization seems to be decreased more rapidly than rate of transport. This result suggests that transport rates could be altered as consequences of decreased overall metabolism. Another very different indication that growth and transport are not closely coupled is that factors that stimulate these two processes can be physically separated (Cunningham and Pardee, 1969).

In summary, altered transport rates have been observed as correlates of growth states of cells and of tumorigenic transformation. These changes are generally only a few fold in magnitude, and they are seen for some nutrients but not others. The attractive idea that such changes actually control growth, by limiting the supply of some essential nutrient, has not received any direct confirmation, and there is evidence against it for some substances (see Pardee, 1982). Most major nutrients—sugars, amino acids, and phosphate—are not growth-limiting except when supplied at extremely low concentrations which are unlikely to be physiologically relevant. It is still possible that transport of some yet untested nutrient could be rate-limiting under physiological conditions, though there is no direct evidence for this possibility.

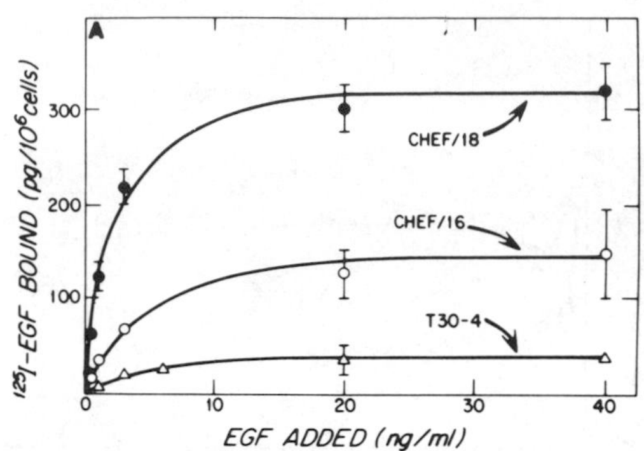

Fig. 3. Diminished binding of EGF by transformed CHEF cells. Final cell densities (in cells/cm^2) in this experiment were $4.1 \pm 0.2 \times 10^4$ for CHEF/18 cells (●), $7.7 \pm 0.06 \times 10^4$ for CHEF/16 cells (o), and $13.9 \pm 0.5 \times 10^4$ for T30-4 cells (▲). Binding of various quantities of EGF to CHEF cells in PBS plus 0.1% BSA; nonspecific binding was subtracted. Bars represent S.E.M. (Cherington and Pardee, 1982).

GROWTH FACTOR RECEPTORS AND GROWTH CONTROL

Growth factors have become increasingly evident as determinants of cell proliferation (Sato, et al., 1982). Whether receptors for these factors could limit or control growth in normal vs. transformed cells is a question that has been approached by cultivating cells in totally defined media (Cherington, et al., 1979). Chinese hamster embryo fibroblast line CHEF/18 can be grown without serum in the presence of the four small proteins insulin, epidermal growth factor, thrombin, and transferrin at a rate comparable to its growth in medium containing 10 percent serum. Various transformed lines related to CHEF/18 also grow well in this medium, and in general have specifically lost their requirement for EGF. After these lines that demonstrate transformed phenotypes in culture are subcultivated in nude mice, and the cells from the tumor put back into culture, other growth factors also are not required. These results show a close correspondence between loss of growth factor requirements, particularly for EGF, and transformed phenotypes in culture as well as tumorigenicity in vivo.

The question to be addressed here is whether such a decreased requirement for growth factors is due to increased activity of receptors for the growth factor, or alternatively depends on some event within the cell that short circuits the requirement for added growth factor, or upon sufficient overproduction by the transformed cells of a factor to satisfy the growth requirement. If receptors are limiting, one expects growing and transformed cells to have higher receptor activities. Measurements made of the number of EGF

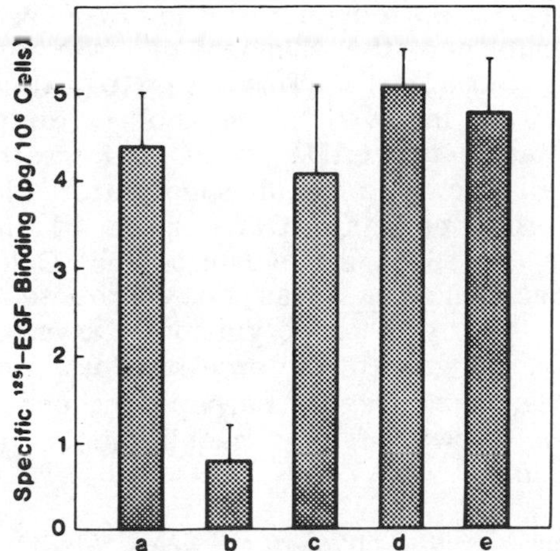

Fig. 4. ^{125}I-labeled EGF binding to CHEF/18 cells pretreated with media conditioned by transformed cell lines. ^{125}I-EGF binding was assayed in PBS + 0.1% BSA using CHEF/18 cells pretreated for 1 hr as follows: (a) with fresh Waymouth's medium; (b) Waymouth's medium conditioned by 3B11-1C; (c) by CHEF/18; (d) by T30-4; (e) nonpretreated cells. Bars represent S.E.M. (Cherington and Pardee, 1982).

receptors on quiescent and growing CHEF/18 cells showed there is little difference. In variously transformed cells (Cherington and Pardee, 1982), surprisingly, the number of receptors per cell decreased after CHEF/18 cells were transformed (a 2X decrease); after passage of these transformed cells through nude mice, a 10-fold decrease was found (Fig. 3). The EGF binding constant remained the same, but the total number of sites per cell diminished. Thus this lost requirement for EGF cannot be attributed to increased activity or quantity of the EGF receptor. Indeed, the results were exactly to the contrary. Furthermore, the decreased number of measured EGF receptors could not be attributed to secretion by the transformed cell of a tumor growth factor, in the manner described by Todaro et al. (1976). As a control, conditioned medium prepared with a Moloney transformed 3T3 cell line that secretes transforming growth factors blocked EGF binding by CHEF/18 cells (Fig. 4). But media similarly prepared with transformed CHEF/18 cells had no effect on this binding, and hence these cells were not secreting an "autocrine" competitive factor.

Alterations of receptors may be very important if they bypass activation by growth factors. The decreased binding that was seen in parallel with loss of the EGF requirement of transformed CHEF/18 cells indicates that the EGF receptor itself was modified. In a very similar way, cholera toxin eliminates the EGF requirement of 3T3 cells, and also diminished the number of available EGF receptors (Wharton et al., 1982). This result also suggests that phosphorylations may be involved, since cholera toxin raises the cAMP level and thereby activates cAMP dependent protein kinases. The EGF receptor is well known to be phosphorylated when it is activated by EGF, and also when cells are transformed by Rous sarcoma virus which provides the src oncogene (Ushiro and Cohen, 1980). Very recently another oncogene, erb, has been proposed to code for the EGF receptor, perhaps by phosphorylation (Downward et al., 1984). Thus the modulation of growth factor receptors, under the influence of transformations and oncogenes, perhaps by phosphorylation, could replace the normal growth factor requirements and provide steps toward transformation.

Growth factors and nutrients are clearly necessary for proliferation of at least normal cells, since total removal of either can stop growth. Specific membrane sites are required for both growth factors and nutrients. But the regulation of proliferation does not appear to be determined by amounts or activities of at least these EGF receptors of CHEF cells. It is too early to generalize this result; we cannot say whether modulations of receptor numbers or activities control growth of other cells under the influence of other factors. Very recently, data have been presented that suggest receptors for Interleukin 2 are important in growth control of T lymphocytes (Cantrell and Smith, 1984).

The cell surface nevertheless may be important for controlling growth, perhaps by processes dependent on chemical modifications as suggested so long ago by Herman Kalckar (1964). Much indirect evidence relates to cytoskeleton, containing microfilaments and microtubules, to growth and transformation. Connections via the cytoskeleton between cytoplasmic membrane and nucleus could be a very important means of signal transmission. Very recent information emphasizing the importance of surface events for growth and cancer have come unexpectedly from a novel line of research. Oncogenes are genetic elements related to tumorigenicity (Land, et al., 1983). Some oncogenes code for proteins located at the inner side of the cytoplasmic membrane; these proteins thus may have membrane related roles in growth. Furthermore, some oncogenes are protein kinases, of the rare kind that phosphorylate tyrosine residues. It will be recalled that EGF and other growth factors have cell membrane receptors that also are tyrosine-protein kinases (Carpenter and Cohen, 1979). These novel clues give us only hazy outlines of processes by which both oncogenes and growth factors may modulate growth by altering specific membrane components.

SUMMARY

The experiments briefly summarized here lead us to the conclusion that membrane receptors for nutrients and for factors are important and necessary for growth, but their quantities are probably not limiting fibroblasts' growth. Current data are contrary to the original hypothesis that growth control and its variation following transformation probably are mediated by these ektobiological properties. The secrets of growth regulation, indeed influenced by external factors that are utilized via ekto systems, must reside within the cell. It is in this domain that our more recent efforts have been focused (Croy and Pardee, 1983).

To conclude this little essay I would like to draw a parallel between these ektobiological studies and the ektobiology of Herman himself. Ekto-Herman has remained admirable throughout the years, as is plain for all to see! But even more remarkable, as I have learned during the past 20 years, is endo-Kalckar—a man of remarkable depth and sympathy. Furthermore, as I got to know Herman, I became better able to interpret his conversation. Many have told me of puzzlement in understanding Herman's oblique style. I have concluded that he carries on an ekto-conversation. His words comprise a surface, within which hides his idea, and it is this idea that he leaves for us listeners to decipher.

EKTO-HERMANESE

Sometimes, Herman's ekto-conversation
Creates in listeners consternation.
Our comprehension approaches nought!
We must then seek his endo-thought
To reach enlightened illumination.

REFERENCES

Cantrell, C.A., and Smith, K.A., 1984, Transient expression of Interleukin 2 receptors: consequences for T cell growth, J. Exp. Med., 158:1895.

Carpenter, G., and Cohen, S., 1979, Epidermal growth factor, Ann. Rev. Biochem., 48:193.

Cherington, P.V., Smith, B.L., and Pardee, A.B., 1979, Loss of epidermal growth factor requirement and malignant transformation, Proc. Natl. Acad. Sci. USA, 76:3937.

Cherington, P.V., and Pardee, A.B., 1982, On the basis for loss of the EGF growth requirement by transformed cells, in: "Growth of Cells in Hormonally Defined Media," Cold Spring Harbor Laboratory, New York.

Croy, R.G., and Pardee, A.B., 1983, Enhanced synthesis and stabilization of M_r 68,000 protein in transformed BALB/c-3T3 cells:

candidate for restriction point control of cell growth, Proc. Natl. Acad. Sci. USA, 80:4699.

Cunningham, D.D., and Pardee, A.B., 1969, Transport changes rapidly initiated by serum addition to "contact inhibited" 3T3 cells, Proc. Natl. Acad. Sci. USA 64:1049.

Cunningham, D.D., 1972, Changes in phospholipid turnover following growth of 3T3 mouse cells to confluency, J. Biol. Chem., 247:2464.

Downward, J., Yarden, Y., Mayes, E., Scrace, G., Totty, N., Stockwell, P., Ullrich, A., Schlessinger, J., and Waterfield, M.D., 1984, Close similarity of epidermal growth factor receptor and v-erb-B oncogene protein sequences, Nature, 307:521.

Dubrow, R., Pardee, A.B., and Pollack, R., 1978, 2-Amino-isobutyric acid and 3-O-methyl-D-glucose transport in 3T3, SV40-transformed 3T3 and revertant cell lines, J. Cell. Physiol., 95:203.

Ephrussi, B., 1972, Hybridization of somatic cells, Princeton University Press, Princeton, N.J.

Foster, D.O., and Pardee, A.B., 1969, Transport of amino acids by confluent and nonconfluent 3T3 and polyoma virus-transformed 3T3 cells growing on glass cover slips, J. Biol. Chem., 244:2675.

Holley, R.W., 1972, A unifying hypothesis concerning the nature of malignant growth, Proc. Natl. Acad. Sci. USA, 69:2840.

Kalckar, H.M., 1964, Aberrations of metabolic patterns of malignant cells and their relevance to cell biology, Natl. Cancer Inst. Monograph, 14:21.

Kamely, D., and Rudland, P., 1976, Nutrient-dependent arrest of fibroblast growth is partially reversed by insulin but not fibroblast growth factor, Nature, 260:51.

Land, H., Parada, L.F., and Weinberg, R.A., 1983, Cellular oncogenes and multistep carcinogenesis, Science, 222:771.

McKeehan, W.L., and McKeehan, K.A., 1980, Serum factors modify the cellular requirement for Ca^{2+}, K^{+}, Mg^{2+}, phosphate ions, and 2-oxocarboxylic acids for multiplication of normal human fibroblasts, Proc. Natl. Acad. Sci. USA, 77:3417.

Nicolson, G.L., 1982, Cancer metastasis—Organ colonization and the cell-surface properties of malignant cells, Biochim. Biophys. Acta, 695:113.

Pardee, A.B., 1964, Cell division and a hypothesis of cancer, Natl. Cancer Inst. Monograph, 14:7.

Pardee, A.B., 1975, The cell surface and fibroblast proliferation some current research trends, Biochim. Biophys. Acta, 417:153.

Pardee, A.B., 1982, Growth regulation by transport into mammalian cells, in "Membranes and Transport," volume 2, A. Martonosi, ed., Plenum Publishing Corp., New York.

Sato, G.H., Pardee, A.B., and Sirbasku, D.A., eds., 1982, "Growth of Cells in Hormonally Defined Media," Cold Spring Harbor Laboratory, New York.

Todaro, G.J., DeLarco, J.E., and Cohen, S., 1976, Transformation by murine and feline sarcoma virus specifically blocks binding of epidermal growth factor to cells, Nature, 264:26.

Ushiro, H., and Cohen, S., 1980, Identification of phosphotyrosine as a product of epidermal growth factor-activated protein kinase in A-431 cell membranes, J. Biol. Chem., 255:8363.

Wharton, W., Leof, E., Pledger, W.J., and O'Keefe, E.J., 1982, Modulation of the epidermal growth factor receptor by platelet-derived growth factor and choleragen: effects of mitogenesis, Proc. Natl. Acad. Sci. USA, 79:5567.

CARBOHYDRATE ANTIGENS OF CELL SURFACES DETECTED BY MONOCLONAL ANTIBODIES

Victor Ginsburg

National Institute of Arthritis, Diabetes, and
Digestive and Kidney Diseases
National Institutes of Health
Bethesda, Maryland

ABSTRACT

To obtain monoclonal antibodies specific for various cancers, mice and rats have been immunized with human tissues in many laboratories. Some of the antibodies derived from spleen cells of the immunized animals have an apparent specificity for certain cancers and are directed against carbohydrates. Out of 325 of these antibodies sent to us from various laboratories in the past three years, 97 are directed against carbohydrates as determined by solid-phase radioimmunoassay and autoradiography. Of the 97 antibodies, 7 are directed against the H type 1 or H type 2 antigens, 4 against the Le^b antigen, 8 against the A antigen, and 3 against the B antigen of the human ABO and Lewis blood group systems. Fifty-five are directed against a sugar sequence found in the human milk oligosaccharide lacto-N-fucopentaose III, 2 against a sialyated Le^a antigen and 18 against unidentified carbohydrate sequences in glycolipids and/or glycoproteins.

The carbohydrates on cell surfaces change during development as they are probably involved in cell recognition (Kalckar, 1967). It is these developmentally regulated changes that allow some antibodies directed against carbohydrates to discriminate among various tissues, both normal and malignant. The most important glycosyl residues as far as immunological specificity is concerned are generally at nonreducing ends of the carbohydrate chains. These "immunodominant" residues determine the antigenic specificity of the molecules in which they occur. Because the terminal sequences of sugars in the sugar chains of glycoproteins and glycolipids are somtimes identical, some antibodies directed against carbohydrates react with both glycoproteins and glycolipids.

Unlike protein antigens, which are primary gene products, carbohydrate antigens are secondary gene products. The primary gene products are glycosyltransferases which add single glycosyl residues to the growing chains, and it is the presence or absence of these enzymes that determine which particular chains or antigens are synthesized. Over 70 different disaccharide sequences occur in the complex carbohydrates of cell surfaces (Dawson, 1978), most of which are products of separate glycosyltransferases. As the acceptor specificity of these enzymes is greater than just single glycosyl residues, the total number of glycosyltransferases that a cell can potentially produce is probably in the hundreds. The expression of these enzymes varies during development, and it is this variation that accounts for the characteristic pattern of carbohydrate structures that occur in different tissues and for the appearance and disappearance of carbohydrate antigens (Hakomori and Kannagi, 1983). In addition, because some glycosyltransferases compete for the same chains, the amount of antigen formed depends on the relative activities of those enzymes that are present (Rohr et al., 1980). Thus, there are several ways that carbohydrate antigens are modulated as illustrated in Fig. 1. Assuming the trisaccharide sequence labeled B is an antigen, the level of antigen B will be maximal if only enzymes a and b are active. Its level will be reduced if enzymes c or d are active, and it will disappear if enzymes a or b disappear.

In attempts to obtain monoclonal antibodies specific for various cancers, mice and rats have been immunized with human tissues in many laboratories. Some of the antibodies derived from spleen cells of the immunized animals have an apparent specificity for certain

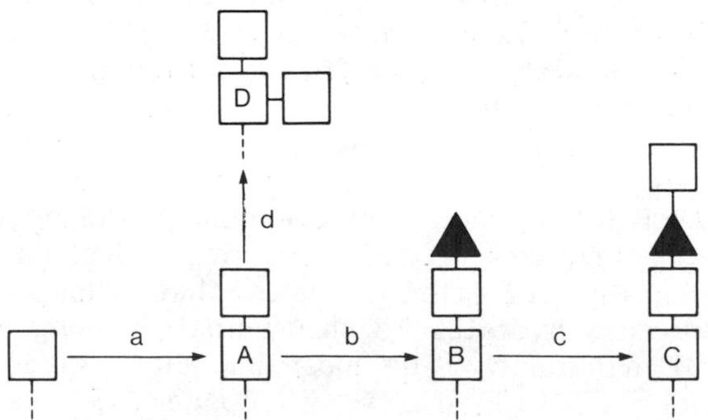

Fig. 1. Modulation of expression of carbohydrate antigens. A, B, C, and D are different carbohydrate antigens: a, b, c, and d are glycosyltransferases (see text).

cancers and are directed against carbohydrates. We have tested for carbohydrate specificity about 325 of these antibodies sent to use from various laboratories in the past three years (Brockhaus et al., 1981; Brockhaus et al., 1982; Magnani et al., 1982; Huang et al., 1983a, 1983b; Fredman et al., 1983; and unpublished results). About one third are directed against carbohydrates as determined by solid-phase radioimmunoassay and autoradiography (Table 1). The first six antigens are associated with the human ABO and Lewis blood group systems. Of these, the antibody against the H type 1 antigen is especially interesting as it specifically precipitates the epidermal growth factor receptor of the human epidermoid carcinoma cell line A431 (a glycoprotein of Mr 170,000) and also reacts with glycolipids containing the H type 1 sequence of sugars (Fredman et al., 1983).

About one sixth of the antibodies tested are directed against a sugar sequence found in the human milk oligosaccharide lacto-N-fucopentaose III (Kobata and Ginsburg, 1969) and which must be extremely immunogenic in mice and rats. A glycolipid containing lacto-N-fucopentaose III was first isolated from a human adenocarcinoma (Yang and Hakomori, 1971). The same sugar sequence minus the glycosyl residue occurs in higher glycolipids (Hakomori et al., 1981) and also in glycoproteins (Lloyd et al., 1968). Antibodies against this sugar sequence detect a stage-specific embryonal antigen (called SSEA-1) of the murine embryo and teratocarcinoma (Hakomori et al., 1981; Gooi et al., 1981); they also detect an antigen (called My-1) that is strongly expressed in human granulocytes and granulocyte precursor cells but not in normal peripheral blood lymphocytes, monocytes, platelets or red cells (Huang et al., 1983); and they also detect an antigen characteristic of human small cell carcinoma, adenocarcinoma, and squamous cell carcinoma of the lung (Huang et al., 1983). Interestingly, the My-1 antigen transiently appears on My-1-negative mouse fibroblasts following transfer of DNA from human myeloblastic or lymphoblastic leukemia cells (Chang et al., 1982). As My-1 is a carbohydrate antigen, it is not a direct gene product (Ginsburg, 1972; Watkins, 1980). Possibly DNA transfer transiently alters the pattern of glycosyltransferases normally present in the recipient cells, and it is these altered glycosyltransferases that catalyze the synthesis of My-1.

The antibodies directed against the sialylated Le^a antigen (Magnani et al., 1982) may be useful diagnostically as they detect antigen in the serum of many patients with gastrointestinal and pancreatic cancer (Heryln et al., 1982). Although the sialylated Le^a sequence of sugars occurs in the gangliosides of pancreatic and gastrointestinal cancers (Magnani et al., 1982; Hansson et al., 1983), the antigen detected in the sera of patients are in high molecular weight carbohydrate-rich glycoproteins known as mucins. The evidence for this is as follows (Magnani, 1983): little antigen is extracted by organic solvents from sera and that which is extracted remains at the origin under conditions of thin-layer chromatography

Table 1: Monoclonal Antibodies Directed Against Carbohydrates

Number of Antibodies (out of 325 tested)	Antigen	Structure
1	H type 1	Fucα1-2Galβ1-3GlcNAc ...
1	H type 2	Fucα1-2Galβ1-4GlcNAc ...
5	H type ?	Fucα1-2Galβ1-?GlcNAc ...
4	Le^b	Fucα1-2Galβ1-3GlcNAc ... 4 | Fucα1
8	A	Fucα1-2Galβ1-4GlcNAc ... 3 | GalNAcα1
3	B	Fucα1-2Galβ1-4GlcNAc ... 3 | Galα1
55	Lacto-N-fucopentaose III (also called SSEA-1, VIM, My-1, Le^x, and X-hapten)	Galβ1-4GlcNAc ... 3 | Fucα1
2	Sialylated Le^a	NeuNAcα2-3Galβ1-3GlcNAc ... 4 | Fucα1
18	Unidentified carbohydrate sequences in glycolipids and/or glycoproteins	
97		

where the ganglioside antigen migrates up the plate. Upon gel filtration of serum on Sephacryl S-400, the antigen elutes in the void volume, indicating an Mr $\geq$ 5 x 10^6. Incubation for 5 hr at 37° C in 0.1 N NaOH destroys the serum antigen but does not affect the ganglioside antigen. The density of the serum antigen as determined in a CsCl gradient is 1.50 g/ml, while in 4 M guanidine HCl its density is 1.43 g/ml. Finally, antigen affinity-purified by anti-sialylated Le^a antibody from the serum of a cancer patients belonging to the Le(a-b+) blood group contains Le^b antigen, consistent with the multiple antigenic specificities exhibited by mucins. The occurrence of mucins in the blood of cancer patients has been reported many times (Race and Sanger, 1975).

In the previous study (Magnani et al., 1982), sialylated Le^a antigen was not detected by solid-phase radioimmunoassay in extracts from normal adult tissues. By immunoperoxidase labeling of normal tissue sections, however, the antigen was found in a layer of ductal cells in normal pancreas and a layer of cells in normal salivary glands and bronchial epithelium that secrete mucins (Atkinson et al., 1982) and by autoradiography low levels of ganglioside antigen were detected in extracts of normal pancreas (Hansson et al., 1983). The antigen is also found in salivary mucins from most normal individuals belonging to the Le(a+b-) or Le(a-b+) blood group and is not found in salivary mucins from normal individuals belonging to the Le(a-b-) blood group (Brockhaus et al., 1984). About 7 percent of the population belong to the Le(-b-) blood group because they lack the fucosyltransferase that catalyzes the synthesis of the sugar sequence Fucα1-4GlcNAc ... (Grollman et al., 1969). As a consequence, cancer patients belonging to the Le(a-b-) blood group cannot synthesize sialylated Le^a antigen (Koprowski et al., 1982). Since the carbohydrate composition of mucins varies in different patients, monoclonal antibodies directed against other carbohydrate sequences found in mucins might be used together with the anti-sialylated Le^a antibody for a more sensitive serum test for gastrointestinal and pancreatic cancer.

The monosaccharide residues commonly found in the complex carbohydrates of cell surfaces include three hexoses (D-glucose, D-galactose and D-mannose), one 6-deoxyhexose (L-induronic acid), one pentose (D-xylose), and three amino sugars (N-acetylneuranimic acid). Fucose and N-acetylneuraminic acid are usually found only at the nonreducing ends of sugar chains, which might partially explain their importance as antigens (see Table 1). The curious distribution of glucose supports the idea that complex carbohydrates function in cell recognition (Gesner and Ginsburg, 1964). This hexose is the most abundant sugar found in nature and occurs in many complex carbohydrates of plants and bacteria. In animals it comprises the reserve polysaccharide glycogen and substantial amounts are found free in body fluids. Yet with rare exceptions, chiefly collagen-like structural proteins, it is not found in mammalian glycoproteins. In glycolipids it occurs as the glycosyl residue closest to the lipid, and not in distal parts of the sugar chains which extend outwards. The exclusion of glucose has a rational basis if the carbohydrate structures of cell surfaces were indeed binding complementary molecules, presumably mammalian lectins (Barondes, 1981): the efficiency of binding sites based on glucosyl residues would be impaired by the free glucose of body fluids, much as haptens inhibit antigen-antibody reactions. Evolutionary selection against the impairment would exclude glucose as a component of these surfaces.

REFERENCES

Atkinson, B.F., Ernst, C.S., Herlyn, M., Steplewski, Z., Sears, H.F., and Koprowski, H., 1982, Gastrointestinal cancer-associated antigen in immunoperoxidase assay, Cancer Res., 42:4820.

Barondes, S.H., 1981, Lectins: their multiple endogenous cellular functions, Ann. Rev. Biochem., 50:207.

Brockhaus, M., Magnani, J.L., Blaszczyk, M., Steplewski, Z., Koprowski, H., Karlsson, K.A., Larsson, G., and Ginsburg, V., 1981, Monoclonal antibodies directed against the human Le^b blood group antigen, J. Biol. Chem., 256:13223.

Brockhaus, M., Magnani, J.L., Herlyn, M., Blaszczyk, M., Steplewski, Z., Koprowski, H., and Ginsburg, V., 1982, Monoclonal antibodies directed against the sugar sequence of lacto-N-fucopentaose III are obtained from mice immunized with human tumors, Arch. Biochem. Biophys., 217:647.

Brockhaus, M., Wysoca, M., Magnani, J.L., Steplewski, Z., Koprowski, H., and Ginsburg, V., 1984, The gastrointestinal and pancreatic cancer-associated antigen detected by monoclonal antibody 19-9 in the serum of mucin of patients also occurs in normal salivary mucin, Vox Sang., submitted.

Chang, L.J.-A., Gamble, C.L., Izaguirre, C.A., Minden, M.D., Mak, T.W., and McCullough, E.A., 1982, Detection of genes coding for human differentiation markers by their transient expression after DNA transfer, Proc. Natl. Acad. Sci. USA, 79:146.

Dawson, G., 1978, Disaccharide units from complex carbohydrates of animals, Methods Enzymol., 50:272.

Fredman, P., Richert, N.D., Magnanai, J.L., Willingham, M.C., Pastan, I., and Ginsburg, V., 1983, A monoclonal antibody that precipitates the glycoprotein receptor for epidermal growth factor is directed against the human blood group H type 1 antigen, J. Biol. Chem., 258:11206.

Gesner, B.M., and Ginsburg, V., 1964, Effect of glycosidases on the fate of transfused lymphocytes, Proc. Natl. Acad. Sci. USA, 52:750.

Ginsburg, V., 1972, Enzymatic basis for blood types in man, in: "Advances in Enzymology and Related Areas of Molecular Biology," A. Meister, ed., John Wiley & Sons, New York.

Gooi, H.C., Feizi, T., Kapadla, A., Knowles, B.B., Solter, D., and

Evans, M.J., 1981, Fucosylated type 2 blood group chains, Nature 292:156.

Grollman, E.F., Kobata, A., and Ginsburg, V., 1969, An enzymatic basis for Lewis blood types in man, J. Clin. Invest. 48:1489.

Hakomori, S., and Kannagi, R., 1983, Glycosphingolipids as tumor-associated and differentiation markers, J. Natl. Cancer Inst., 71:231.

Hakomori, S., Nudelman, E., Levery, S., Solter, D., and Knowles, B.B., 1981, The hapten structure of a developmentally regulated glycolipid antigen (SSEA-1) isolated from human erythrocytes and adenocarcinoma: A preliminary note, Biochem. Biophys. Res. Commun., 100:1578.

Hansson, G.C., Karlsson, D.A., Larsson, G., McKibbin, J.M., Blaszczyk, M., Herlyn, M., Steplewski, A., and Koprowski, H., 1983, Mouse monoclonal antibodies against human cancer cell lines with specificities for blood group and related antigens. Characterization by antibody binding to glycosphingolipids in a chromatogram binding assay, J. Biol. Chem., 258:4091.

Herlyn, M., Sears, H.F., Steplewski, A., and Koprowski, H., 1982, Monoclonal antibody detection of a circulating tumor-associated antigen. I. Presence of antigen in sera of patients with colorectal, gastric, and pancreatic carcinoma, J. Clin. Immunol., 2:135.

Huang, L.C., Brockhaus, M., Magnani, J.L., Cuttitta, F., Rosen, S., Minna, J.D., and Ginsburg, V., 1983a, Many monoclonal antibodies with an apparent specificity for certain lung cancers are directed against a sugar sequence found in lacto-N-fucopentaose III, Arch. Biochem. Biophys., 220:318.

Huang, L.C., Civin, C.I., Magnani, J.L., Shaper, J.H., and Ginsburg, V., 1983b, My-1, the human myeloid-specific antigen detected by mouse monoclonal antibodies, is a sugar sequence found in lacto-N-fucopentaose III, Blood, 61:1020.

Kalckar, H.M., 1967, Galactose metabolism and cell sociology, Science, 150:305.

Kobata, A., and Ginsburg, V., 1969, Oligosaccharides of human milk. II. Isolation and characterization of a new pentasaccharide, lacto-N-fucopentaose III, J. Biol. Chem., 244:5496.

Koprowski, H., Blaszczyk, M., Steplewski, Z., Brockhaus, M., Magnani, J.L., and Ginsburg, V., 1982, Lewis blood type may affect the incidence of gastrointestinal cancer, Lancet 1:1332.

Lloyd, K.O., Kabat, E.A., and Licerio, E., 1968, Immunochemical studies on blood groups. XXXVIII. Structures and activities of oligosaccharides produced by alkaline degradation of blood-group Lewis^{a} substance. Proposed structure of the carbohydrate chains of human blood-group A, B, H, Le^{a} and Le^{b} substances, Biochemistry, 7:2976.

Magnani, J.L., Nilsson, B., Brockhaus, M., Zopf, D., Steplewski, Z., Koprowski, H., and Ginsburg, V., 1982, A monoclonal antibody-defined antigen associated with gastrointestinal cancer is a ganglioside containing sialylated lacto-N-fucopentaose II, J. Biol. Chem. 257:14365.

Magnani, J.L., Steplewski, Z., Koprowski, H., and Ginsburg, V., 1983, The gastrointestinal and pancreatic cancer-associated antigen detected by monoclonal antibody 19-9 in the sera of patients is a mucin, Cancer Res., 43:5489.

Race, R.R., and Sanger, R., 1975, Blood Groups in Man, 6th Ed., Blackwell, Oxford.

Rohr, T.E., Smith, D.F., Zopf, D.A., and Ginsburg, V., 1980, Le^b-active glycolipid in human plasma: measurement by radioimmunoassay, Arch. Biochem. Biophys., 199:265.

Watkins, W.M., 1980, Biochemistry and genetics of the ABO, Lewis, and P blood group systems, in: "Advances in Human Genetics," vol. 10, H. Harris and K. Hirschorn, ed., Plenum Press, New York.

Yang, H.-J., and Hakomori, S., 1971, A sphingolipid having a novel type of ceramide and lacto-N-fucopentaose III, J. Biol. Chem., 246:1192.

STUDIES ON A CANCER ASSOCIATED GLYCOPROTEIN

Kurt J. Isselbacher and Daniel K. Podolsky

Harvard Medical School
Boston, Massachusetts 02114

ABSTRACT

Cancer-associated galactosyltransferase acceptor (CAGA) is a small glycoprotein purified from human malignant effusions that selectively kills transformed cells. CAGA was tritiated by reductive methylation in the presence of NaB^3H_4. CAGA-glycoprotein-sensitive cells [(baby-hamster kidney cells transformed by polyoma virus (BHKpy) and chick-embryo fibroblasts infected with Ts68 temperature-sensitive mutant of Rous sarcoma virus (CEF-RSV)] grown at 37° C, (the permissive temperature) bound 3- to 5-fold more ^{3}H-labelled CAGA glycoprotein than did their CAGA-glycoprotein-resistant nontransformed counterparts. The CEF-RSV grown at 41° C (nonpermissive temperature) bound an intermediate amount of ^{3}H-labelled CAGA glycoprotein; however, this intermediate amount appeared to be sufficient to induce inhibition of cell growth when the infected CEF treated at 41° C were switched to 37° C. ^{3}H-Labelled CAGA glycoprotein bound to intact cells could be removed by trypsin treatment up to 4 hr after addition of the glycoprotein but not thereafter. This time course paralleled the decreasing reversibility of growth inhibition. Thus, binding appears to be a necessary but not a sufficient condition to induce cell killing. Growth inhibition appears to depend on internalization of the glycoprotein and the presence of a transformation-specific cell process. The effect of CAGA on incorporation of a variety of macromolecular precursors was also studied in transformed and nontransformed cells. Incorporation of [^{3}H]mannose, [^{3}H]galactose, and [^{3}H]glucosamine into acid precipitable material after one-hour pulse was inhibited more than 70

This work was supported by a grant from the National Cancer Institute (CA-31277) and a Searle Scholar Award (to Dr. Podolsky) percent within four to eight hours in transformed cells. These data suggest that the action of CAGA on transformed cells may include inhibition of glycoconjugate synthesis.

The role of glycoconjugates in the maintenance of cellular integrity and regulation of growth and differentiation remains uncertain. Despite extensive documentation of widespread glycosylation of both intracellular and surface membrane proteins, the biological functions of the oligosaccharide moieties are largely unproven. The ubiquity of glycosylation supports the notion that these components are essential to normal cellular function. Clearly some structural features of oligosaccharide chain composition may play a role in targeting of intracellular enzymes [e.g., mannose-6-phosphate found in lysosomal hydrolases (Gabel et al., 1982)], and uptake of extracellular glycoproteins [e.g., asialofetuin (Tolleshaug et al., 1979)]. Still other studies suggest that carbohydrate side chains may provide recognition sites at the cell surface and facilitate intercellular interactions [e.g., sperm-egg binding (Shur and Hall, 1982)].

In light of these wide-ranging putative functions, it is not surprising that considerable attention has been directed to the potential role of glycoprotein structures in mediating the altered behavior of transformed and malignant cells. To examine these issues, many investigators have compared the glycopeptide content of transformed cell lines and their nontransformed counterparts. Although these studies have failed to demonstrate alterations unique to transformed cells, collectively they indicate the association of the transformed state with some general alterations in cellular glycopeptide composition. Thus, a large number of reports have noted an increase in larger, sialylated complex-type glycopeptides and a reduction in smaller, intermediate, and high mannose glycopeptides (Buck et al., 1970; Buck et al., 1971; Lai and Duesberg, 1972; Rieber and Irwin, 1974; Leonard et al., 1978; Glick, 1979; Litin and Grimes, 1983).

Alternatively, glycoprotein products of malignant cells may be examined as shed material in the growth media. We have previously described the purification from human malignant effusions of a low-molecular-weight glycoprotein (Mr approximately 3600) termed cancer-associated galactosyltransferase acceptor (CAGA); initial studies suggested that this glycoprotein may serve as an acceptor substrate for a transformation-associated galactosyltransferase (Podolsky, et al., 1978; Podolsky and Weiser, 1979). This preparation was found to cause profound inhibition of cell growth, leading to cell death, when added to media of transformed cells in culture (Podolsky et al., 1978; Podolsky and Isselbacher, 1980). In contrast, nontransformed cells were unaffected by addition of CAGA glycoprotein. In addition, administration of CAGA glycoprotein to hamsters inoculated

Table 1: Effect of ^{3}H-Labelled CAGA Glycoprotein on Cell Growth

Cells	Temperature (° C)	^{3}H-Labelled CAGA glycoprotein	10^{-6} x No. of cells	[^{3}H]-Thymidine uptake (cpm/ 100 μl)
CEF	41	-	5.7 ± 1.1	2830
		+	6.2 ± 1.3	2140
CEF	37	-	4.6 ± 0.8	1760
		+	4.3 ± 0.7	1840
CEF-RSV	41	-	4.8 ± 1.2	2150
		+	3.9 ± 0.8	1920
CEF-RSV	37	-	5.1 ± 1.4	2170
		+	0.5 ± 0.3	150
CEF-RSV	41 → 37*	-	4.6 ± 1.6	2060
		+	0.9 ± 0.4	270

Cells were seeded at 1 x 10^{6}/100 mm dish; 100 μl of CAGA glycoprotein solution (5 μg/ml) or phosphate-buffered saline was added on day 3 after seeding and incubation was continued for 24 hr. Before harvesting of the cells, [^{3}H]-Thymidine (10 μCi; specific radioactivity 80 Ci/mmol) was added; 2 hr after this addition, cells were harvested and resuspended in 1.0 ml, and radioactivity was determined in a 100 μl sample dissolved in 10.0 ml of Aquasol. Determinations were made in duplicate on separate experiments.
*At 6 hr after the addition of CAGA glycoprotein, the medium was changed and cells were shifted to incubation at 37° C.

with tumorigenic cells was found to prevent tumor development or cause actual tumor regression.

The mechanism whereby the glycoprotein leads to selective killing of transformed cells in vitro and in vivo has been the subject of further studies. In these studies, the effects of CAGA glycoprotein on cell growth, morphology, and ^{3}H-thymidine incorporation were examined with chick embryo fibroblasts (CEF), as well as CEF infected with a temperature-sensitive RSV mutant (Ts68) grown at 37° C and 41° C (Podolsky and Isselbacher, 1982). As shown in Table 1, CAGA glycoprotein selectively killed the infected (transformed) cells grown at 37° C (permissive temperature) but did not affect either uninfected CEF cells or infected CEF cells at 41° C. However, when RSV infected CEF, initially treated with CAGA glycoprotein at 41° C, were subsequently shifted to 37° C, cell killing was observed at 24 hr (the time necessary for expression of transformation), despite a change to non-CAGA-glycoprotein containing media before the shifting of the temperature (Table 1). Uninfected CEF pretreated with CAGA glycoprotein were not affected by a shift to 37° C.

Evaluation of the binding of ^{3}H-labelled CAGA glycoprotein in CEF, RSV-CEF (41° C) and RSV-CEF (37° C) cells disclosed differential binding similar to that observed with transformed BHKpy and nontransformed BHK cell lines. As shown in Table 2, the transformed cells bound approximately five times more glycoprotein than did the uninfected CEF cells. The time course and saturation concentration for binding of ^{3}H-labelled CAGA glycoprotein to CEF, RSV-CEF (41° C) and RSV-CEF (37° C) cells were similar. Determination of binding affinity constants by equilibrium dialysis of whole cells indicates a homogeneous set of receptors in each of the cell types studied, and Scatchard-plot analysis suggests equivalent affinity for all cells types (K_a = 1.7 x 10^6 $\pm$ 0.4 x 10^6 mol). It is noteworthy that the infected nontransformed RSV-CEF (41° C) cells bound an intermediate amount of ^{3}H-labelled CAGA glycoprotein, i.e., about twice that found with CEF but half that observed for transformed RSV-CEF (37° C) cells. It is especially important to note that this intermediate amount of binding of CAGA glycoprotein appeared to be sufficient to initiate cell death, in view of the cytotoxic effect of CAGA glycoprotein observed when these cells were shifted to 37° C without any further addition to the CAGA glycoprotein. Thus, specific binding of CAGA glycoprotein appears to be necessary, but not in itself sufficient, to make a cell susceptible to the CAGA-glycoprotein-induced cytotoxic effect. The similarity of calculated association constants for both CAGA-glycoprotein-sensitive and CAGA-glycoprotein-insensitive cells suggests that differential binding may indeed be due primarily to quantitative differences.

Although CAGA glycoprotein appears to bind initially at the cell surface, as determined by sensitivity to limited protease treatment, there is a progressive decline in the amount of CAGA glycoprotein

Table 2: Binding of ^{3}H-Labelled CAGA Glycoprotein to Transformed and Nontransformed Cells

Cells	10^{-3} x [^{3}H]-CAGA glycoprotein bound (cpm/10^6 cells)	10^{-4} x CAGA glycoprotein bound (molecules/cell)
BHK	1.2 ± 0.5	1.52
CEF	1.8 ± 0.4	2.28
CEF-RSV (41° C)	3.8 ± 0.6	4.83
BHKpy	4.8 ± 0.7	6.10
CEF-RSV (37° C)	8.7 ± 0.8	11.05

Cells (10^6/100 μl) were incubated with 100 μl of [^{3}H]-CAGA glycoprotein solution (20 μg/ml; 3 x 10^7 cpm/mg) at 37° C for 30 min.

solubilized by such treatment suggesting that bound glycoprotein may be internalized over a period of 4 hr, thus making it inaccessible to an exogenous protease. Alternatively, the progressive resistance to solubilization by trypsin could result from a change in the conformation or distribution of the glycoprotein an its receptor in the cell membrane rendering CAGA glycoprotein resistant to release by the protease. It is noteworthy that the 4 hr period leading to resistance to trypsin-induced solubilization of CAGA glycoprotein is comparable with the time during which the cytotoxic effect of CAGA glycoprotein remained reversible (Fig. 1).

In more recent studies, the effect of CAGA preparations on the uptake and incorporation of a variety of macromolecular precursors by both transformed and nontransformed cells was examined in order to gain some insight into the nature of its cytotoxic effect (Podolsky et al., 1983). In initial studies, precursors of protein ([^{3}H]-serine and [^{14}C]-leucine), RNA ([^{3}H]-uridine, and DNA ([^{3}H]-thymidine) were added to media of baby hamster kidney fibroblasts (BHK) and polyoma transformed BHK (BHKpy) monolayers at varying times after addition of CAGA. Total uptake was defined as the total cellular content of label after one hour of incubation; incorporation was defined as the amount of label appearing into acid-precipitable material during that same time period. Similar experiments were undertaken after addition of labelled carbohydrate precursors of glycoconjugates ([^{3}H]-glycosamine, [^{3}H]-mannose, and [^{3}H]-galactose).

The addition of CAGA (1.0 μg/ml) caused a rapid decrease in incorporation of labelled glucosamine into acid-precipitable material

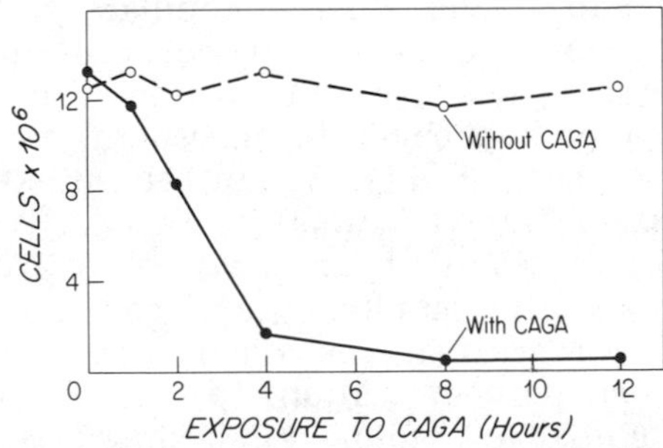

Fig. 1. Effect of Duration of Exposure to CAGA Glycoprotein on Reversibility of Growth Inhibition. Semi-confluent dishes (35 mm) of BHKpy cells were incubated with 100 μl of phosphate-buffered saline (o) or CAGA glycoprotein (10 μg) (o). After various lengths of time, the medium was aspirated off and replaced with non-CAGA-glycoprotein-containing media. After further 24 hr incubation at 37 ° C, cell counts were determined with a Coulter counter, and viability was assessed by Trypan Blue exclusion.

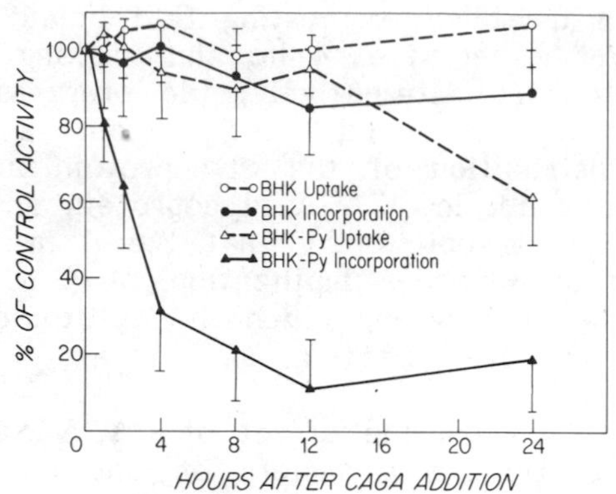

Fig. 2. Effect of CAGA on Uptake and Incorporation of [^{3}H]-Glucosamine in BHK and BHKpy Cells. CAGA (1.0 μg) was added to semiconfluent 35 mm dishes of BHK and BHKpy cells. One hour prior to harvesting, D-[1-^{3}H]-glucosamine (10 μCi/dish) was added to dishes in duplicate. Total uptake and incorporation into acid-precipitable material were determined. Activity is expressed as percentage compared with untreated controls per 10^6 cells ± S.E. (100% cpm BHK uptake = 12,370, BHK incorporation = 2,470, BHKpy uptake = 15,320, BHKpy incorporation = 3,360).

in BHKpy monolayers (Fig. 2). Incorporation was found to be diminished 35% ± 8% within two hours and 75% ± 12% within four hours. Decreased incorporation was not due to diminished monosaccharide uptake since total cellular content of labelled monosaccharide was comparable to controls up to 12 hours; only after 12 hours was any apparent decline in long-term uptake observed. Inhibition of carbohydrate incorporation was found only in the BHKpy cells. As seen in Fig. 2, neither uptake nor incorporation of the monosaccharide tested was significantly inhibited in the nontransformed cells despite exposure to the CAGA for up to 24 hours. Experiments with labelled galactose and mannose yielded similar results. Incorporation was diminished 40-50 percent within two hours and 70-90 percent within four hours without significant dimunition in total uptake of radiolabelled sugar.

The relatively rapid inhibition of incorporation of hexoses into acid precipitable material was compared with the effects of CAGA on incorporation of precursors into protein, RNA, and DNA. CAGA caused significant reduction in the amount of labelled leucine into acid-precipitable material in BHKpy cells but not in nontransformed BHK cells. However, in contrast to the results with labelled carbohydrates, less than 25 percent inhibition of incorporation was observed within 4 to 12 hours after exposure to CAGA. Maximal inhibition of leucine incorporation was observed only after 12 hours, reaching 90 percent after 24 hours; this was associated with only a small decrease in amino acid uptake (Fig. 3). CAGA also caused a temporally similar inhibition of serine incorporation; but even at 24 hours, serine uptake was reduced by only one third. Uptake of [^{3}H]-AIB, a nonmetabolizable amino acid analogue, by BHKpy cells was reduced less than 10 percent up to 12 hours following exposure to CAGA. These data suggest that the decrease in acid-precipitable activity was related to a decrease in protein synthesis.

The effects of CAGA on incorporation of [^{3}H]-thymidine into DNA were comparable to those on protein synthesis. Incorporation of thymidine was diminished in BHKpy but not in BHK cells. The time course of inhibition was similar to that observed with labelled amino acids and again in contrast to that noted with labelled monosaccharides. However, it should be noted that the effects of CAGA on nucleic acid precursor uptake and incorporation differed in at least two respects from that observed with monosaccharides and amino acids. Thus, in contrast to results with hexoses, where there was a predominant inhibition of incorporation only, there was a parallel decrease in both uptake and incorporation of uridine and thymidine. Indeed, inhibition of uridine uptake was greater than inhibition of incorporation up to 12 hours following exposure to CAGA. However, neither was inhibited to the extent observed with carbohydrate incorporation over this 12-hour period. In addition, the early inhibitory effect of CAGA on uridine uptake and incorporation was observed in both BHKpy and BHK cells.

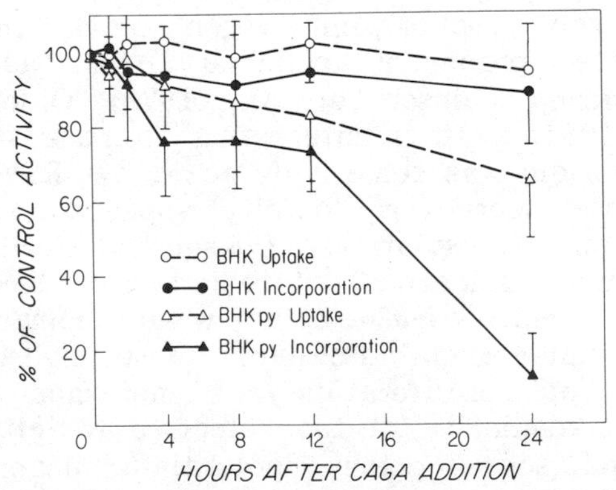

Fig. 3. Effect of CAGA on Uptake and Incorporation of [^{3}H]-leucine in BHK and BHKpy Cells. Uptake and incorporation were determined after addition of L-[4,5-^{3}H]-leucine (10 μCi/35 mm dish) one hour prior to harvesting (100% cpm, BHK uptake = 15,860, BHK incorporation = 6,270, BHKpy uptake = 18,590, BHKpy incorporation 7,140).

The effects of CAGA were similarly assessed in secondary cultures of CEF, RSV-CEF (41° C), and RSV-CEF (37° C). The addition of labelled monosaccharides (mannose, galactose, or glucosamine) to monolayers of RSV-CEF (37° C) resulted in a decrease in incorporation into acid-precipitable material after exposure to CAGA, analogous to that observed in BHKpy cells. However, the decrease in glucosamine incorporation produced by CAGA was more delayed than that observed in BHKpy cells; approximately half-maximal inhibition was seen at approximately six hours (cf two hours in BHKpy), and maximal inhibition required 8-12 hours (cf four hours in BHKpy). Uptake of monosaccharide was not significantly reduced during this 12-hour period. In corresponding experiments, RSV-CEF and uninfected CEF at 41° C showed no significant decrease in either uptake or incorporation of the tested carbohydrate in 12 hours. Identical results were obtained when incorporation of labelled galactose and mannose were studied. Less than 20 percent inhibition of incorporation of each monosaccharide was observed at 24 hours for both uninfected CEF and RSV-CEF (41° C).

Collectively, these data suggest that CAGA preparations may be related to disruption of glycoconjugate biosynthesis. The present findings are compatible with disruption of this process at any of these stages, although the uniform inhibition of incorporation of each of the monosaccharides tested suggests that the site of CAGA action may be early in the glycosylation process. While these experiments do not assess the effect of CAGA on incorporation of carbohydrate into glycolipids (which would not be measured by these techniques), it is possible that this process is also inhibited.

It will be interesting to compare the effect of CAGA with those of tunicamycin and amphomycin, which block the formation of dolichol phosphate precursor necessary for addition of core carbohydrate residues (Gibson et al., 1979; Olden et al., 1978; Banerjee et al, 1981). It is interesting to note that Olden et al. found that tunicamycin was selectively toxic for RSV-CEF (37° C) and not CEF (41° C). Moreover, toxicity appeared to correlate with inhibition of protein glycosylation. Duksin and Bornstein (1977) have also noted relatively selective cytotoxicity for SV40 and polyoma-transformed 3T3 and WI-38 cells when compared with their nontransformed counterparts. However, these investigators did note some inhibition of proliferation of nontransformed lines by tunicamycin. The similarity of the selective toxicity of tunicamycin and CAGA underscores the potential important role of glycoconjugates in processes related to transformation. If indeed CAGA selectively inhibits glycoprotein formation in transformed cells, it may prove to be an especially useful probe to examine alterations in glycoprotein biosynthesis associated with transformation in view of its apparent specificity for the transformed state. Earlier workers have noted marked alteration in cellular structure of transformed cells grown in media supplemented by various monosaccharides

(Molnar and Bekesi, 1972). If the observed decline in incorporation of labelled carbohydrate into acid-precipitable material is related to subsequent growth inhibition and cell death, elucidation of the effects of CAGA may provide further insight into the role of glycosylation in maintaining the integrity of cellular processes.

While the observed decrease in monosaccharide incorporation may reflect specific interference in glycoconjugate biosynthesis, the data might also reflect an action of CAGA on intermediate metabolism and possibly on the shunting of monosaccharides away from pathways leading to their incorporation into acid precipitable material. Such a disruption of intermediate metabolism would need to be highly selective in view of the relatively well preserved incorporation of amino acid, uridine, and thymidine early after exposure to CAGA. Recent experiments to define the structure of newly synthesized glycoproteins in CAGA-treated cells in order to delineate the nature of the apparent interference with glycosylation indicate substantial alterations in the carbohydrate side chain structure of glycoprotein isolated from CAGA-treated cells. Additional studies are needed to determine the possible effect of CAGA on glycolysis and other reactions of intermediate metabolism.

In summary, studies demonstrate that the glycoprotein CAGA is initially bound at the cell surface after which it may undergo simple rearrangement or internalization making it inaccessible to exogenous proteases. It appears that these processes are followed by a corresponding decline in the incorporation of carbohydrates into glycoconjugates; this in turn may lead to disruption of other cellular processes and ultimately to cell death through as yet undetermined mechanisms. The possible importance of the more moderate early effect on uridine and thymidine incorporation remains unclear. It is intriguing that a glycoprotein selectively toxic for transformed cells is present in malignant effusions. Further work is necessary to elucidate this apparent paradox. Finally, the relationship of these biological activities to glycopeptide acceptor activity remains unknown and it is possible they are indeed independent. Further purification and characterization will be necessary to answer these important issues.

REFERENCES

Banerjee, D.K., Scher, M.G., and Waechter, C.J., 1981, Amphomycin: Effect of the lipopeptide antibiotic on the glycosylation and extraction of dolichyl monophosphate in calf brain membranes, Biochemistry, 20:1561.

Buck, C.A., Glick, M.C., and Warren, L., 1970, A comparative study of glycoproteins from the surface of control and rous sarcoma virus transformed hamster cells, Biochemistry, 9:4567.

Buck, C.A., Glick, M.C., and Warren, L., 1971, Effect of growth on the glycoproteins from the surface of control and rous sarcoma virus transformed hamster cells, Biochemistry, 10:2176.

Duksin, D., Bornstein, P., 1977, Changes in surface properties of normal and transformed cells caused by tunicamycin, an inhibitor of protein glycosylation, Proc. Natl. Acad. Sci. USA, 74:3433.

Gabel, C.A., Goldberg, D.E., and Kornfeld, S., 1982, Lysosomal enzyme oligosaccharide phosphorylation in mouse lymphoma cells: Specificity and kinetics of binding to the mannose 6-phosphate receptor in vivo, J. Cell Biol., 95:536.

Gibson, R., Schlesinger, S., and Kornfeld, S., 1979, The nonglycosylated glycoprotein of vesicular stomatitis virus is temperature-sensitive and undergoes intracellular aggregation at elevated temperature, J. Biol. Chem., 254:3600.

Glick, M.C., 1979, Membrane glycopeptides from virus-transformed hamster fibroblasts and the normal counterpart, Biochemistry, 18:2525.

Lai, M.M.C., and Duesberg, P.H., 1972, Differences between the envelope glycoproteins and glycopeptides of avian tumor viruses released from transformed and from nontransformed cells, Virology, 50:359.

Leonard, S.G., Hale, A.H., Roll, D.E., Conrad, H.E., and Weber, M.J., 1978, Turnover of cellular carbohydrates in normal and rous sarcoma virus-transformed cells, Cancer Res., 38:185.

Litin, B.S., and Grimes, W.J., 1983, Heterogeneity of sialoglycoproteins purified from normal and malignant cells, Cancer Res., 43:2131.

Molnar, Z., and Bekesi, J.G., 1972, Effects of D-glucosamine, D-mannosamine, and 2-deoxy-D-glucose on the ultrastructure of ascites tumor cells in vitro, Cancer Res., 32:380.

Olden, K., Pratt, R.M., and Yamada, K.M., 1978, Role of carbohydrates in protein secretion and turnover: Effects of tunicamycin on the major cell surface glycoprotein of chick embryo fibroblasts, Cell, 13:461.

Podolsky, D.K., Weiser, M.M., and Isselbacher, K.J., 1978, Inhibition of growth of transformed cells and tumors by an endogenous acceptor of galactosyltransferase, Proc. Natl. Acad. Sci. USA, 75:4426.

Podolsky, D.K., and Weiser, M.M., 1979, Detection, purification, and characterization of a human cancer-associated galactosyltransferase acceptor, Biochem. J., 178:279.

Podolsky, D.K., and Isselbacher, K.J., 1980, Cancer-associated galactosyltransferase acceptor, Cancer, 45:1212.

Podolsky, D.K., and Isselbacher, K.J., 1982, Transformation-specific cell killing by a cancer-associated galactosyltransferase acceptor and cellular binding, Biochem. J., 208:249.

Podolsky, D.K., Fournier, D., and Isselbacher, K.J., 1983, Inhibition of carbohydrate incorporation in transformed cells by a cancer-associated galactosyltransferase acceptor (CAGA), J. Cell. Physiol., 115:23.

Rieber, C., and Irwin, J.C., 1974, Possible correlation of growth rate and expression transformation with temperature-dependent modification high-molecular-weight membrane glycoproteins in mammalian cells transformed by a wild-type and by a thermosensitive mutant of avian sarcoma virus, Cancer Res., 34:3469.

Shur, B.D., and Hall, N.G., 1982, A role for mouse sperm surface galactosyltransferase in sperm binding to the egg zona pellucida, J. Cell Biol., 95:574.

Tolleshaug, H., Berg, T., Frolich, W., and Norum, K.R., 1979, Intracellular localization and degradation of asialofetuin in isolated rat hepatocytes, Biochim. Biophys. Acta, 585:71.

GLYCOSPHINGOLIPIDS AS MARKERS FOR DEVELOPMENT AND DIFFERENTIATION AND AS REGULATORS OF CELL PROLIFERATION

Sen-itiroh Hakomori

Biochemical Oncology/Membrane Research
Fred Hutchinson Cancer Research Center
Departments of Pathobiology, Microbiology, and Immunology
University of Washington
1124 Columbia Street
Seattle, Washington 98104

ABSTRACT

Dramatic changes in glycolipid synthesis from one series to another series have been observed to be associated with early mouse embryogenesis and with differentiation of mouse leukemia cells. Decreased synthesis of the extended globoseries* recognized by anti-SSEA-3 and anti-SSEA-4 antibodies, and a concomitant increase in synthesis of lactoseries (including those recognized by the anti-SSEA-1 antibody) have been observed in developing mouse preimplantation embryos and in human teratocarcinoma. Similar shifts of glycolipid synthesis from ganglio- through lacto- to globoseries have been found to be associated with differentiation of mouse M1 leukemia cells.

Drastic changes of glycolipid synthesis have also been found to occur in a number of experimental tumors and human cancers. Three

---

*Three series of glycolipids are defined as follows: ganglioseries contain Galβ1→3GalNAcβ1→4Galβ1→4Glc, lactoseries contain Galβ1→4GlcNAcβ1→3Galβ1→4Glc, and globoseries contain GalNAcβ1→3Galα1→4Galβ1→4Glc as the core structure (see Hakomori and Kannagi, 1983).

classes of changes, i.e., incomplete synthesis, neosynthesis, and loss of crypticity, can lead to the formation of tumor-associated antigens. Examples of tumor-associated antigens with defined structure are presented in Tables 2-4. Monoclonal antibodies directed to these structures should prove useful in the diagnosis and treatment of human cancer.

The biological significance of incomplete synthesis of gangliosides was studied by addition of gangliosides to cell culture. The results, summarized in the third section, strongly suggest that some gangliosides can modulate growth factor receptors and can affect PDGF-stimulated tyrosine-phosphorylation of the Mr=170,000 receptor protein. A possible consequence of transformation-dependent, blocked synthesis of murine cell gangliosides is discussed.

INTRODUCTION

One of the most lucid documents addressing the possible role of cell surface membranes and sugar metabolism in cellular recognition is the provocative review written by Herman Kalckar in the October 1965 issue of Science (Kalckar, 1965). My personal interest, as a carbohydrate biochemist, in a possible role of cell surface carbohydrates in animal cell recognition was greatly stimulated by this paper. However, no solid data was available at that time on the molecular basis for differences in social behavior between normal and oncogenically transformed cells. Changes in blood group ABH antigens in human cancer had been previously ascribed to modified blood group activities and changes in sugar composition (Masamune and Hakomori, 1960). Subsequently, large quantities of a new type of fucose-containing glycolipid were found to accumulate in some human adenocarcinomas (Hakomori and Jeanloz, 1964). These fucolipids were later determined to be a mixture of lacto-fucopenta-osyl(II)ceramide and lacto-fucopentaosyl(III)ceramide (Hakomori and Andrews, 1970; Yang and Hakomori, 1971). Through encouraging discussions with Doctors Hiroshi Nikaido and Herman Kalckar concerning these earlier observations, I became aware of the emerging concept of "ektobiology" which was essentially based on microbiological phenomena. Only a few of these phenomena, such as glycosidase-sensitive lymphocyte homing (Gesner and Ginsburg, 1964) and contact-inhibition related to cell surface sialic acid density (Abercrombie and Ambrose, 1962), described carbohydrates responsible for differences in animal cell phenotypes. These phenomena influenced me to pursue the possibility that blood group changes and abnormal glycolipids in human cancer could be related to the aberrant social behavior of tumor cells. Consequently, the major effort of my research at Brandeis University focused on the differences in glycolipid composition and metabolism between normal and oncogenically transformed cells in vitro. Results of these studies clearly indicated a pattern of incomplete synthesis with the accumulation of precursor glycolipids in oncogenically transformed cells (Hakomori and Murakami, 1968). These initial studies were performed with the strong and sustained support of Dr. Kalckar. With these recollections and admiration, it is my honor and pleasure to be invited to this symposium in commemoration of Dr. Kalckar's seventy-fifth birthday.

The major research currently in progress in my laboratory will be summarized in this paper under three topics: (1) characterization of the changes in carbohydrate chains associated with development and differentiation, (2) identification of glycolipid tumor-associated antigens, and (3) regulation of cell growth by gangliosides.

THE MAJOR CARBOHYDRATE CHANGES ASSOCIATED WITH DIFFERENTIATION IN EARLY EMBRYO AND IN MOUSE MYELOGENEOUS LEUKEMIA M1 CELLS

Perhaps the most remarkable example of cell social changes mediated by the change of cell surface membranes has been found in the processes of embryonic development and cellular differentiation. Such membrane changes have been described through the orderly appearance or disappearance of cell surface markers detected by various antibodies, such as blood group ABH (Szulman, 1977), Forssman (Willison and Stern, 1978), and blood group Ii (Kapadia et al., 1981). More recently, the monoclonal antibody directed to embryonal carcinoma F9 cells (Solter and Knowles, 1978) and those directed to 4-8 cell stage mouse embryos (Shevinsky et al., 1982) have been shown to define clear stage-specific changes in certain carbohydrate molecules in the preimplantation embryo. The antigen defined by the former antibody is called stage-specific embryonic antigen 1 (SSEA-1), which is expressed maximally at the morula and in the inner cell mass of blastocysts, and declines at later stages of differentiation (see Fig. 1). The molecule defined by the latter antibody is expressed maximally at 4-8 cell stage mouse embryos, disappears at later stages of differentiation, and is called stage-specific embryonic antigen 3 (SSEA-3) (see Fig. 1). A monoclonal antibody directed to human teratocarcinoma 2102 defines a similar molecule expressed maximally at 4-8 cell stage embryos, which is called stage-specific embryonic antigen 4 (SSEA-4).

We have characterized the chemical structure of the molecules that are defined by these antibodies, as shown in the footnote to Figure 1. SSEA-1 has been identified as having X-hapten structure (Hakomori et al., 1981; Gooi et al., 1981). A series of glycolipid antigens in erythrocyte membranes that are reactive to SSEA-1 antibody have been isolated and characterized as y_2, z_1, and z_2 molecules, as shown in Table 1 (Kannagi et al., 1982a). The glycolipid antigen reactive with SSEA-3 antibody has been isolated and characterized from a large quantity (more than 100 ml of packed

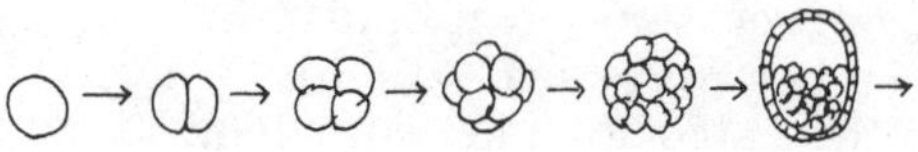

	○	→ ◯◯	→ 4-cell	→ 8-cell	→ morula	→ blastocyst	→ later stage
SSEA-1[1]	-	-	-	+	+++	++	±
SSEA-3&4[2]	±	+	+++	+++	-	-	-

1. Structure: X-hapten/or LFP III
Galβ1→4GlcNAcβ1→R
3
↑
Fucα1
Hakomori et al. (1981) BBRC 100, 1578
Gooi et al. (1981) Nature 292, 156

2. Structure: Extended globo-series
NeuAcα2→3Galβ1→3GalNAcβ1→3Galα1→4Gal
4 3
Kannagi et al. (1983) JBC 258,8334
Kannagi et al. (1983) EMBO, in press

Fig. 1. Two major carbohydrate determinants, SSEA-1 and SSEA-3, are expressed at different stages of preimplantation mouse embryo. SSEA-3 is maximally expressed at the 2-4 cell stage, in contrast to SSEA-1 which is expressed maximally at the morula and the inner cell mass of the blastocyst. The structures of the epitopes defined by the monoclonal antibodies directed to SSEA-1 and SSEA-3 are shown in the footnote. A third antigen, SSEA-4, defined by its monoclonal antibody, constitutes the terminal structure of the same molecule as SSEA-3. The structure defined by the anti-SSEA-4 antibody is NeuAcα2→3Galβ1→3GalNAc.

cells) of human teratocarcinoma 2101. Interestingly, these glycolipids reactive with anti-SSEA-3 antibody belong to a new class of extended globoseries[1], as shown in Table 2 (Kannagi et al., 1983a), and the epitope structure is located at the internal GalNAcβ1→3Galα1→4Gal chain. Galβ1→3 substitution to the GalNAc residue of globoside is essential to enhance the reactivity of the SSEA-3 antibody. The antibody to SSEA-4 has been identified as recognizing the terminal NeuAcα2→3Galβ1→3GalNAc residue. Thus, a new globoseries ganglioside (GL-6) can be identified by the antibodies to SSEA-3 and SSEA-4.

Since SSEA-3 defines an earlier stage in early mouse embryos than SSEA-1, a drastic shift from globo- to lactoseries may take place during the process of differentiation from the 4-8 cell stage embryo, through morula, to blastocyst, as shown in Figure 2. Since undifferentiated human teratocarcinoma 2102 cells express SSEA-3 and 4, but not SSEA-1, and differentiated cells express SSEA-1, but a

Table 1: Carbohydrate Structures Defined by SSEA 1, 3, and 4

SSEA-1

y_2 Galβ1→4GlcNAcβ1→3Galβ1→4GlcNAcβ1→3Galβ1→4Glc
(Fucα1→3 on the first GlcNAc)

z_1 Galβ1→4GlcNAcβ1→3Galβ1→4GlcNAcβ1→3Galβ1→4GlcNAcβ1→3Galβ1→4Glcβ1→1Cer
(Fucα1→3 on the first GlcNAc)

z_2 Galβ1→4GlcNAcβ1→3Galβ1→4GlcNAcβ1→3Galβ1→4GlcNAcβ1→3Galβ1→4Glcβ1→1Cer
(Fucα1→3 on the first GlcNAc; Fucα1→3 on the second GlcNAc)

SSEA-3 (Extended globoseries)

GL4 GalNAcβ1→3Galα1→4Galβ1→4Glcβ1→1Cer

GL5 Galβ1→3GalNAcβ1→3Galα1→4Galβ1→4Glcβ1→1Cer

GL7 NeuAcα2→3Galβ1→3GalNAcβ1→3Galα1→4Galβ1→4Glcβ1→1Cer

SSEA-4 (NeuAcα2→3Galβ1→3GalNAcβ1)

HexCer→LacCer
a: Gb3 → Gb4 → βGal-Gb4; Gb4 → αGalNAc-Gb4; βGal-Gb4 → αFuc-βGal-Gb4; βGal-Gb4 → αNeuAc-βGal-Gb4
Globo-series, SSEA-3 active, maximally expressed at 4 to 8 cell stage.
b: Lc3 → nLc4 → nLc6 → nLc8; nLc4 → III³αFuc-nLc4; nLc6 → V³αFuc-nLc6; nLc8 → VII³αFuc-nLc8
Neolacto-series, SSEA-1 active, maximally expressed at morula stage

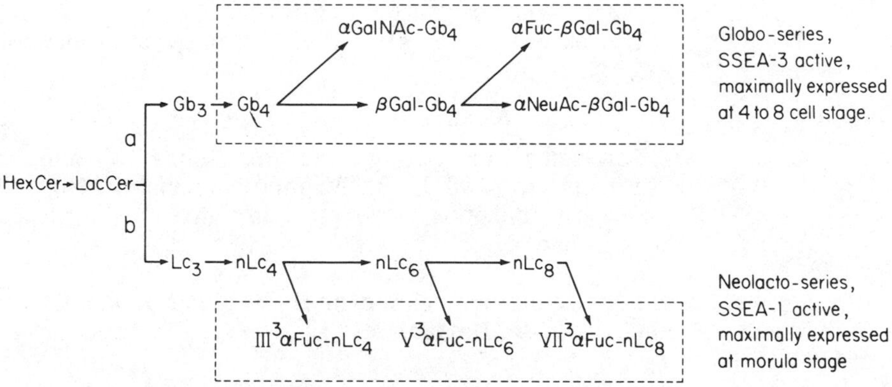

Fig. 2. Synthetic pathways for SSEA-1 and SSEA-3. Route a represents globoseries synthesis, which leads to new extended globoseries glycolipids, including SSEA-3 determinants. The extended globoseries glycolipids and globoside (Gb_4) are shown within the upper dotted circle. Route b indicates the synthetic pathway of lactoseries glycolipids Lc_3, nLc_4, nLc_6, and nLc_8, which contain repeated N-acetyllactosamine units. Fucosylation of the subterminal GlcNAc results in a series of SSEA-1-active glycolipids, which are shown within the lower dotted circle.

Table 2: Glycolipid Tumor Antigens Established by Classical Chemical Analysis Followed by Immunochemical Characterization

---

1. Lactoneotetraosylceramide (nLc_4) in hamster NILpy tumor (Gahmberg & Hakomori, 1975; Sundsmo & Hakomori, 1976)

Structure: Galβ1→4GlcNAcβ1→3Galβ1→4Glcβ1→1Cer

Observation: a) NILpy cells and derived tumors had nLc_4 in appreciable amount (Gahmberg & Hakomori, 1975; Watanabe, K., Matsubara, T., & Hakomori, S., unpublished observation)

b) Sera of hamsters bearing NILpy tumors had antibodies to nLc_4 (Sundsmo & Hakomori, 1976)

2. Gangliotriaosylceramide (Gg_3) in mouse sarcoma (KiMuSV sarcoma in Balb/C) and mouse lymphoma (L-5178 in DBA/2) (Rosenfelder et al., 1977; Young & Hakomori, 1981)

Structure: GalNAcβ1→4Galβ1→4Glcβ1→1Cer

Observation: a) KiMuSV sarcoma (in Balb/C) (Rosenfelder et al., 1977) and L5178 lymphoma (in DBA/2) (Young & Hakomori, 1981) contain a large quantity of Gg_3; normal tissues and organs of these mice do not contain Gg_3 in appreciable amount (Rosenfelder et al., 1977; Young & Hakomori, 1981)

b) Monoclonal IgG_3 antibody to Gg_3 inhibits the lymphoma growth in DBA/2 mice (Young & Hakomori, 1981). Syngeneic immunization by L-5178 lymphoma results in anti-Gg_3 antibody response

3. Blood group A-like antigen in human cancer of blood group O or B individuals (Breimer, 1980; Hattori et al., 1981; Yokota et al., 1981)

Structure: GalNAcα1→3Hex→HexNAc→Hex→Hex→Cer (Yokota et al., 1981)

GalNAcα1→3(Fucα1→2)Hexβ1→4(Fucα1→2) HexNAc→Hex→Hex→Cer (Breimer, 1980)

Observation: a) Gastric, colonic, and hepatic adenocarcinomas of blood group O or B individuals contain glycolipids that weakly inhibit A-hemagglutination

(continued)

Table 2 (continued)

b) Antigens were partially identified as having atypical A-determinants

4. Forssman antigen in human cancer (Hakomori et al., 1977; Young et al., 1979; Yoda et al., 1980; Taniguchi et al., 1981; Yokota et al., 1981)

Structure: GalNAcα1→3GalNAcβ1→Hex→Cer (Yokota et al., 1981)

GalNAcα1→3GalNAcβ1→3Galα1→4Galβ1→4Glcβ1→1Cer (Hakomori et al., 1977)

Observation: Various human cancer tissues derived from Forssman-negative tissue contain Forssman glycolipid antigen (Hakomori et al., 1977)

5. Blood group P_1 and P-like antigen in the tumor of a patient with pp-genotype (Levine et al., 1951; Levine, 1978; Kannagi et al., 1982b)

Structure: Galα1→4Galβ1→4GlcNAcβ1→3Galβ1→4Glc→Cer (P1)

GalNAcβ1→3Galβ1→4GlcNAcβ1→3Galβ1→4Glc→Cer (P-like) (Kannagi et al., 1982b)

Observation: a) Background, see text (Levine et al., 1951; Levine, 1978)

b) The tumor tissue of the blood group pp individual had P-like and P1 glycolipid, the structures of which are identified as above

lower quantity of SSEA-3 and 4, a similar shifting from globo- to lactoseries structures may take place during differentiation of human teratocarcinoma, and perhaps in human embryos as well (Kannagi et al., 1983b).

A similar change of glycolipids from ganglio- through lacto- to globoseries has been noticed during differentiation of mouse myelogeneous leukemia M1 cells. The major glycolipids in undifferentiated myelogeneous leukemia M1 cells have been characterized as gangliotriaosylceramide (Gg3) and i-active lacto-$\overline{\text{nor}}$octaosylceramide (nLc_6). On differentiation, synthesis of ganglioser$\overline{\text{ies}}$ glycolipids declines and lactoseries synthesis is enhanced, which leads to the synthesis of I-active branched lacto-$\overline{\text{iso}}$octaosyl structure. On further differentiation into macrophage-l$\overline{\text{ike}}$ $M1^+$ cells, synthesis of globoseries is

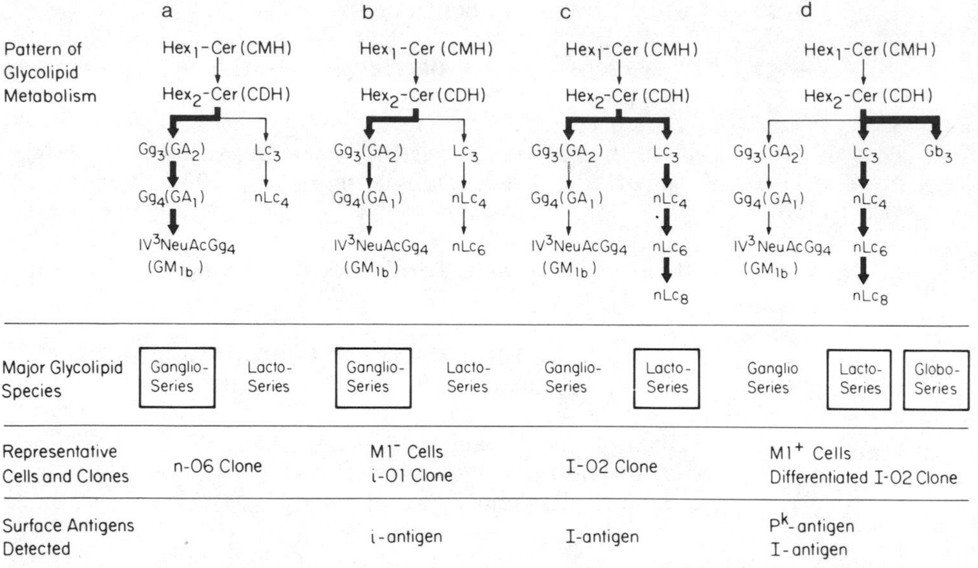

Fig. 3. A sequential shifting of glycolipid synthesis which occurs during differentiation of mouse leukemia $M1^-$ cells to macrophage-like $M1^+$ cells. Undifferentiated $M1^-$ cells contain the major glycolipid series of gangliotriaosylceramide (Gg_3 or GA_2), gangliotetraosylceramide (Gg_4 or GA_1), and GM_{1b} ganglioside. At an early stage of differentiation, this pathway declines and the synthesis of lacto-nor-hexoaosylceramide (nLc_6), which is i antigen, is induced (stage b). Upon further differentiation, the synthesis of extended ganglioseries ceases and synthesis of lactoseries glycolipids, including a branched structure (nLc_8), greatly increases. This event leads to the expression of I antigen. At the terminal stage of differentiation at which phagocytotic activity and the Fc receptor appear, ganglioseries synthesis completely ceases and synthesis of globotriaosylceramide (Gb_3 or p^k antigen) is induced.

initiated, which leads to the formation of p^k positive cells (Kannagi et al., 1983c; see also Fig. 3).

At present, the biological significance of such a drastic change from one series to another series of carbohydrate structure during the processes of differentiation and development is not known. The process of cell recognition may be dramatically altered and cell adhesion may be greatly modified, which are important bases for cell differentiation and development. Much study remains to be done on the biochemical events following such carbohydrate changes.

GLYCOLIPID CHANGES ASSOCIATED WITH ONCOGENIC TRANSFORMATION AND TUMOR-ASSOCIATED GLYCOLIPID ANTIGENS

The glycolipid changes observed in a number of transformed cells in vitro and tumor cells in vivo can be classified into two types: (1) incomplete synthesis with a frequent accumulation of precursor glycolipids, and (2) activation of the synthesis of a new glycolipid which is absent in progenitor cells. In addition, glycolipids in tumor cells are more exposed than in normal cells due to organizational changes to tumor cell membranes (Hakomori et al., 1968; Gahmberg and Hakomori, 1975). These changes have been reviewed repeatedly in various articles (Hakomori, 1973; Brady and Fishman, 1974; Hakomori, 1975; Critchley and Vicker, 1977; Hakomori and Kannagi, 1983). Either of these processes or a combination of mechanisms leads to the formation of tumor-associated antigens. With a systematic chemical analysis using polyclonal antibodies, glycolipid tumor antigens have been well established in a few experimental tumor systems, such as lacto-neotetraosylceramide in NILpy tumors in hamsters (Gahmberg and Hakomori, 1975; Sundsmo and Hakomori, 1976); gangliotriaosylceramide in Kirsten sarcoma in Balb/c mice (Rosenfelder et al., 1977); an unidentified glycolipid antigen in SV40-induced tumors in hamsters (Ansel and Huet, 1980); Forssman antigen in human cancer derived from Forssman negative tissue (Hakomori et al, 1977); A-like antigen in human gastrointestinal tumors derived from blood group O or B individuals (Yokota et al., 1981; see other earlier papers in review by Hakomori and Kannagi, 1983); and P_1 and P-like antigen in the tumor of a blood group p individual (Levine et al., 1951; Kannagi et al., 1982b) (for structures of these glycolipid tumor antigens, see Table 2).

Since the monoclonal antibody approach was introduced in tumor immunology, a number of "tumor-specific" monoclonal antibodies have been described which define certain glycolipids. Some of them define the precursor glycolipids accumulating in certain types of human cancer, as shown in Table 3. Others define "neoglycolipids," which are essentially absent in normal cells or tissues and are newly synthesized in neoplastic cells and tissues. Typical examples of neoglycolipids are the sialosyl-Le^a antigen (Magnani et al., 1982) defined by the antibody N-19-9 (Koprowski et al., 1981) and the poly-X antigen (Hakomori et al., in press) as shown in Table 4. Particular interest has been aroused by an accumulation of di- or trifucosylated type 2 chain. Only tumors which accumulate lacto-fucopentaosyl(III)ceramide accumulate the di- and trifucosylated derivates. A few monoclonal antibodies have been isolated which define these structures. Some of these monoclonal antibodies, such as FH4 which recognizes internal difucosyl structure (see Fig. 4), show higher specificity and more restricted reactivity for certain types of human cancer cells (Fukushi et al., in press). Application of these monoclonal antibodies, which define specific cell surface carbohydrates, to diagnosis and treatment of human cancer is

Table 3: Precursor Glycolipids Accumulating in some Human Cancers Recognized by Monoclonal "Tumor-Specific" Antibodies

---

1. Globotriaosylceramide (Gb3) in Burkitt lymphoma (Nudelman et al., 1983)

   Structure: Galα1→4Galβ1→4Glcβ1→1Cer

   Observation: a) Rat hybridoma IgM antibody (38-13) was selected for specific reactivity to Burkitt lymphoma

   b) The antibody defined globotriasylceramide (P^k antigen)

2. GD_3 ganglioside in human melanoma (Nudelman et al., 1982; Pukel et al., 1982)

   Structure: NeuAcα2→8NeuAcα2→3Galβ1→4Glcβ1→1Cer (GD_3)

   Observation: Mouse hybridoma antibodies (R24, 4.2) selected by the specific reactivity with human melanoma were identified as being directed to GD_3. Another monoclonal antibody (AH) was identified as being directed to GD_2

3. GM_2 in human breast cancer, brain tumor, melanoma, and GD_2 in human melanoma and other neuroectodermal tumors (Irie et al., 1982; Cahan et al., 1982; Watanabe et al., 1982; Tai et al., 1983)

   Structure: NeuAcα2→3[GalNAcβ1→4]Galβ1→4Glcβ1→Cer (GM_2)

   NeuAcα2→8NeuAcα2→3[GalNAcβ1→4]Galβ1→4Glcβ1 →1Cer (GD_2)

   Observation: Antigens defined by human serum and human monoclonal antibodies which show specificities to various human cancers were identified as GM_2 and GD_2

promising, but further extensive studies are needed.

BIOLOGICAL SIGNIFICANCE OF GLYCOLIPID CHANGES IN TUMORS

The biological significance of glycolipid changes associated with oncogenic transformation is essentially unknown, although a strong

Table 4: Complex Glycolipids Accumulating in Human Cancer Defined by Monoclonal Antibodies

1. Poly-X antigen in various human adenocarcinoma (Hakomori et al., in press) as defined by monoclonal antibody "FH4" (Fukushi et al., in press)

Structure:

```
Galβ1→4GlcNAcβ1→3Galβ1→4GlcNAcβ1→3Galβ1→4Glcβ1→1Cer
          3                 3
          ↑                 ↑
        Fucα1             Fucα1            III³V³Fuc₂nLc6

Galβ1→4GlcNAcβ1→3Galβ1→4GlcNAcβ1→3Galβ1→4GlcNAcβ1→3Galβ1→4Glcβ1→1Cer
          3                 3                 3
          ↑                 ↑                 ↑
        Fucα1             Fucα1             Fucα1          III³V³VII³Fuc₃nLc8
```

Background: Detected originally by anti-X monoclonal antibody; accumulating in colonic and liver cancer; absent in normal colonic mucosa, normal liver, and normal granulocytes

2. Novel fucoganglioside of human colonic cancer defined by monoclonal antibody (IB9) (Hakomroi et al., 1983)

Structure:

```
Galβ1→4GlcNAcβ1→3Galβ1→4GlcNAcβ1→3Galβ1→4Glcβ1→1Cer
 6                          3
 ↑                          ↑
NeuAcα2                   Fucα1         III³FucVI⁶NeuAcnLc6
```

Background: Detected by IB9 antibody which is directed to NeuAcα2→6Gal residue; present in colonic adenocarcinomas and absent in normal liver and normal colonic mucosa

possibility exists that these changes may inhibit cell adhesion and disrupt normal cell recognition. On the other hand, the loss of a particular ganglioside, which occurs in a few transformed cell lines, can be correlated with a loss of growth control, as shown in Table 5 in which evidence for such an association is listed. All the phenomena previously described, however, are indirect evidence. Glycolipids exogenously added in cell culture are slowly incorporated into plasma membranes, inhibit cell growth, and modify growth behavior (Laine and Hakomori, 1973; Keenan et al., 1975). With the availability of purified growth factors and serum-free culture conditions in recent years (Barnes and Sato, 1980), we have been able to examine this phenomenon in mouse Swiss 3T3 cells in greater detail with the following results (Bremer and Hakomori, 1982; Bremer et al., manuscript submitted for publication): (1) Cell growth (cell number increase) in serum-free medium was specifically inhibited by the presence of GM_1 and to a lesser extent by GM_3, but not by $NeuAcnLc_4$, although the gangliosides were incorporated equally well into cell membranes. GM_3 inhibited both PDGF- and EGF-stimulated mitogenesis determined by thymidine incorporation, while GM_1 could only inhibit PDGF-stimulated mitogenesis. $NeuAcnLc_4$ had no effect on mitogen-stimulated thymidine incorporation. Both GM_1 and GM_3 inhibited PDGF-dependent DNA synthesis (see Fig. 5). (2) The con-

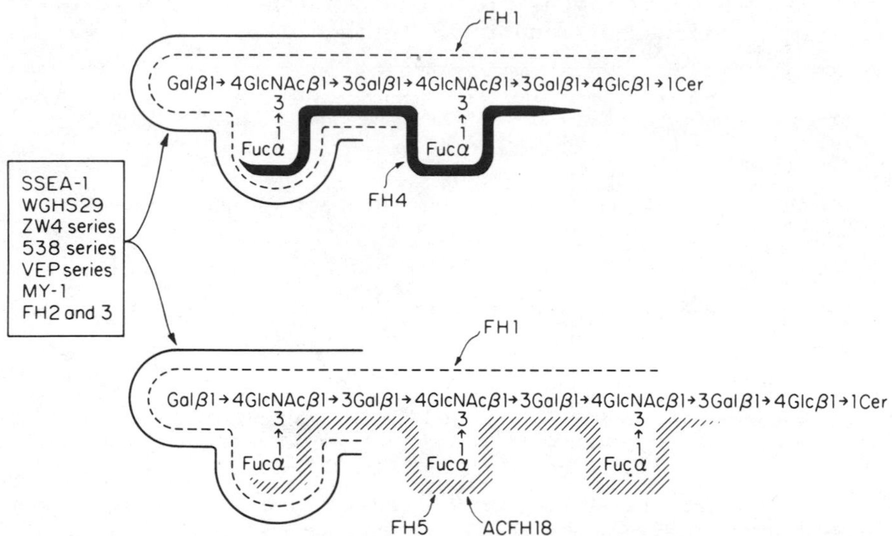

Fig. 4. The epitope structure of various antibodies directed to fucosylated type 2 chain. A number of monoclonal antibodies directed to teratocarcinoma and various human cancers are directed towards Galβ1→4[Fucα1→3]GlcNAc structure. These antibodies include WGHS and ZW series (gastrointestinal tumors), 538 series (lung cancer), and VEP series and My-1 (myelogeneous leukemia). In contrast, a monoclonal antibody (FH4) prepared by immunization with a pure difucosyl type 2 chain (upper structure) is directed towards difucosyl structure. A possible epitope structure of GH4 is shown by a bold solid line. Other monoclonal antibodies (FH5 and ACFH18) are directed towards a trifucosylated structure and their epitope structure is shown by a shadow.

centration-dependent binding of $[^{125}I]$-PDGF to cells indicated that cells whose growth was inhibited by GM_1 or GM_3 showed an increased affinity for PDGF as compared to cells grown without addition of ganglioside, while the total number of receptors stayed the same (see Fig. 6). This indicates that gangliosides that induce growth inhibition alter the affinity of the receptor to PDGF. Addition of ganglioside did not affect the binding of $[^{125}I]$-EGF. (3) No direct interaction was observed between gangliosides and growth factors, as evidenced by the lack of competition by ganglioside-containing liposomes for cellular binding of $[^{125}I]$ growth factors (data not shown). (4) GM_1 and GM_3, but neither $NeuAcnLc_4$ nor Gb_4, inhibited the PDGF-stimulated tyrosine phosphorylation by membrane preparations of a 170,000 molecular weight protein, which was identified as the PDGF receptor (see Fig. 7). Thus, the level of gangliosides GM_1 and GM_3

Table 5: Evidence that Glycolipids May Regulate Cell Proliferation

---

1. Contact inhibition of cell growth accompanies change of glycolipid synthesis (Hakomori, 1970; Sakiyama et al., 1972; Critchley and MacPherson, 1973).

2. Various glycolipids are more highly exposed by G1 phase, and some at G2 phase (Gahmberg and Hakomori, 1975; Lingwood and Hakomori, 1977).

3. Butyrate induces cell growth inhibition and enhances GM_3 synthesis (Fishman et al., 1974; Simmons et al., 1975).

4. Retinoids induce contact inhibition, enhnace GM_3 synthesis and glycolipid response (Patt et al., 1978).

5. Antibodies to GM_3, but not to globoside, inhibit 3T3 and NIL cell growth and enhance GM_3 syntehsis (Lingwood and Hakomori, 1977).

6. Exogenous addition of glycolipids incorporated into cell membranes inhibits cell growth through extension of G1 phase (Laine and Hakomori, 1973; Keenan et al., 1975).

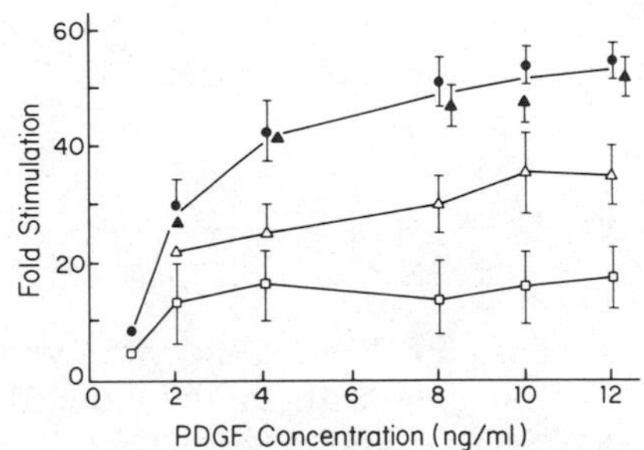

Fig. 5. Inhibition of PDGF-dependent [^{3}H]-thymidine incorporation by gangliosides. 3T3 cells were cultured in DME/F12 medium with or without the addition of gangliosides. The medium was replaced by DME/F12 without serum, and cells were cultured with various concentrations of PDGF for 48 h in the continuous presence or absence of gangliosides. Measurement of [^{3}H]-thymidine incorporation into trichloroacetic acid-insoluble material was made after 18 h. ●, no ganglioside; □, 50 nmol/ml GM_1; △, 50 nmol/ml GM_3; ▲, 50 nmol/ml sialosyl paragloboside.

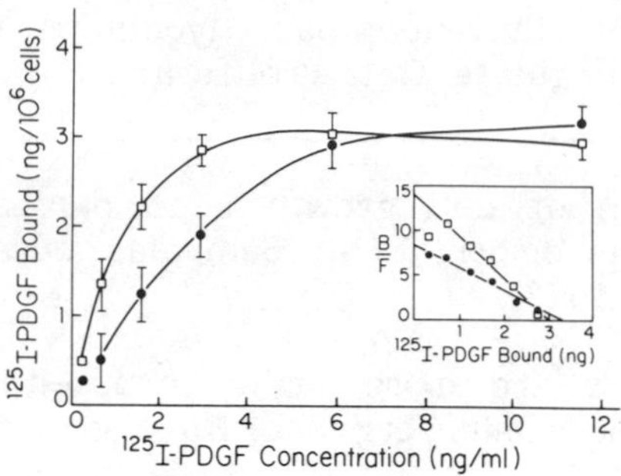

Fig. 6. Binding of [^{125}I]-PDGF to Swiss 3T3 fibroblasts. Cells were grown in DME/F12 medium in the presence or absence of 50 nmol/ml of GM_1 (see legend for Fig. 5). After 4 days, the medium was replaced with fresh medium without serum, with or without ganglioside. [^{125}I]-PDGF binding was determined by incubation with 1 ml of binding medium containing increasing amounts of [^{125}I]-PDGF. The Scatchard plot is shown in the figure insert.

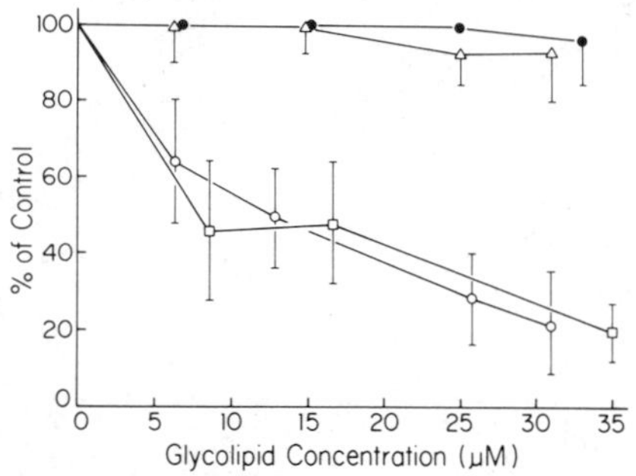

Fig. 7. The inhibition of PDGF-dependent tyrosine-phosphorylation in Mr-170,000 phosphoprotein by gangliosides. Swiss 3T3 cell membranes were incubated with 10 nM [γ^{32}]ATP and 60 nmol of PDGF with various concentrations of glycolipids as indicated for 30 min at 30° C. Proteins were separated by polyacrylamide gel electrophoresis base-treated to eliminate serine or threonine phosphates. After visualization of [γ^{32}]-labeled proteins by autoradiography, the receptor Mr-170,000 protein was cut from the gel and counted in a scintillation counter. Results are expressed as percent of maximum PDGF-dependent response. The results shown are the average of at least three determinations. o, GM_1; □, GM_3; ●, globoside; △, sialosyl paragloboside.

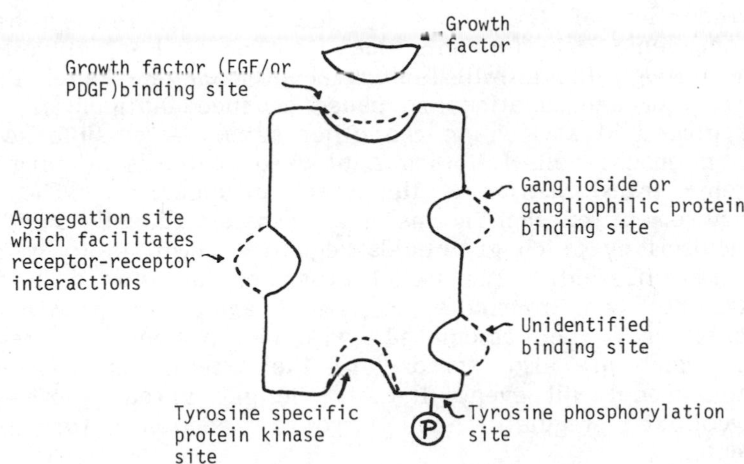

Fig. 8. "Allosteric regulator" model of growth factor receptor adapted from Schlessinger et al (1983). The binding of the growth factor to the receptor may affect 1) tyrosine-specific protein kinase, which is independent of cyclic AMP-dependent kinase, 2) receptor-receptor aggregation, which facilitates clustering and internalization of the receptor-growth factor complex, and 3) direct binding to gangliosides or indirect binding through a gangliophilic protein. This model is based on the fact that PDGF-dependent tyrosine-phosphorylation of the receptor protein (Mr-170,000) as well as PDGF-dependent cell growth and mitogenesis were inhibited by exogenous enrichment of membrane gangliosides. This model also explains how ganglioside enrichment of membranes alters the affinity of cells to growth factors.

in membranes may modulate PDGF receptor function by affecting the degree of tyrosine-phosphorylation and may alter the affinity of the receptor for PDGF (Bremer et al., submitted for publication).

Recently Schlessinger et al. (1983) proposed an "allosteric regulator" model for the EGF receptor in which several binding sites on the receptor protein in addition to the EGF binding site were postulated. Since the chemical level and organization of gangliosides in plasma membranes may affect the allosteric configuration of the receptor, it is possible that one of the additional sites on the growth hormone receptor may bind GM_1 or GM_3 either directly or through an intermediate gangliophilic protein (see Fig. 8). Organization of gangliosides with such receptors may alter the receptor-receptor interaction which is necessary for internalization of the receptor-growth hormone complex. The model may be useful for further studies of cell growth regulation through membrane gangliosides. A loss or reduction of GM_3 or GM_1 due to a blocked synthesis in various oncogenic transformants may cause a loss of allosteric regulation through the growth factor receptor on one hand, and may induce precursor accumulation and cause enhanced antigenicity of the precursor glycolipid antigen on the other hand. A possible sequence of events in mouse cells following a blocked synthesis of gangliosides by oncogenic transformation is illustrated in Figure 9. The growth factor requirement of each normal and transformed cell is different. The mechanisms by which glycolipids regulate growth factor receptors may also be different. The model described above (and in Fig. 9) represents only one example. A large variety of growth factor requirements may be associated with a number of regulator glycolipids which are also "sensors" of the external environment. I believe this model will eventually fill the gap in our knowledge on how glycolipids regulate cell growth adapted to external environments.

ACKNOWLEDGEMENTS

The authors thank for general support the National Institutes of Health (Grants CA20026, CA19224, GM23100), the American Cancer Society (BC9M), and the Otsuka Research Foundation. This work has been accomplished in collaboration with Doctors Reiji Kannagi, Eric G. Bremer, Eric Holmes, Ed Nudelman, Steven B. Levery, Nancy

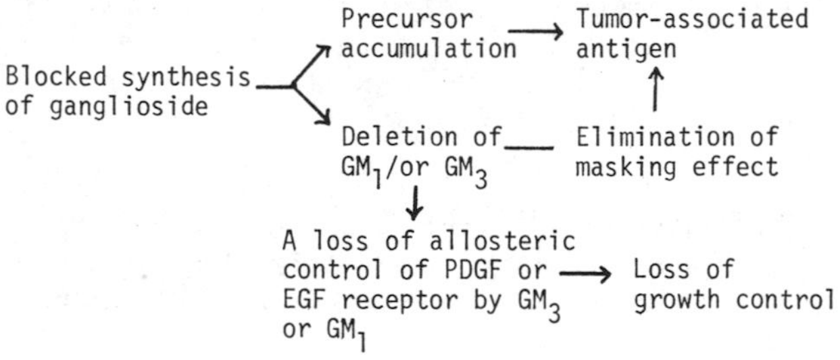

Blocked synthesis of ganglioside → Precursor accumulation → Tumor-associated antigen

Blocked synthesis of ganglioside → Deletion of GM_1/or GM_3 → Elimination of masking effect → Tumor-associated antigen

Deletion of GM_1/or GM_3 → A loss of allosteric control of PDGF or EGF receptor by GM_3 or GM_1 → Loss of growth control

Fig. 9. A blocked synthesis of ganglioside results in accumulation of precursors which are recognized by antibodies and by the immune system as tumor-associated antigens. On the other hand, a blocked synthesis of ganglioside results in deletion of a specific ganglioside which causes a loss of allosteric control of the growth factor receptor and consequently results in loss of growth control. The deletion of higher gangliosides may eliminate the masking effect of precursor antigenicity, thus enhancing the antigenicity of glycolipid tumor antigens. The masking effect of tumor antigens by longer chain glycolipids was recently described (Kannagi et al, 1983d).

Cochran, William W. Young, Jr., Davor Solter, and Barbara B. Knowles.

REFERENCES

Abercrombie, M., and Ambrose, E.J., 1962, The surface properties of cancer cells: A review, Cancer Res., 22:525.

Ansel, S., and Huet, C., 1980, Specific glycolipid antigen in SV30-transformed cell membranes, Int. J. Cancer, 25:797.

Barnes, D., and Sato, G., 1980, Methods for growth of cultured cells in serum-free medium, Anal. Biochem., 102:255.

Brady, R.O., and Fishman, P., 1974, Biosynthesis of glycolipids in virus-transformed cells, Biochim. Biophys. Acta, 335:121.

Breimer, M.E., 1980, Adaptation of mass spectrometry for the analysis of tumor antigens as applied to blood group glycolipids of a human gastric carcinoma, Cancer Res., 40:897.

Bremer, E.G., and Hakomori, S., 1982, GM_3 ganglioside induces hamster fibroblast growth inhibition in chemically-defined medium: Ganglioside may regulate growth factor receptor function, Biochem. Biophys. Res. Commun., 106:711.

Bremer, E.G., Hakomori, S., Bowen-Pope, D.F., Raines, E., and Ross, R., submitted for publication, Ganglioside-mediated cell growth modulation: the change of cell surface affinity to growth factors and the inhibition of growth factor-dependent Mr-170,000 protein phosphorylation by gangliosides, J. Biol. Chem.

Cahan, L.D., Irie, R.I., Singh, R., Cassidenti, A., and Paulsen, J.C., 1982, Identification of human neuroectodermal tumor antigen (IA-I-2) as ganglioside GD_2, Proc. Natl. Acad. Sci. USA, 79:7629.

Critchley, D.R., and MacPherson, I., 1973, Cell density-dependent glycolipids in NIL_2 hamster cells derived from malignant and transformed cell lines, Biochim. Biophys. Acta, 296:145.

Critchley, D.R., and Vicker, M.G., 1977, Glycolipids as membrane receptors important in growth regulation and cell-cell interactions, in: "Dynamic Aspects of Cell Surface Organization," G. Poste and G.L. Nicolson, eds., Elsevier/North-Holland Biomedical Press.

Fishman, P.H., Simmons, J.L., Brady, R.O., and Freese, E., 1974, Induction of glycolipid synthesis by sodium butylate in Hela cells, Biochem. Biophys. Res. Commun., 59:292.

Fukushi, Y., Hakomori, S., Nudelman, E., and Cochran, N., in press, Novel fucolipids accumulating in human adenocarcinoma. II. Selective isolation of hybridoma antibodies that differentially recognize mono-, di-, and trifucosylated type 2 chain, J. Biol. Chem.

Gahmberg, C.G., and Hakomori, S., 1975, Surface carbohydrates of hamster fibroblasts. I. Chemical characterization of surface-labeled glycosphingolipids and a specific ceramide tetrasaccharide for transformants, J. Biol. Chem., 250:2438.

Gesner, B.M., and Ginsburg, V., 1964, Effect of glycosidases on the fate of transfused lymphocytes, Proc. Natl. Acad. Sci. USA, 52:750.

Gooi, H.C., Feizi, T., Kapadia, A., Knowles, B.B., Solter, D., and Evans, J.M., 1981, Stage-specific embryonic antigen involves $\alpha 1 \to 3$ fucosylated type 2 blood group chains, Nature, 292:156.

Hakomori, S., 1970, Cell density-dependent changes of glycolipid concentrations in fibroblasts, and loss of this response in virus-transformed cells, Proc. Natl. Acad. Sci. USA, 67:1741.

Hakomori, S., 1973, Glycolipids of tumor cell membranes, in: "Advances in Cancer Research," S. Weinhouse, ed., Academic Press, N.Y.

Hakomori, S., 1975, Structure and organization of cell surface glycolipids and glycoproteins, dependency on cell growth and malignant transformation, Biochim. Biophys. Acta, 417:55.

Hakomori, S., and Andrews, H.D., 1970, Sphingoglycolipids with Le^b activity and the co-presence of Le^a and Le^b glycolipids in human tumor tissue, Biochim. Biophys. Acta, 202:225.

Hakomori, S., and Jeanloz, R.W., 1964, Isolation of a glycolipid containing fucose, galactose, glucose, and glucosamine from human cancerous tissue, J. Biol. Chem., 239:3606.

Hakomori, S., and Kannagi, R., 1983, Glycosphingolipids as tumor-associated and differentiation markers, J. Natl. Cancer Inst., 71:231.

Hakomori, S., and Murakami, W.T., 1968, Glycolipids of hamster fibroblasts and derived malignant transformed cell lines, Proc. Natl. Acad. Sci. USA, 59:254.

Hakomori, S., Nudelman, E., Levery, S., Solter, D., and Knowles, B.B., 1981, The hapten structure of a developmentally regulated glycolipid antigen (SSEA-1) isolated from human erythrocytes and adenocarcinoma: A preliminary note, Biochem. Biophys. Res. Commun., 100:1578.

Hakomori, S., Nudelman, E., Levery, S.B., Kannagi, R., in press, Novel frucolipids accumulating in human adenocarcinoma. I. Glycolipids with di- or trifucosylated type 2 chain, Biol. Chem.

Hakomori, S., Teather, C., and Andrews, H.D., 1968, Organizational difference of cell surface hematoside in normal and virally transformed cells, Biochem. Biophys. Res. Commun., 33:563.

Hakomori, S., Wang, S.-H., and Young, W.W., Jr., 1977, Isoantigenic expression of Forssman glycolipid in human gastric and colonic mucosa: its possible identity with "A-like antigen" in human cancer, Proc. Natl. Acad. Sci. USA, 74:3023.

Hattori, H., Uemura, K., and Taketomi, T., 1981, Glycolipids of gastric cancer. The presence of blood group A-active glycolipids in cancer tissues from blood group O patients, Biochim. Biophys. Acta, 666:361.

Irie, R.F., Sze, L.L., and Saxton, R.E., 1982, Human antibody to OFA-I, a tumor antigen, produced in vitro by Epstein-Barr virus-transformed human B-lymphoid cell lines, Proc. Natl. Acad. Sci. USA, 79:5666.

Kalckar, H.M., 1965, Galactose metabolism and cell sociology, Science, 150:305.

Kannagi, R., Nudelman, E., Levery, S.B., and Hakomori, S., 1982a, A series of human erythrocyte glycosphingolipids reacting to the monoclonal antibody directed to a developmentally regulated antigen, SSEA-1, J. Biol. Chem., 257:14865.

Kannagi, R., Levine, P., Watanabe, K., and Hakomori, S., 1982b, Recent studies of glycolipid and glycoprotein profiles and characterization of the major glycolipid antigen in gastric cancer of a patient of blood group genotype pp (Tj^{a-}) first studied in 1951, Cancer Res., 42:5249.

Kannagi, R., Levery, S.B., Ishigami, F., Hakomori, S., Shevinsky, L.H., Knowles, B.B., and Solter, D., 1983a, New globoseries glycosphingolipids in human teratocarcinoma reactive with the monoclonal antibody directed to a developmentally regulated antigen, SSEA-3, J. Biol. Chem., 258:8934.

Kannagi, R., Cochran, N.A., Ishigami, F., Hakomori, S., Andrews, P.W., Knowles, B.B., and Solter, D., 1983b, Stage-specific embryonic antigens (SSEA-3 and -4) are epitopes of a unique globoseries ganglioside isolated from human teratocarcinoma cells, EMBO J. in press.

Kannagi, R., Levery, S.B., and Hakomori, S., 1983c, Sequential change of carbohydrate antigen associated with differentiation of murine leukemia cells: Ii-antigenic conversion and shifting of glycolipid synthesis, Proc. Natl. Acad. Sci. USA, 80:2844.

Kannagi, R., Stroup, R., Cochran, N.A., Urdal, D.L., Young, W.W., Jr., and Hakomori, S., 1983d, Glycolipid tumor antigen in cultured murine lymphoma cells and factors affecting its expression at the cell surface, Cancer Res., 43:4997.

Kapadia, A., Feizi, T., and Evans, M.J., 1981, Changes in expression and polarization of blood group I and i antigens in post-implantation embryos and teratocarcinomas of mouse associated with cell differentiation, Exp. Cell Res., 131:185.

Keenan, T.W., Schmid, E., Franke, W.W., and Wiegandt, H., 1975, Exogenous gangliosides suppress growth rate of transformed and untransformed 3T3 mouse cells, Exp. Cell Res. 92:259.

Koprowski, H., Herlyn, M., Steplewsksi, Z., and Sears, H.F., 1981, Specific antigen in serum of patients with colon carcinoma, Science, 212:53.

Laine, R.A., and Hakomori, S., 1973, Incorporation of exogenous glycosphingolipids in plasma membranes of cultured hamster cells and concurrent change of growth behavior, Biochem. Biophys. Res. Commun., 54:1039.

Levine, P., 1978, Blood group and tissue genetic markers in familial adenocarcinoma: Potential specific immunotherapy, Semin. Hematol., 5:28.

Levine, P., Bobbit, O.B., Waller, R.K., and Kuhmichel, A., 1951, Isoimmunization by a new blood factor in tumor cells, Proc. Soc. Exp. Biol. Med., 77:403.

Lingwood, C., and Hakomori, S., 1977, Selective inhibition of cell growth and associated changes in glycolipid metabolism induced by monovalent antibody to glycolipids, Exp. Cell Res., 108:385.

Magnani, J.L., Nilsson, B., Brockhaus, M., Zopf, D., Steplewski, Z., Koprowski, H., and Ginsburg, V., 1982, A monoclonal antibody-defined antigen associated with gastrointestinal cancer is a ganglioside containing sialylated lacto-N-fucopentaose II, J. Biol. Chem., 257:14365.

Masamune, H., and Hakomori, S., 1960, On the glycoprotein and mucopolysaccharides of cancer tissue, in particular references on "cancer blood group substances," Symposia Cell Chemistry, 10:37.

Nudelman, E., Hakomori, S., Kannagi, R., Levery, S., Yeh, M.-Y., Hellström, K.E., and Hellström, I., 1982, Characterization of a human melanoma-associated ganglioside antigen defined by a monoclonal antibody, 4.2, J. Biol. Chem., 257:12752.

Nudelman, E., Kannagi, R., Hakomori, S., Parsons, M., Lipinski, M., Wiels, J., Fellous, M., and Tursz, T., 1983, A glycolipid antigen associated with Burkitt lymphoma defined by a monoclonal antibody, Science, 220:509.

Patt, L.M., Itaya, K., and Hakomori, S., 1978, Retinol induces density-dependent growth inhibition and changes in glycolipids and LETS, Nature, 273:379.

Pukel, C.S., Lloyd, K.O., Trabassos, L.R., Dippold, W.G., Oettgen, H.F., and Old, L.J., 1982, GD_3, a prominent ganglioside of human melanoma: Detection and characterization by mouse monoclonal antibody, J. Exp. Med., 155:1133.

Rosenfelder, G., Young, W.W., Jr., and Hakomori, S., 1977, Association of the glycolipid pattern with antigenic alterations in mouse fibroblasts transformed by murine sarcoma virus, Cancer Res., 37:1333.

Sakiyama, H., Gross, S.K., and Robbins, P.W., 1972, Glycolipid synthesis in normal and virus-transformed hamster cell lines, Proc. Natl. Acad. Sci. USA, 69:872.

Schlessinger, J., Schreiber, A.B., Levi, A., Lax, I., Liberman, T., and Yarden, Y., 1983, Regulation of cell proliferation by epidermal growth factor, CRC Rev. Biochem., 14:93.

Shevinsky, L.H., Knowles, B.B., Damjanov, I., and Solter, D., 1982, A stage-specific embryonic antigen (SSEA-3) defined by monoclonal antibody to murine embryos, expressed on mouse embryos and on human teratocarcinoma cells, Cell, 30:697.

Simmons, J.L., Fishman, P.H., Freese, E., and Brady, R.O., 1975, Morphological alterations and ganglioside sialyltransferase activity induced by small fatty acids in Hela cells, J. Cell Biol., 66:414.

Solter, D., and Knowles, B.B., 1978, Monoclonal antibody defining a stage-specific embryonic antigen (SSEA-1), Proc. Natl. Acad. Sci. USA, 75:5565.

Sundsmo, J., and Hakomori, S., 1976, Lacto-N-neotetraosylceramide ("paragloboside") as a possible tumor-associated surface antigen of hamster NILpy tumor, Biochem. Biophys. Res. Commun., 68:799.

Szulman, A.E., 1977, The ABH and Lewis antigens of human tissues during prenatal and postnatal life, in: "Human Blood Groups," 5th Int. Convoc. Immunol., Karger, Basel.

Tai, T., Paulson, J.C., Cahan, L.D., and Irie, R.F., 1983, Human tumor associated antigen: Ganglioside GM_2 and DG_2, in "Glycoconjugates," M.A. Chester, D. Heinegard, A., Lundblad, and S. Svensson, eds., 7th International Glycoconjugate Symposium, Lund.

Taniguchi, N., Yokosawa, N., Narita, M., Mitsuyama, T., and Makita, A., 1981, Expression of Forssman antigen synthesis and degradation in human lung cancer, J. Natl. Cancer Inst., 67:577.

Watanabe, T., Pukel, C.S., Takeyama, H, Lloyd, K.O., Shiku, H., Li, L.T.C., Trabassos, L.R., Oettgen, H.F., and Old, L.J., 1982, Humn melanoma antigen AH is an autoantigen ganglioside related to GD_2, J. Exp. Med., 156:1884.

Willison, K.R., and Stern, P.L., 1978, Expression of a Forssman antigenic specificity in the preimplantation mouse embryo, Cell, 14:785.

Yang, H.J., and Hakomori, S., 1971, A sphingolipid having a novel type of ceramide and lacto-N-pentaose III, J. Biol. Chem., 246:1192.

Yoda, Y., Ishibashi, T., and Makita, A., 1980, Isolation, characterization, and biosynthesis of Forssman antigen in human lung and lung carcinoma, J. Biochem. (Tokyo), 88:1887.

Yokota, M., Warner, G.A., and Hakomori, S., 1981, Blood group A-like glycolipid and a novel Forssman antigen in the hepatocarcinoma of a blood group O individual, Cancer Res., 41:4185.

Young, W.W., Jr., and Hakomori, S., 1981, Therapy of mouse lymphoma with monoclonal antibodies to glycolipid: Selection of low antigenic variants in vivo, Science, 211:487.

Young, W.W., Jr., Hakomori, S., and Levie, P., 1979, Characterization of anti-Forssman (anti-Fs) antibodies in human sera: Their specificity and possible changes in patients with cancer, J. Immunol., 123:92.

PROBING THE MACROPHAGE SURFACE:

Alterations in Specific Surface Carbohydrates Accompany Macrophage Activation

Arthur M. Mercurio*, Gerald A. Schwarting†, and Phillips W. Robbins*

*Department of Biology and Center for Cancer Research
Massachusetts Institute of Technology
Cambridge, Massachusetts 02139

† Department of Biochemistry
E. K. Shriver Center
Waltham, Massachusetts 02254

ABSTRACT

We have begun to analyze the carbohydrates on the surface of mouse peritoneal macrophages with the long-range intent of implicating specific carbohydrates in macrophage surface function. As an initial approach to this problem, populations of resident and activated macrophages were surface labeled by the galactose oxidase and neuroaminidase/NaB^3H_4 method. Surface glycoproteins labeled by this method were digested with pronase and the resultant glycopeptides analyzed by gel-filtration chromatography. The results indicate that the macrophage surface is enriched in high MW lactosaminoglycan-containing glycopeptides; upon activation a marked increase in the amount of labeling and degree of branching of these lactosaminoglycans is observed.

Surface glycolipids were separated into neutral glycolipid and ganglioside fractions, and analyzed by both chemical and radiochemical methods. The results demonstrate that macrophage activation results in alterations in the chemical amounts and in the surface exposure of specific glycolipids. These alterations include an increase in the surface exposure of asialo GM_1 on the surface of activated macrophages as well as alterations in the amount and surface exposure of specific moieties of the ganglioside GM_1 upon activation.

INTRODUCTION

The macrophage surface participates in an impressive array of specialized functions including phagocytosis, secretion, interaction with T cells, and the recognition and destruction of tumor cells (for general reviews see Van Furth, 1980). Many of these functions are enhanced, or are only expressed, in populations of macrophages that have been "activated" with specific stimuli (Cohn, 1978; Karnovsky and Lazdins, 1978). Consequently, it is of considerable interest to characterize those components of the macrophage surface that are altered upon activation and, ideally, to relate such alterations to specific macrophage function. The protein components of the macrophage surface--specific receptors, antigens, and ectoenzymes--have been most intensively studied in this regard, and the results obtained have provided substantial insight into the dynamics of the macrophage surface. However, despite the fact that most macrophage surface proteins are glycosylated, little attention has been directed toward the structure and function of these oligosaccharides, or toward other surface carbohydrates. Given the increasing number of cellular functions in which carbohydrates are being implicated, it seems likely that they participate, at least to some extent, in the affairs of the macrophage surface.

For the above reasons, we have begun to characterize the carbohydrates on the surface of the mouse peritoneal macrophage, focusing on the possibility that macrophage activation results in alterations in specific surface carbohydrates. In this report we demonstrate that a general consequence of macrophage activation is (1) an enrichment and a structural change in a specific class of protein-bound surface carbohydrates, the lactosaminoglycans, and (2), an alteration in both the amount and surface exposure of specific glycolipids.

GENERAL METHODOLOGY

Macrophages

All macrophages were obtained by lavage of the peritoneal cavity of C57/BL6 mice. Resident macrophages were obtained from untreated mice. Inflammatory (TG macrophages) were obtained from mice which had been injected four days previously with thioglycollate-broth (TG). Tumoricidal (BCG macrophages) were obtained from mice that had been injected 4-6 weeks previously with live Bacillus Calmette-Guerin (BCG).

Harvested peritoneal cells (approximately 1-2 x 10^6 macrophages) were plated in MEM containing 10 percent fetal bovine serum in 35 mm tissue culture plates. The dishes were incubated at 37° C for 4 hr and then washed vigorously to remove the nonadherent cells.

Surface Labeling

Macrophage monolayers, washed free of nonadherent cells, were surface labeled by the galactose oxidase ± neuraminidase/NaB^3H_4 method as described by Gahmberg and Hakomori (1973) with the exception that the entire procedure was carried out at 4° C instead of 37° C. After labeling, the labeled glycoproteins were exhaustively digested with pronase and the resultant glycopeptides were analyzed by gel filtration chromatography.

For the glycolipid experiments, surface labeled macrophages were solubilized in chloroform-methanol (2:1). Neutral glycolipids were separated and analyzed as described by Schwarting (1980) and gangliosides as described by Schwarting and Gajewski (1983).

RESULTS

Analysis of the surface glycopeptides obtained from resident, BCG, and TG macrophages labeled by the neuraminidase/galactose oxidase/NaB^3H_4 on Bio-Gel P-10 (Fig. 1) reveals the presence of two broad peaks of labeled glycopeptides (designated Peak I and Peak II). No label was present in either Peak I or II in control experiments in which the neuraminidase/galactose oxidase incubation was omitted, establishing that the observed labeling is specifically in carbohydrates. Metabolic labeling studies with ^3H-fucose indicate that both Peak I and Peak II glycopeptides contain fucose (data not shown).

Peak II contains typical complex-type glycopeptides as evidenced by the fact that complex glycopeptides purified from Sindbis virions (gift of Dr. Peggy Hsieh, M.I.T.) migrate in the position of Peak II on P-10 columns. Peak II most likely also contains hybrid-type and O-linked glycopeptides given the method of labeling and the size range of this peak.

Peak I consists of glycopeptides of much higher molecular weight than typical complex glycopeptides, i.e., larger than 4000 daltons. These glycopeptides are not complex glycopeptides that have been incompletely digested with pronase, as evidenced by the fact that redigestion of Peak I with pronase does not alter its migration on P-10. As seen in Figure 1 and Table 1, marked differences exist in the labeling of Peak I glycopeptides relative to the labeling of Peak II glycopeptides among the three macrophage populations. Labeling of Peak I is most evident in glycopeptides obtained from TG macrophages and least in those from resident macrophages. Indeed,

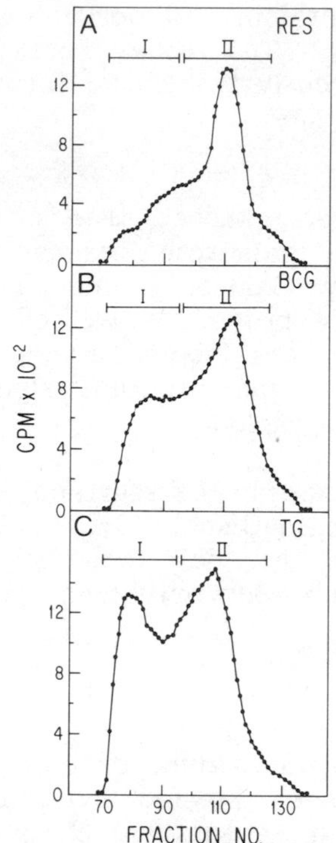

Fig. 1. Gel filtration profiles of macrophage surface glycopeptides. Glycopeptides were obtained from pronase digests of macrophage surface glycoproteins labeled by the neuraminidase/galactose oxidase/^{3}H-borohydride method and chromatographed on Bio-Gel P-10. The exclusion and inclusion volumes of the column are at tractions 70 and 165, respectively. Glycopeptides obtained from A) resident macrophages; B) BCG macrophages; C) TG-elicited macrophages. Peaks I and II were pooled as indicated, lyophilized, and used in subsequent analyses.

in the case of TG macrophages, the labeling in Peak I accounts for approximately 40 percent of the total surface glycopeptides labeled by this method. Relative to resident macrophages, the labeling of Peak I from BCG macrophages is also substantially increased.

In an attempt to determine the nature of the glycopeptides in Peak I, this peak was pooled and treated with endo-β-galactosidase. Susceptibility to endo-β-galactosidase is diagnostic for a class of high molecular weight carbohydrates, the lactosaminoglycans which are characterized by the repeating disaccharide $(Gal\beta1 \rightarrow 4GlcNAc\beta1 \rightarrow 3)_n$ (Fukuda et al., 1979; see Fig. 4 for structure); endo-β-galactosidase cleaves after internal galactose residues in this sequence. Peak I from all three populations of macrophages is susceptible to endo-β-galactosidase (Fig. 2). In contrast, Peak II is more resistant to the action of this enzyme. Less than 12 percent of the material in Peak II from all three populations of macrophages was altered in mobility by treatment with this enzyme.

Considerable differences were observed in the endo-β-galactosidase digestion profile of Peak I among the three macrophage populations (Fig. 2 and Table 1). Material obtained from resident macrophages was most susceptible to the action of this enzyme; almost 60 percent of the labeled material from Peak I was degraded to a peak of lower molecular weight material (designated Peak III). Peak III consists of two oligosaccharides: the tetrasaccharide Galβ1→4(Fucα1→3)GlcNAcβ1→3Gal and the trisaccharide Galβ1→4GlcNAc β1→3Gal as determined by their comigration with purified standard oligosaccharides on P-4. These two oligosaccharides are the smallest labeled oligosaccharides expected to be generated by endo-β-galactosidase given the method of labeling and the specificity of the enzyme (Fukuda, 1981; see Fig. 4). In contrast to resident macrophages, only 24 percent of the labeled material from TG macrophages was degraded to Peak III oligosaccharides, and a more

Table 1: Effect of Macrophage Activation on the Sensitivity of Peak I Surface Glycopeptides to Endo-β-Galactosidase*

Source of Glycopeptides	Total Surface Label in Peak I (%)	Peak I Radioactivity Degraded to Peak III by Endo-β-galactosidase (%)
Freshly explanted resident macrophages	20.6	57.0
BCG macrophages	32.2	32.2
TG-elicited macrophages	42.2	24.8
Cultured (24 hr) resident macrophages	43.6	24.3
Lymphokine-treated (24 hr) resident macrophages	50.1	17.4

*Based on the Bio-Gel P-10 column profiles of neuraminidase/galactose oxidase/^{3}H-borohydride labeled glycopeptides, the amount (%) of total label in Peak I was quantitated and the Peak I material was pooled, lyophilized, and incubated with endo-β-galactosidase. The resulting degradation products were rechromatographed on P-10 and the amount (%) of total label in Peak III was quantitated. Peak III consists of two oligosaccharides: Galβ1→4(Fucα1→3)GlcNacβ1→3Gal and Galβ1→4GlcNAcβ1→3Gal.

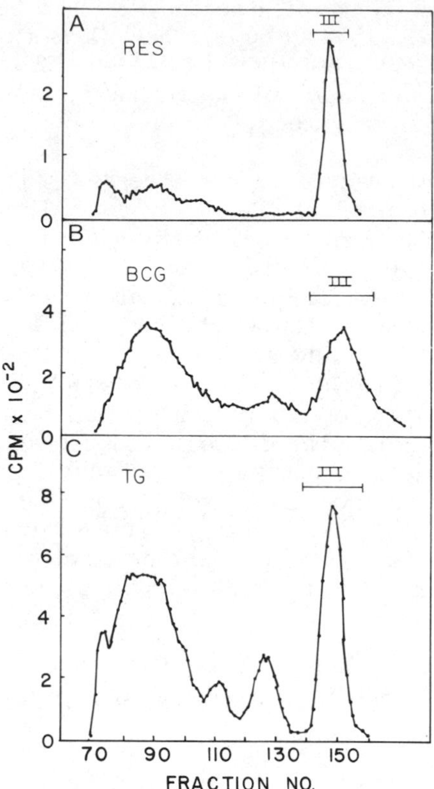

Fig. 2. Gel filtration profiles of the degradation products obtained after treatment of macrophage surface glycopeptides with endo-β-galactosidase. Peak I glycopeptides (see Fig. 1) were incubated in the presence of endo-β-galactosidase (2 additions of 70 milliunits each for 48 hr at 37° C) and rechromatographed on Bio-Gel P-10. Endo-β-galactosidase degradation profiles of Peak I glycopeptides from A) resident macrophages; B) BCG macrophages; C) TG-elicited macrophages. The exclusion and inclusion volumes of the column are at fractions 70 and 165, respectively. Peak III is observed only after treatment of macrophage surface glycopeptides with endo-β-galactosidase and consists of the tetrasaccharide Galβ1→4(Fucα1→3)GlcNAcβ1→3Gal and the trisaccharide Galβ1→4GlcNAcβ1→3Gal.

heterogeneous mixture of degradation products was observed (Fig. 2C). Similarly, 32 percent of the labeled Peak I material from BCG macrophages was degraded to Peak III, and the degradation pattern resembling that obtained for TG macrophages, although not as heterogeneous. As established by Fukuda et al (1979), increased resistance to degradation by endo-β-galactosidase (resulting in a more heterogeneous degradation profile) is indicative of increased branching in the lactosaminoglycan structure (see Fig. 4). Thus, the lactosaminoglycans on TG and BCG macrophages are considerably more branched than those on resident macrophages.

Higher concentrations of endo-β-galactosidase (150 munits/ml) did not alter the profiles shown in Figure 2. In addition, control experiments (i.e., DEAE chromatography and precipitation with cetyl-

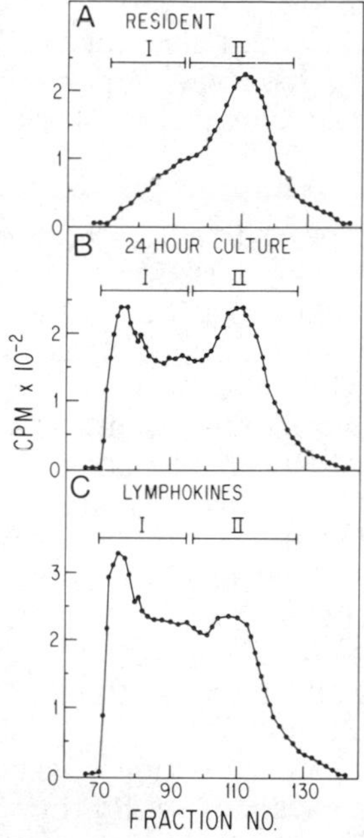

Fig. 3. Gel filtration profiles of neuraminidase/galactose oxidase/^{3}H-borohydride labeled glycopeptides obtained from A) freshly explanted resident macrophages (defined as macrophages that have been incubated in normal medium for 4 hours; B) resident macrophages maintained in culture for 24 hours in normal medium; C) resident macrophages cultured for 24 hours in the presence of lymphokines (a 1:4 dilution of the supernatant of con A-stimulated Balb/c spleen cells). The exclusion and inclusion volumes of the column are at fractions 70 and 165, respectively. Resident macrophages cultured for only 4 hours in the presence of lymphokines yielded a gel filtration profile identical to that of freshly explanted resident macrophages (A). As in Fig. 1, Peaks I and II were pooled, lyophilized, and used in subsequent analyses.

pyridinium chloride) excluded the possibility that Peak I contained proteoglycan-type material.

We also examined the possibility that the macrophage surface glycopeptide profile could be altered as a consequence of in vitro activation of resident macrophages. Resident macrophages when cultured in the presence of lymphokines (present in the supernatant of con A stimulated spleen cells) acquire some of the characteristics of activated macrophages such as increased secretion of plasminogen activator and increased endocytic activity (Cohn, 1978; Karnovsky and Lazdins, 1978). Similarly, resident macrophages, when cultured in normal medium, also begin to express some properties of activated macrophages. For example, Bianco et al. (1975) reported that resident macrophages maintained in culture for 24 hours drastically increase their rate of phagocytosis of antibody-coated erythrocytes. Presumably, this activation can be attributed to components in serum.

Figure 3 compares the surface glycopeptide profile of freshly explanted resident macrophages, resident macrophages cultured for 24 hours in normal medium, and resident macrophages cultured for 24 hours in the presence of lymphokines. Morphologically, those macrophages cultured in the presence of lymphokines, and to a lesser extent those cultured in normal medium, are considerably more spread than freshly explanted resident macrophages. These three populations of macrophages also differ markedly in their surface glycopeptide profiles. The labeling of Peak I glycopeptides obtained from lymphokine-treated and cultured macrophages is substantially increased relative to Peak I glycopeptides from freshly explanted macrophages; however, culturing in the presence of lymphokines is more effective in increasing the labeling in Peak I than culturing in normal medium alone (compare Figs. 3B and 3C). A striking feature of these profiles is that the increased labeling observed in lymphokine-treated and cultured macrophages is specifically in Peak I glycopeptides; the labeling in Peak II glycopeptides remains relatively constant. Resident macrophages when cultured in the presence of lymphokines for only 4 hours did not exhibit this increased labeling in Peak I glycopeptides, and yielded a profile identical to that of freshly explanted macrophages.

Similar to the results we obtained with in vivo activated macrophages, the increase in Peak I labeling observed upon in vitro activation is accompanied by a substantial increase in resistance to degradation of endo-β-galactosidase to Peak III oligosaccharides (Table 1). Only 17 percent of the Peak I glycopeptides from lymphokine-treated macrophages are degraded to Peak III, compared with 57 percent for freshly explanted resident macrophages. These data are consistent with our observation that a general consequence of macrophage activation is an increased branching in surface lactosaminoglycans.

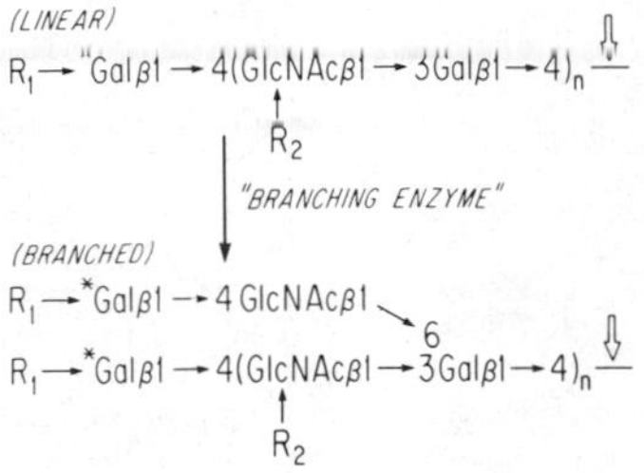

(LINEAR)

$R_1 \rightarrow {}^{*}Gal\beta 1 \rightarrow 4(GlcNAc\beta 1 \rightarrow 3Gal\beta 1 \rightarrow 4)_n$ ⇩

↑ R_2

"BRANCHING ENZYME"

(BRANCHED)

$R_1 \rightarrow {}^{*}Gal\beta 1 \rightarrow 4\,GlcNAc\beta 1 \searrow 6$

$R_1 \rightarrow {}^{*}Gal\beta 1 \rightarrow 4(GlcNAc\beta 1 \rightarrow 3Gal\beta 1 \rightarrow 4)_n$ ⇩

↑ R_2

Fig. 4. Minimal outer structures of linear and branched lactosaminoglycans present on the mouse peritoneal macrophage surface based on the data presented in this paper. Adapted from Fukuda et al (1979). R_1 = sialic acid $\alpha 2 \rightarrow 3/6$, or H; R_2 = fucose $\alpha 1 \rightarrow 3$ or H. Asterisks indicate the galactose residues presumed labeled by the neuraminidase/galactose oxidase/^{3}H-borohydride method. Open arrows indicate the cleavage sites of endo-β-galactosidase. The minimum structural requirement for endo-β-galactosidase is GlcNAc β1⟶3Galβ1→4GlcNAc (Fukuda, 1981).

Surface Glycolipids

Glycolipids on the surface of resident, BCG, and TG macrophages were labeled by the galactose oxidase/NaB^3H_4 method, and chloroform-methanol (2:1) extracts were separated into neutral glycolipid and ganglioside fractions and subjected to both chemical and radiochemical analyses.

Chemical analyses (Fig. 5B) indicate that the major neutral glycolipids present in the mouse macrophage are CMH, CDH, and asialo GM_1. Quantitative differences in the amount of these glycolipids were not observed between resident and BCG (tumoricidal) macrophages. Interestingly, however, asialo GM_1 is labeled on the surface of BCG macrophages but not on resident macrophages (Fig. 5A). This suggests that resident and tumoricidal macrophages differ markedly in their surface exposure of asialo GM_1 as determined by accessibility to galactose oxidase. Compared to resident and tumoricidal macrophages, an increase in both the amount and surface labeling of asialo GM_1 was observed in TG macrophages.

As shown in Fig. 6, GM_1 is the most prominent ganglioside in mouse macrophages and is labeled on the surface of all three populations of macrophages by the galactose oxidase/NaB^3H_4 method. We attribute four bands to GM_1 in our TLC separations (Fig. 6) based on their comigration with standard GM_1 and on their ability to bind cholera toxin. These multiple GM_1 bands can be attributed to fatty acid heterogeneity as well as the presence of both N-acetyl and N-glycolyl neuraminic acid in macrophage gangliosides. Close

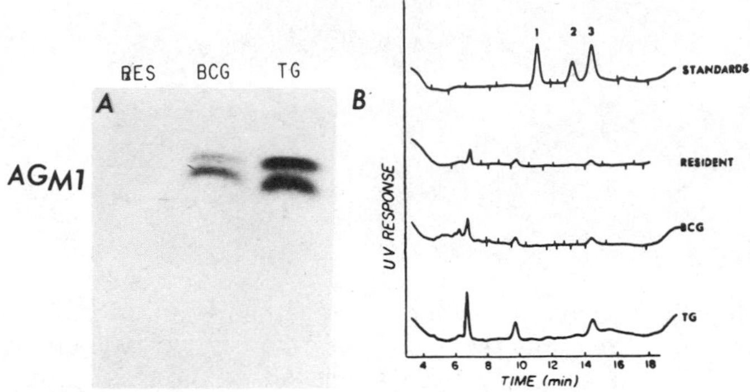

Fig. 5. Autoradiogram (A) and HPLC (B) of galactose oxidase/^{3}H-borohydride labeled neutral glycolipids from mouse macrophages. Each preparation represents 150 mg of cellular protein. Glycolipids in A were detected by exposure to LKB ultrafilm for 7 days. Glycolipids in B were detected by UV at 205 nm as perbenzoyl derivatives. Standards are the derivatives of 1, globotriaosylceramide; 2, globoside; 3, asialo-Gm$_1$. The peak at 7 min is glucosylceramide (CMH), and the peak at 10 min is lactosylceramide (CDH).

inspection of Fig. 6 reveals distinct differences in both the relative chemical amounts and in the surface labeling of these four GM$_1$ bands among the three macrophage populations. Most notable is the increased labeling of third GM$_1$ band on BCG macrophages compared to TG macrophages, even though chemically, TG macrophages contain more of this GM$_1$ moiety. Also, the first GM$_1$ band is labeled on the surface of BCG and TG macrophages but not on resident macrophages; chemically, this band is only detectable on TG macrophages. These observations suggest that macrophage activation alters the surface exposure of specific moieties of individual gangliosides.

DISCUSSION

The results presented above constitute the first demonstration that macrophage activation results in alterations in specific surface carbohydrates. These alterations include a marked increase in both the amount of labeling and degree of branching (determined by resistance to endo-β-galactosidase digestion) of lactosaminoglycan-containing glycopeptides, as well as differences in the chemical amounts and surface exposure (determined by accessibility to galactose oxidase) of both neutral glycolipids and gangliosides. Indeed, the glycolipid data suggest that macrophage activation results in a substantial reorganization of the three-dimensional structure of the cell surface. It will be of considerable interest to elucidate the mechanism which controls glycolipid exposure on the macrophage surface, and to determine if such exposure is correlated with alterations in specific surface polypeptides.

The results we have obtained may be of functional importance. For example, it is known that tumoricidal macrophages are capable of binding to and lysing most types of neoplastic cells but not normal cells (Fink, 1977). The surface structures involved in this interaction have not been elucidated although preliminary evidence suggests that surface carbohydrates may play an important role (Weir and Ogmundsdottir, 1980). In this direction, therefore, we are actively studying those alterations in surface carbohydrates which accompany the acquisition of tumoricidal capacity macrophages. Clearly, our data to date suggest that branched lactosaminoglycans and differences in glycolipid exposure may differentiate the tumoricidal macrophage from the resident macrophage. The recent availability of recombinant γ-interferon (a lymphokine which is capable of activating macrophages to a tumoricidal state in vitro; Schreiber et al., 1983) will enable us to characterize in more detail these and other carbohydrate alterations under well-defined conditions.

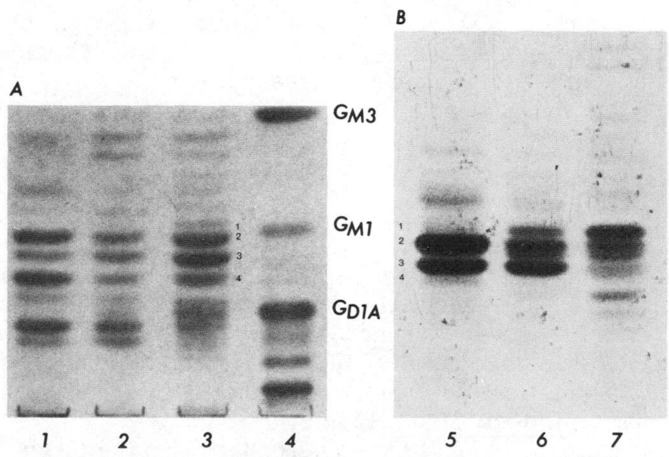

Fig. 6. Thin layer chromatogram (A) and corresponding autoradiogram (B) of galactose oxidase/^{3}H-borohydride labeled gangliosides from mouse macrophages. The amount of sample applied to each lane was normalized to total cellular protein. Lanes 1 and 5, resident macrophages; lanes 2 and 6, BCG macrophages; lanes 3 and 7, TG macrophages; standards are in lane 4. Gangliosides in A were detected with resorcinol spray. Gangliosides in B were detected by exposure to LKB ultrafilm for 7 days. The numeral 1,2,3,4 denote four distinct moieties of GM_1 as evidenced by their ability to bind cholera toxin.

One obvious question raised by the lactosaminoglycan data is which surface glycoproteins carry these carbohydrates? In preliminary studies (Mercurio, Springer, and Robbins), we have shown that the macrophage surface antigen Mac-3 (Ho and Springer, 1983) contains lactosaminoglycan chains as well as both complex and high-mannose-type oligosaccharides. The apparent MW of Mac-3 obtained from activated macrophages is considerably higher than that obtained from resident macrophages, and our data indicate that this MW increase can be attributed, at least in part, to alterations in the lactosaminoglycan structure. It seems likely that other surface glycoproteins also contain lactosaminoglycans. We are now in the process of identifying these proteins and determining the extent to which their carbohydrate moieties are altered upon activation. From a functional perspective, this information will enable us to study the possibility that alterations in the expression and function of macrophage surface glycoproteins known to occur upon activation (e.g., Ezekowitz et al., 1981) may be the result of alterations in their glycosylation.

ACKNOWLEDGEMENTS

This work was supported by NIH Postdoctoral Fellowship EY05487 (AMM), NIH Grant CA26712 (PWR), and NIH Grant CA25532. Doctors Michiko and Minoru Fukuda generously provided endo-β-galactosidase purified standard oligosaccharides and many helpful discussions. We thank Devon Young for typing the manuscript.

REFERENCES

Bianco, C., Griffin, F.M., Jr., and Silverstein, S.C., 1975, Studies of the macrophage complement receptor, J. Exp. Med., 141:1278.

Cohn, Z.A., 1978, The activation of mononuclear phagocytes: Fact, fancy, and future, J. Immunol., 121:813.

Ezekowitz, R.A.B., Austyn, J., Stahl, P.D., and Gordon, S., 1981, Surface properties of bacillus calmette-Guerin-activated mouse macrophage, J. Exp. Med., 154:60.

Fink, M.A. (Editor), 1977, "The Macrophage in Neoplasia," Academic Press, New York.

Fukuda, M.N., 1981, Purification and characterization of endo-β-galactosidase from Escherichia freundii induced by hog gastric mucin, J. Biol. Chem., 256:3900.

Fukuda, M., Fukuda, M.N., and Hakomori, S., 1979, Developmental change and genetic defect in the carbohydrate structure of Band 3 glycoprotein of human erythrocyte membrane, J. Biol. Chem., 254:3700.

Gahmberg, C.G., and Hakomori, S., 1973, External labeling of cell surface galactose and galactosamine in glycolipid and glycoprotein of human erythrocytes, J. Biol. Chem., 248:4311.

Ho, M.K., and Springer, T.A., 1983, Tissue distribution, structural characterization and biosynthesis of Mac-3, a macrophage surface glycoprotein exhibiting molecular weight heterogeneity, J. Biol. Chem., 258:636.

Karnovsky, M.L., and Lazdins, J.K., 1978, Biochemical criteria for activated macrophages, J. Immunol., 121:809.

Schreiber, R.D., Pace, J.L., Russell, S.W., Altman, A., and Katz, D.H., 1983, Macrophage-activating factor proudced by a T cell hybridoma: physiochemical and biosynthetic resemblance to γ-interferon, J. Immunol., 826.

Schwarting, G.A., 1980, Quantitative analysis of neutral glycosphingolipids from human lymphocyte subpopulations, Biochem. J., 189:407.

Schwarting, G.A., and Gajewski, A., 1983, Glycolipids of murine lymphocyte subpopulations. Structured characterization of thymus gangliosides, J. Biol. Chem., 258:5893.

Van Furth, R. (Editor), 1980, "Mononuclear Phagocytes. Functional Aspects," Parts I and II, Martinus Nijhoff, The Hague, Netherlands.

Weir, D.M., and Ogmundsdottir, H.M., 1980, in: "Mononuclear Phagocytes. Functional Aspects," Van Furth, R., (ed.), Martinus Nijhoff, The Hague, The Netherlands, pp. 865-880.

HARDERIAN GLAND AS A MODEL ORGAN FOR STUDY OF LIPID METABOLISM

Yousuke Seyama

Department of Physiological Chemistry and Nutrition
Faculty of Medicine
The University of Tokyo
Bunkyo-ku
Tokyo, 113, Japan

ABSTRACT

The Harderian gland is well developed in certain animals, for example, rodents. It occupies a considerable part of the orbit and is located around the posterior half of the orbit. This gland has several functions as an extraretinal photoreceptor, regulation of the circadian rhythm and its pheromonal effect, in addition to the excretion of lipids as lubricants. The constituents of the lipids are different from species to species. This gland is an interesting model organ, which has specific functions together with specific membrane constituents. The fatty acid compositions of excretory lipid, 1-alkyl-2,3-diacylglycerol, and the membrane lipids, phosphatidyl choline and phosphatidyl ethanolamine, of the guinea pig Harderian gland were examined. A large amount of methyl-branched fatty acids was detected. The absence of essential fatty acids was also noticed. The fatty acid synthetase of this gland was different from that of liver; the former enzyme produced many odd-numbered and methyl-branched fatty acids in the presence of methylmalonyl-CoA, but the latter enzyme was strongly inhibited by methylmalonyl-CoA. These results indicated that the Harderian gland is like an "Independent Factory" performing a unique lipid metabolism isolated from surroundings. The natural labeling of fatty acids with methyl-branchings is thought to be a useful tool to analyze the lipid metabolism, and it will undoubtedly offer a clue in the near future to elucidate the physiological roles of these lipids in the special functions of Harderian gland.

INTRODUCTION

Every cell of every organism consists of nucleus, mitochondria, endoplasmic reticulum, and cytoplasm, which are enwrapped in a plasma membrane. The plasma membrane is much more than just a sac. This membrane controls the osmotic pressure and the flow of materials into and out of cells. Kalckar (1965) called these cellular interactions with the environment through cell-surface structures "ektobiology." These functions are the manifestations of the constituents of each plasma membrane. Although all plasma membranes contain both protein and lipid in essentially similar natures, different plasma membranes show their specific functions according to the minute difference of chemical compositions, and physical architectures of each membrane, as well as their different metabolisms.

To understand the nature of these functions it is essential to know the chemical compositions of plasma membranes. It is also very important to select an adequate model organ that has specific functions together with specific membrane constituents. The Harderian gland is an interesting organ in this sense. This gland has several functions as an extraretinal photoreceptor (Wetterberg et al., 1970a,b), regulation of the circadian rhythm (Pevet et al., 1980), and a pheromonal effect (Thiessen et al., 1976; Payne, 1977) in addition to the excretion of lipid as lubricants. The constituents of the lipids are different from species to species. We investigated this gland from two standpoints. First is the determination of the chemical constituents of lipids in this gland, and second is the characterization of the lipid metabolism in the cell.

HARDERIAN GLAND

The Harderian gland was first described by Swiss anatomist Harder in 1694. He found this gland in the orbit of deer and follow deer and named it as "Glandula nova lachrymalis," which was different from the nictitans gland. After its discovery by Harder, many investigations have been made to find the same gland in many animals, not only in mammals but also in birds, reptiles, and amphibians. Gradually, however, the distinction between the Harderian gland and the nictitans gland was forgotten, and the ocular glands in the inner canthus of the eye became generally regarded as the Harderian gland. In the late 19th century, the distinction between the two glands was recognized again. Recently, the mammalian Harderian gland was defined as follows (Sakai, 1981): *The mammalian Harderian gland are those ocular glands that have tubuloalveolar endpieces (tubular alveioli) and secrete lipid by a merocrine mechanism.* Morphologically, the mammalian Harderian gland has a single layer of columnar epithelial cells surrounding the lumen and a single secretory duct opens near the third eyelid in the inner canthus. The gland is covered with an endothelium of the well-developed orbital venous sinus. The Harderian gland is well-developed in rodents, lagomorphs and the big-clawed shrew, but this gland presumably may have been lost secondarily in chiroptera, primates, terrestrial carnivores, perissodactyls, etc.

The Harderian gland is known to have several functions. It is natural to suppose this gland as a source of lubrication for the eye (Cohn, 1955; Kennedy, 1970). However, this may not be the main function in terrestrial animals, because the Harderian gland mainly secretes lipid, which is not the main lubricant of the eye in these animals. The second function ascribed to this gland is its pheromonal effects (Thiessen and Rice, 1976; Payne, 1977). Buschke (1933) reported a sexually excitatory effect of its secretory product. The homogenate of female Harderian gland inhibited the aggressive response of a male golden hamster towards an opponent (Payne, 1977). Sexual dimorphism in both morphology and their chemical composition is also pointed out by several investigators (Hoffman, 1971); the gland in the female is highly pigmented, whereas the gland in the male is almost devoid of porphyrins (Christensen and Dam, 1953). The third function may be the photoreception in relation to the regulation of circadian rhythm. Removal of the Harderian gland abolishes the effect of light on the pineal hydroxy-O-methyltransferase and serotonin levels in blinded 12-day-old rats (Wetterberg et al., 1970a,b). The Harderian gland itself contains a large amount of melatonin, the concentration of which demonstrates diurnal rhythms (Bubenik et al., 1978).

The Harderian gland also shows a peculiar phenomenon. Injection of rats with the neurotransmitter acetylcholine produced a quick excretion of "bloody tears" from the Harderian gland (Tashiro et al., 1940). This phenomenon was used in the assay of acetylcholine at that time. This may be the result of contraction of myoepithelial cells in response to acetylcholine, squeezing out the contents of the secretory cells into the lumen (Chiquoine, 1958).

Biochemical studies have revealed the chemical composition of the Harderian gland, and the metabolism related to the secretory products like porphyrin, melatonin and lipids. The main secretory material of the mammalian Harderian gland is lipid. The chemical nature of these lipids differed from species to species (Fig. 1). The major lipid component of the pink portion of the rabbit Harderian gland is alkyldiacylglycerols, containing esterified, hydroxy-substituted alkyl chains instead of saturated or unsaturated chains (Kasama et al., 1973; Rock and Snyder, 1975). The white portion of this gland contains mainly a mixture of 2-(O-acyl)hydroxy fatty acid esters (Rock et al., 1976). The major lipid in the rat Harderian gland is wax ester (Murawski and Jost, 1974). The major components of both fatty acid and fatty alcohol moieties in this lipid were a homologous series of n-7 monoenes. Alkyldiacylglyerol is also the main lipid

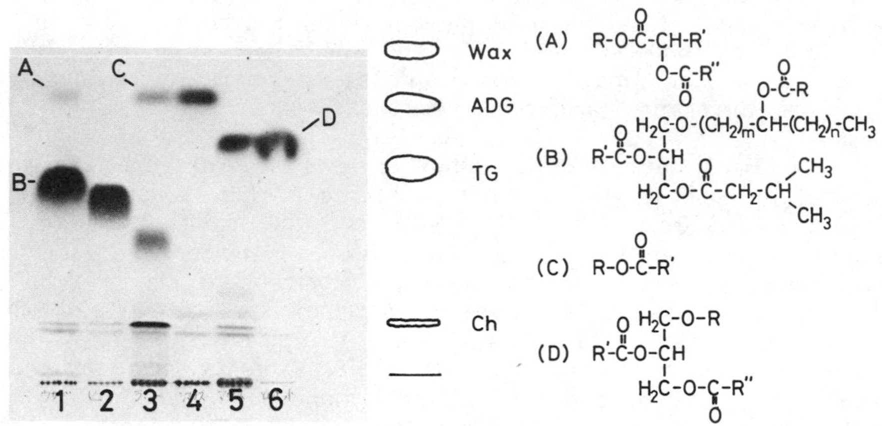

Fig. 1. Thin layer chromatogram on a silica gel plate with hexane-diethylether-acetic acid (80:20:1, v/v). Samples: the lipids in the Harderian glands of rabbit (lane 1), rat (lane 3), squirrel (lane 4), mouse (lane 5), and guinea pig (lane 6), and human orbital fat (lane 2). Standards: Wax, wax ester; ADG, alkyldiacylglycerol; TG, triacylglycerol; Ch, cholesterol.

components of the mouse (Kasama et al., 1970; Watanabe, 1980) and golden hamster (Lin and Nadakavukaren, 1981). The excretory lipids of the Mongolian gerbil (Meriones unguiculatus) are 2,3-alkenediol diacyl esters (2,3-diol waxes) (Otsuru et al., 1983).

EXCRETORY LIPIDS OF GUINEA PIG HARDERIAN GLAND

The lipid components in the mammalian Harderian glands are species-specific as shown in Figure 1. As to the lipids from the guinea pig Harderian gland, Valeri et al. (1973) reported the presence of large amounts of triacylglycerol and small amounts of cholesterol and fatty acids. However, our experiment demonstrated that the main lipid class was 1-alkyl-2,3-diacylglycerol (Yamazaki et al., 1981). We were unable to detect triacylglycerol in this gland. Valeri et al (1973) noticed the presence of this lipid on the TLC plate, but they could not characterize it and reported as an unidentified lipid. The total lipids in this gland accounted for about 23 percent of the fresh tissue. About 92 percent of the total lipids consisted of 1-alkyl-2,3-diacylglyerols. Trace amounts of free cholesterol, phospholipids, and glycolipids were also present.

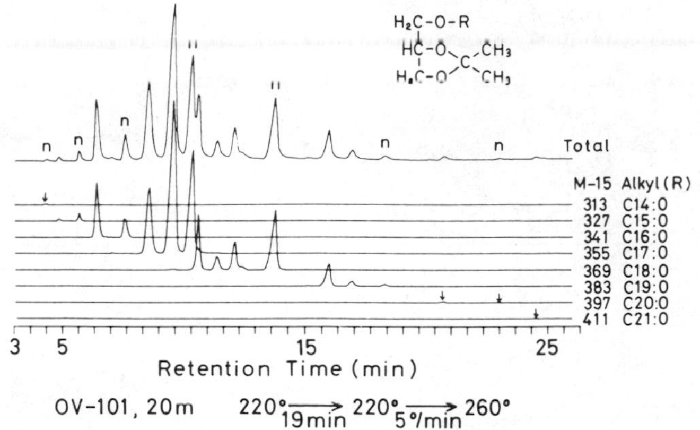

Fig. 2. Analysis of alklglycerols. Mass chromatogram of isopropylidene derivatives of alkylglycerols. The intensities of the signals of total ions (Total) and M-15 ions are plotted separately against retention time. "n" indicates that the alkyl moiety is a straight chain.

Analysis of Alkyl Moieties

Alkylglycerols were obtained by methanolysis of alkyldiacylglycerols followed by Iatrobeads column chromatography, separating alkylglycerols and fatty acid methyl esters. The NMR spectrum of alkylglycreols indicated the presence of methyl branchings. No signals attributable to protons on double bands were observed. The isopropylidene derivatives of alkylglycerols were analyzed by gas chromatography-mass spectrography (GC-MS) (Fig. 2). The alkylmoieties of these components consisted of saturated chains ranging from 14 to 21 carbon number, and the presence of branched chains was also indicated. Methyl branches occurred at the 4,8,10, and terminal anteiso positions. Iso type chains were not found. 8-Methyl-16:0, 14-methyl-16:0 (anteiso), and n-17:0 aliphatic chains were the major chains. No unsaturated chain was found in the alkyl moieties.

Analysis of Acyl Moieties

The total fatty acids from alkyldiacylglycerols and the fatty acids at the 2-position and those at the 3-position produced by lipase hydrolysis were analyzed as methyl esters, respectively, by GC-MS and gas liquid chromatography (GLC) (Fig. 3). The acyl moieties consisted of only saturated fatty acids ranging from 15 to 26 in carbon number and contained 61 mol% of methyl branched fatty acids. The fatty acid composition at the 2 position and that at 3 positoin were quite different. The 2-acyl moieties contained larger amounts of short chain fatty acids and imethyl branched chain fatty

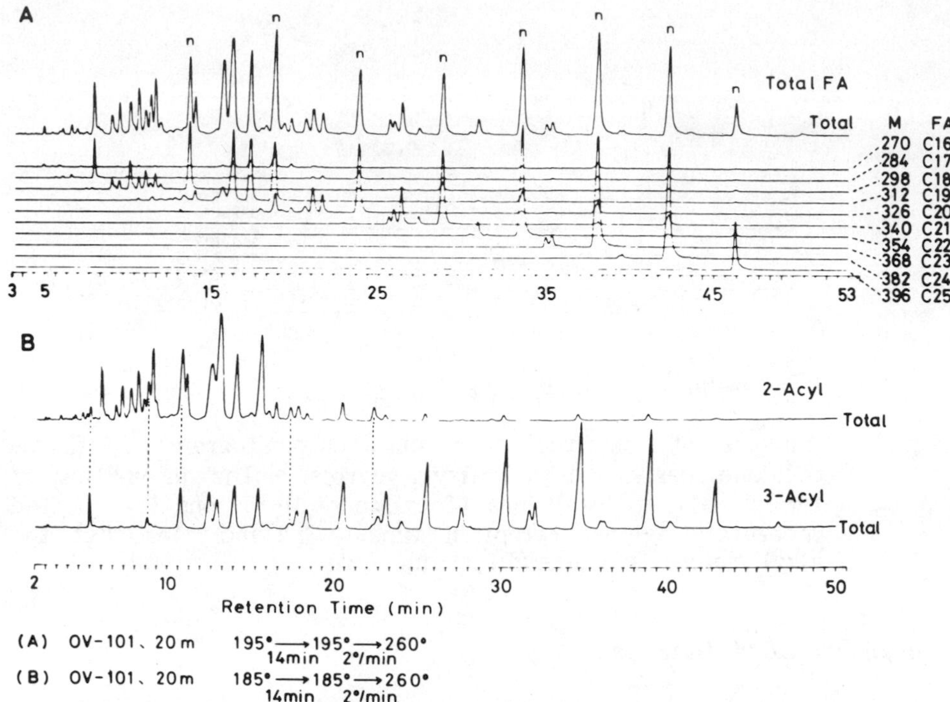

Fig. 3. Analysis of acylmoieties of alkyldiacylglycerols. Mass chromatograms of fatty acid methyl esters. (A): alkyldiacylglycerols (Total FA). (B): Fatty acid at the 2-position (2-Acyl) and 3-position (3-Acyl). The intensities of signals of total ions (Total) and molecular ions (M) (only in A) are plotted separately against retention time. "n" indicates a normal fatty acid.

acids than the 3-acyl moieties (Table 1). All of the methyl branches were located at the even numbered carbon atoms. The methyl branches between the carboxyl group and 10th carbon atom tended to migrate to the methyl terminal side with chain elongation.

Generally speaking, the distribution of fatty acids in glycerolipids is not random and is controlled by the biosynthetic regulation. Saturated fatty acids are esterified predominantly at the 1- or 3-position and unsaturated fatty acids at the 2-position (Tattrie, 1959; Hanahan et al., 1960). In the alkyldiacylglycerol in the guinea pig Harderian gland, 82 mol% of fatty acids at the 2-position are saturated methyl branched ones. Low melting branched fatty acids probably take the place of unsaturated fatty acids at the 2-position. In higher animals, branched chain fatty acids have been found in the lipids of human sebaceous gland (Nicolaides and Apon,

1976), the uropygial gland of waterfowl (Jacob, 1978), the adipose tissue of sheep (Smith et al., 1979), and so on. The contents and structures of the branched chain fatty acids are variable among these tissues. All of the methyl branches in these fatty acids are located at the even numbered carbon atoms except for some terminal iso methyl groups. It has been demonstrated that fatty acids with methyl branches at the even-numbered carbon atoms are synthesized by the condensation of methylmalonyl-CoA by fatty acid synthetase preparations from the uropygial gland and the liver of goose (Buckner and Kolattukudy, 1975; Buckner et al., 1978), the liver of rat (Buckner et al., 1978) and chicken (Scaife et al., 1978), and the adipose tissue of sheep (Scaife et al., 1978). The methyl branched fatty acids in the Harderian gland of guinea pig may thus be synthesized from methylmalonyl-CoA, as shown in the later section.

MEMBRANE LIPIDS OF GUINEA PIG HARDERIAN GLAND

Phosphatidyl choline and phosphatidyl ethanolamine are the predominant membrane lipids. These phospholipids were detected in the lipids extracted from the guinea pig Harderian gland by Valeri et al. (1973), but they did not identify the fatty acid compositions of these phospholipids. About 6.9 percent of the total lipids consisted of phospholipids; 41.2 percent was phosphatidyl ethanolamine and 47.7 percent was phosphatidyl choline (Seyama et al., 1983). Commonly, these phospholipids bear two fatty acid chains, one saturated and one unsaturated to varying degree. However, our investigations revealed that both phospholipids from guinea pig Harderian gland contained a lot of methyl-branched fatty acids but, on the other hand, linolenic and arachidonic acids were not detected.

Table 1: Chain Types of 1-alkyl-2,3-diacylglycerols. Mol% Composition of Aliphatic Chains

Chain Type	1-Alkyl		2-Acyl		3-Acyl	
Straight						
Even	15.49	31.99	13.96	18.01	32.67	60.06
Odd	16.50		4.05		27.39	
Monomethyl branched	58.45	68.01	59.20	82.00	37.77	39.93
Dimethyl branched	7.52		21.20		2.16	
Unknown (branched)	2.04		1.60		trace	
Totals	100.00		100.01		99.99	

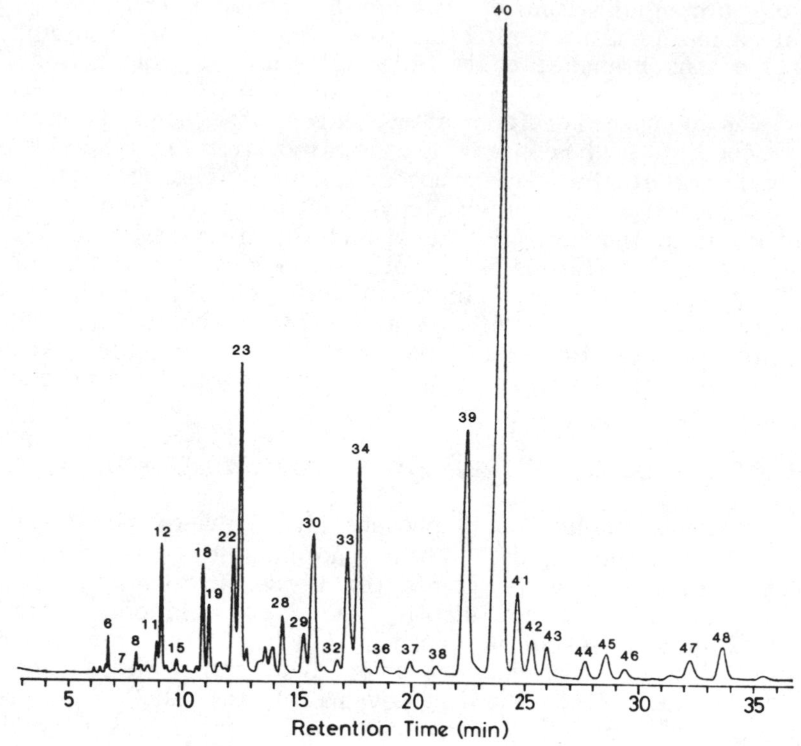

Fig. 4. Gas chromatogram of fatty acid methyl esters of phosphatidyl choline from the Harderian gland of guinea pig.

Fatty Acid Composition of Phosphatidyl Choline

Phosphatidyl choline and phosphatidyl ethanolamine were separated and purified by column chromatography. The fatty acids at the 1-position and 2-position were analyzed after treatment with phospholipase A_2. The total fatty acids derived from phosphatidyl choline are shown in Figure 4. At least 48 peaks were detected, and the peak area was determined with an integrator. Oleic acid (peak No. 40) was the most prominent component, and accounted for 35.8 percent of the total fatty acids. Straight chain saturated fatty acids like palmitic, heptadecanoic, stearic, and nonadecanoic acids were also detected, but occupied only 20 percent. The others, 43 percent, were branched chain fatty acids (Table 2). All of the methyl branches were located at the even-numbered carbon atoms. The most abundant branched chain fatty acid in phosphatidyl choline was 4-methyl-16:0. Arachidonic acid was not detected (less than 0.01 percent). Oleic acid was enriched at the 2-position, and stearic acid was mainly detected at the 1-position. Methyl branched fatty acids resided at both the 1- and 2-positions. Some fatty acids like

2-methyl-16:0 or 8-methyl-17:0 were found mainly at the 1-position, but 4-methyl-15:0 or 6-methyl-16:0 were strictly located at the 2-position of phosphatidyl choline.

Fatty Acid Composition of Phosphatidyl Ethanolamine

Almost the same components as those of phosphatidyl choline were detected, but the composition was slightly different. Oleic acid was also the most prominent fatty acid (45.8 percent). Saturated straight chain fatty acids accounted for 18.8 percent (Table 2). Branched chain fatty acids occupied 34.5 percent of the total fatty acids.

Fatty Acid Composition of Phospholipids from Liver, Brain, Plasma, and Erythrocyte Stroma

The total fatty acids obtained from phosphatidyl choline and phosphatidyl ethanolamine from liver, cerebrum, cerebellum, erythrocyte stroma and plasma were analyzed by GLC. Palmitic, palmitoleic, stearic, oleic, linoleic, linolenic, and arachidonic acids were found as the major constituents (Fig. 5). The compositions of these fatty acids differed according to the source of phospholipids, but the fatty acid compositions in Table 3 resemble those of the phospholipids of mammalian tissues reported (White, 1973). Odd numbered straight chain fatty acids (15:0, 17:0, and 19:0) and branched chain fatty acids (marked with an * in Fig. 5) were also detected as minor constituents.

Table 2: Fatty Acid Compositions of Phosphatidyl Choline and Phosphatidyl Ethanolamine (Mol%)

	Straight Chain		Branched Chain
	Saturated	Unsaturated	
Phosphatidyl choline			
Total	18.7	38.4	42.9
1-position	31.4	14.3	54.3
2-position	16.5	47.3	36.2
Phosphatidyl ethanolamine			
Total	18.8	46.8	34.4
1-position	34.2	24.9	40.9
2-position	13.8	55.0	31.2

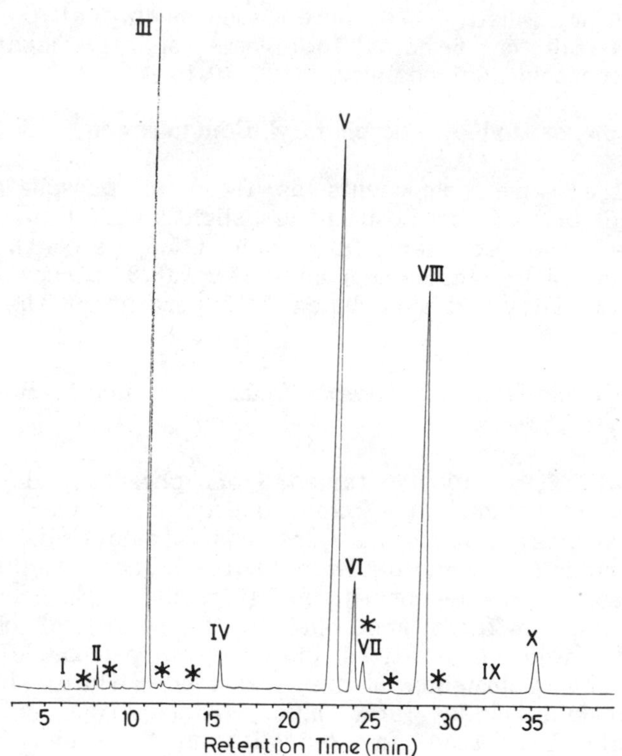

Fig. 5. Gas chromatogram of fatty acid methyl esters of phosphatidyl choline from liver of guinea pig. I, 14:0; II, 15:0; III, 16:0; IV, 17:0; V, 18:0; VII, 19:0 (branched); VIII, 18:2; IX, 19:0; X, 18:3. *Branched chain fatty acid.

FATTY ACID SYNTHESIS IN THE HARDERIAN GLAND OF GUINEA PIG

Guinea pig Harderian gland secretes a large amount of lipid, 1-O-alkyl-2,3-diacylglycerol (Yamazaki et al., 1981), and contains phosphatidyl choline and phosphatidyl ethanolamine as membrane constituents (Seyama et al., 1983). In these lipids, methyl branched fatty acids were the most abundant fatty acids. This finding implies that the fatty acid synthetase from this gland must be able to synthesize methyl-branched fatty acids, like the enzyme from the uropygial gland of birds (Buckner and Kolattukudy, 1975), differing from the usual mammalian fatty acid synthetases that catalyze the production of saturated straight-chain fatty acids. Previously we established a new assay method for fatty acid synthetase with mass fragmentography (Seyama et al., 1978). This method showed effectiveness in the analysis of the character of the fatty acid synthetase from the Harderian gland of guinea pig.

Assay Method for Fatty Acid Synthetase by Mass Fragmentography

The overall activity of fatty acid synthetase can be determined by two methods: spectrophotometric assay of the disappearance of NADPH (and/or NADH) (Lynen, 1969), and the other is radioactivity assay measuring the incorporation of ^{14}C-labeled acetyl-CoA (Hsu et al., 1965). The first method is simple and quick, but the products cannot be determined separately. Moreover, it is not suitable for crude enzyme preparations, because of the high rate of nonspecific degradation of NADPH (NADH). The second method can determine the amounts of individual fatty acids when used in combination with thin layer chromatography or gas chromatography with a radioactivity detector, but the accuracy is not satisfactory.

The m/z 74 fragment of fatty acid methyl esters in gas chromatography-mass spectrometric analysis shifted to m/z 77 when the enzyme reaction proceeded in deuterated water. This enabled us to differentiate the newly synthesized fatty acids from the endogenous ones by monitoring the intensities of m/z 77 and 74 fragments with mass fragmentography. If a known amount of heptadecanoic acid was added as an internal standard, it was possible to determine the absolute amounts of newly synthesized fatty acids. A typical mass fragmentogram is shown in Figure 6. The m/z 74

Table 3: Fatty Acid Compositions of Phosphatidyl Choline and Phosphatidyl Ethanolamine in Guinea Pig (Mol%)

	Phosphatidyl Choline				Phosphatidyl Ethanolamine		
FA	Liver	Brain	RBC	Plasma	Liver	Cerebellum	Cerebrum
14:0	0.2	1.0	0.4	0.3	n.d.	0.5	0.1
15:0	0.5	0.2	0.4	0.3	n.d.	0.3	0.2
16:0	16.8	6.5	19.7	10.7	11.1	37.9	4.9
16:1	0.3	n.d.	0.1	0.2	0.1	0.3	9.0
17:0	1.4	0.8	1.2	1.2	2.1	0.4	1.1
18:0	34.7	23.6	26.2	36.0	62.1	17.1	34.7
18:1	6.8	19.9	10.2	7.4	3.3	23.5	12.8
18:2	26.3	2.5	14.4	24.0	4.7	4.0	0.4
18:3	2.9	2.7	1.4	1.6	n.d.	n.d.	2.2
20:0	0.4	1.6	1.5	0.5	1.3	0.4	0.8
20:4	4.7	3.6	4.6	7.0	3.6	3.9	2.9
Others	5.0	37.6	19.9	10.8	11.7	11.7	30.9
Totals	100.	100.	100.	100.	100.	100.	100.

RBC, red blood cell; n.d., not detected.

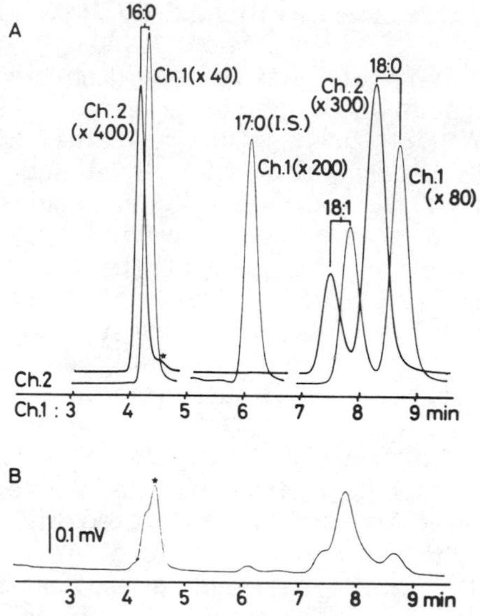

Fig. 6. Mass fragmentogram of fatty acids obtained from incubation with fatty acid synthetase from rat liver. Channels 1 and 2 were set for m/z 74 and 77, respectively. The amplifications of the channels are shown in the parentheses. The peak area of methyl heptadecanoate corresponded to 0.44 µg. *Phthalic acid ester came from n-hexane as a contaminant.

fragment was monitored on channel 1, reflecting the amount of internal standard (17:0) and those of endogenous fatty acids (16:0, 18:1, 18:0). Newly synthesized fatty acids were monitored by m/z 77 in channel 2. The newly synthesized fatty acids eluted earlier than that protium one due to the incorporation of deuterium atoms (McClosky, 1975). To examine the relationship between spectrophotometric and mass fragmentographic assays, an aliquot from the spectrophotometric assay was analyzed by mass fragmentography. The optical density at 340 nm decreased by 0.093, which was equivalent to 277 ng palmitic acid production in 1 ml of reaction mixture. Mass fragmentographic analysis indicated the production of 217 ng palmitic acid and 55.8 stearic acid in this reaction mixture. The correlation of two assay procedures was found to be 0.99. The standard error of the latter was 3.55 percent (n = 5), and linearity was observed from 0.02 µg up to 4 µg. This procedure is more sensitive and specific than the conventional spectrophotometric method (Seyama et al., 1981).

Products of Fatty Acid Synthetase from the Harderian Gland

Fatty acid synthetase was isolated from the particle-free supernatant fraction (cytosol) of the Harderian gland. Under standard incubation conditions with acetyl-CoA and malonyl-CoA as substrate, a large amount of palmitic acid together with small amounts of myristic and stearic acid was synthesized (Fig. 7a). Acetyl-CoA could be omitted, although the amount of newly synthesized fatty acids was reduced to about 90 percent of that in Figure 7a (Fig. 7b). The relative ratio of the three fatty acids observed in Figure 7a did not change under these conditions. When acetyl-CoA was replaced by propionyl-CoA, both odd-numbered fatty acids (13:0, 15:0, 17:0) and even-numbered fatty acids were also produced (Fig. 7d). When methylmalonyl-CoA was added together with malonyl-CoA, more than 19 kinds of fatty acids were produced (Fig. 7e). Most of the product were methyl-branched fatty acids, but straight chain acids with even and odd numbers of carbon atoms were also produced. When malonyl-CoA was completely replaced by methylmalonyl-CoA, practically no fatty acids were produced by this enzyme.

Products of Fatty Acid Synthetase form Liver

Experiments identical with those in Figure 7 were also carried out with fatty acid synthetase from guinea pig liver. Under standard incubation conditions with acetyl-CoA and malonyl-CoA, the main product was palmitic acid, but myristic acid was also produced in a fair amount. Stearic acid was produced only to a minor extent under these incubation conditions. Palmitic, myristic and stearic acids were also produced in the same ratio even upon omission of acetyl-CoA. When acetyl-CoA was replaced by propionyl-CoA, the production of palmitic acid was strongly suppressed, and pentadecanoic acid was the major product together with heptadecanoic acid. When both acetyl-CoA and propionyl-CoA were added as primers, myristic, pentadecanoic, palmitic and heptadecanoic acids were produced. The addition of methylmalonyl-CoA almost completely inhibited the production of fatty acids, and only small amounts of four kinds of fatty acids (14:0, 15:0, 16:0, 17:0) were detected. Branched chain fatty acids were not produced in this incubation mixture.

Effect of Methylmalonyl-CoA Concentration on Fatty Acid Synthesis

When methylmalonyl-CoA was added together with malonyl-CoA, fatty acid synthetase from the Harderian gland produced many different methyl-branched fatty acids along with odd- and even-numbered straight chain fatty acids as shown in Figure 7e. The effect of methylmalonyl-CoA concentration on this feature was examined by changing the amount of methylmalonyl-CoA relative to malonyl-CoA (Fig. 8). Production of even-numbered fatty acids was strongly inhibited by methylmalonyl-CoA, but the formation of odd-

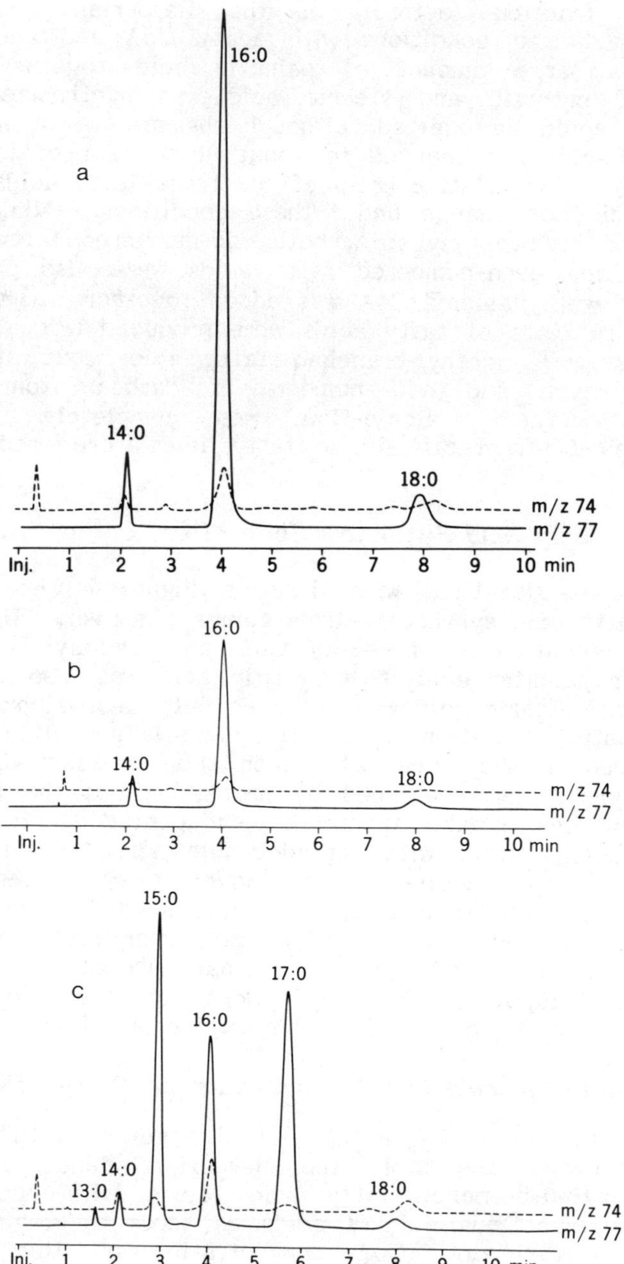
a
16:0
14:0
18:0
m/z 74
m/z 77
Inj. 1 2 3 4 5 6 7 8 9 10 min
b
16:0
14:0
18:0
m/z 74
m/z 77
Inj. 1 2 3 4 5 6 7 8 9 10 min
c
15:0
17:0
16:0
14:0
13:0
18:0
m/z 74
m/z 77
Inj. 1 2 3 4 5 6 7 8 9 10 min

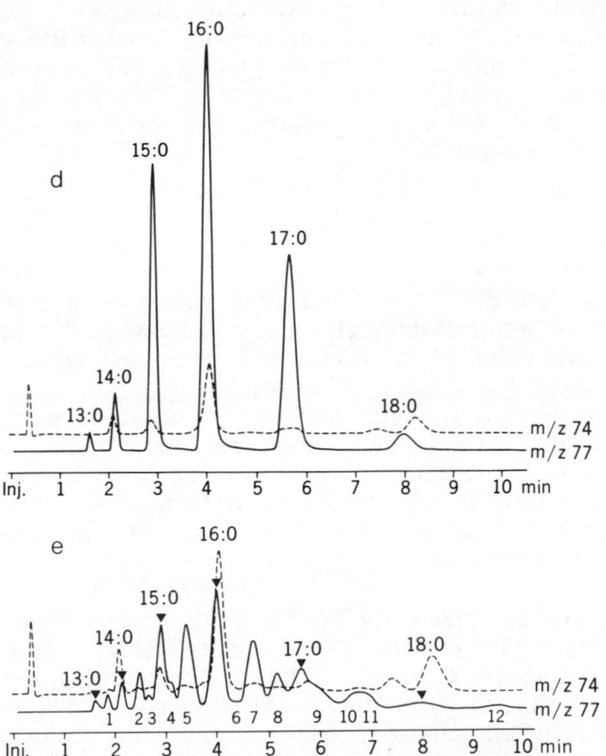

Fig. 7. Fatty acids produced by fatty acid synthetase from guinea pig Harderian gland. The substrates used in experiments (a)-(e) were as follows: (a) acetyl-CoA and malonyl-CoA, (b) malonyl-CoA, (c) propionyl-CoA and malonyl-CoA, (d) acetyl-CoA, propionyl-CoA, and malonyl-CoA, and (e) acetyl-CoA, malonyl-CoA, and methylmalonyl-CoA. Identification of each peak is shown in each figure (a) to (d). In figure (e), together with six kinds of straight chain fatty acids (13:0 to 18:0), at least 12 methyl-branched fatty acids were detected as follows: 1) 4-methyl-13:0, 2) 4-methyl-14:0, 3) 12-methyl-14:0, 4) 4,13-dimethyl-15:0, 5) 8-methyl-15:0, 6) 4,14-dimethyl-16:0, 7) 4-methyl-16:0, 8) 14-methyl-16:0, 9) 6,15-dimethyl-17:0 or 8,15-dimethyl-17:0, 10) 8-methyl-17:0, 11) 4-methyl-17:0, 12) 4-methyl-18:0.

numbered and methyl-branched fatty acids increased to some extent with increasing amount of methylmalonyl-CoA. Thus the availability of methylmalonyl-CoA regulates the composition of fatty acids produced by the fatty acid synthetase from the Harderian gland, as has been shown in the study with the uropygial gland of birds (Buckner et al., 1978). In the presence of appropriate concentrations of acetyl-CoA, propionyl-CoA, malonyl-CoA, and methylmalonyl-CoA, the enzyme can produce the proper fatty acids needed for the synthesis of the 1-O-alkyl-2,3-diacylglycerol, especially for the acylation at the 2-position.

DISCUSSION

The Harderian gland of guinea pig excretes a lipid, which was identified as 1-alkyl-2,3-diacylglycerol (Yamazaki et al., 1981). All of the aliphatic chains in this lipid were saturated chains; no unsaturated chain was found. The aliphatic chains were composed of straight chains with both even and odd numbers of carbon atoms and methyl branched chains. The three aliphatic chains bound to the glycerol differ significantly in chain length and type of methyl branches. This lipid is an excretory product of Harderian gland, and its function is thought to be a lubricant or the solvent of pheromones and porphyrin.

We also characterized the fatty acid compositions of membrane lipids, phosphatidyl choline and phosphatidyl ethanolamine of Harderian gland (Seyama et al., 1983). These lipids are the common constituents of plasma membrane. At least 48 kinds of fatty acids were detected in phosphatidyl choline, and the same was true for phosphatidyl ethanolamine. Most of these fatty acids were methyl-branched, as in the case of 1-alkyl-2,3-diacylglycerol. On the other hand, though linoleic acid was found in phosphatidyl choline (0.79%), linolenic and arachidonic acids were not detectable in both phosphatidyl choline and phosphatidyl ethanolamine. Linoleic acid was also not detected in phosphatidyl ethanolamine. The fatty acid compositions phosphatidyl choline and phosphatidyl ethaholamine in several tissues (liver, brain, red blood cells, and plasma) of the same animal were also examined as a control experiment. These compositions resemble those of the usual phospholipids of mammalian tissues (White, 1973). There is remarkable difference between Harderian gland and the outer environment, for example, blood plasma; the presence of methyl branched fatty acids and the absence of essential fatty acids in the former, and the reverse is true in the latter.

These characteristics of the lipids of Harderian gland are endorsed by the metabolic study of this gland (Seyama et al., 1981). The fatty acid synthetase of the Harderian gland catalyzes the production of methyl branched fatty acids, but that of liver cannot.

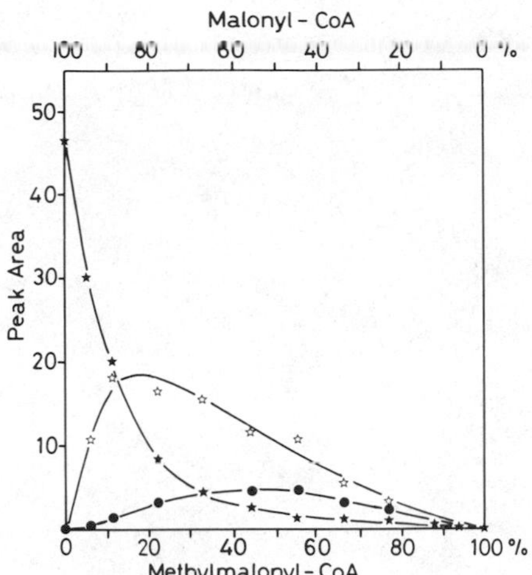

Fig. 8. Effect of methylmalonyl-CoA concentration on fatty acid synthesis. Peak areas (arbitrary unit) of even-numbered fatty acids (★), odd-numbered fatty acids (☆), and methyl-branched fatty acids (●) were plotted against the concentration ratio of malonyl-CoA and methylmalonyl-CoA.

These findings indicate that the phosphatidyl choline and phosphatidyl ethanolamine in this gland were synthesized in the cell itself and not transported from the surroundings. The fatty acids found in these phospholipids of plasma membrane of Harderian gland are supplied by the fatty acid synthetase of this gland itself (Fig. 9). Harderian gland is like an "Independent Factory" performing a unique lipid metabolism isolated from surroundings.

At present the relationship between the functions of Harderian gland and the lipids of plasma membrane is still obscure, but the natural labeling of these lipids with methyl branchings will undoubtedly offer a clue in the near future to elucidate the physiological roles of these lipids in the special functions of the Harderian gland.

SUMMARY

The fatty acid compositions of excretory lipid, 1-alkyl-2,3-diacylglycerol, and the membrane lipids, phosphatidyl choline and phosphatidyl ethanolamine, of the guinea pig Harderian gland were examined. A large amount of methyl-branched fatty acids was detected. The absence of essential fatty acids was also noticed. The fatty acid synthetase of this gland was different from that of liver; the former enzyme produced many odd-numbered and methyl-

Fig. 9. The situation of lipid metabolism in the guinea pig Harderian gland. Hgl, Harderian gland; FAS, fatty acid synthetase.

branched fatty acids in the presence of methylmalonyl-CoA, but the latter enzyme was strongly inhibited by methylmalonyl-CoA. These results indicated that the Harderian gland is like an "Independent Factory" performing an unique lipid metabolism isolated from surroundings. The natural labeling of fatty acids with methyl-branchings is thought to be a useful tool to analyze the lipid metabolism in the Harderian gland.

ACKNOWLEDGEMENTS

The author would like to acknowledge Dr. Tamio Yamakawa of Tokyo Metropolitan Institute of Medical Science, and Prof. Shigenobu Okuda of The University of Tokyo for their helpful advice and encouragement. The author wishes to express his gratitude to Dr. Akihiko Kawaguchi of The University of Tokyo, and Dr. Hideaki Otsuka of Hiroshima University for their collaborations. The author is also grateful to Mr. Takeshi Kasama of his department for his technical assistance in mass spectrometry. Some parts of the present work were done with Doctors Hideko Ogawa of Sagami Women's University and Toshiharu Imamura, Kazumasa Ohashi of his department.

REFERENCES

Bubenik, G.A., Purtill, R.A., Brown, G.M., and Grota, L.J., 1978, Melatonin in the retina and the Harderian gland. Ontogeny, diurnal variations and melatonin treatment, Exp. Eye Res., 323-333.

Buckner, J.S., Kolattukudy, P.E., 1975, Lipid metabolism in the sebaceous glands: synthesis of multibranched fatty acids from methylmalonyl-CoA in cell-free preparations from the uropygial gland of goose, Biochemistry, 14:1774.

Buckner, J.S., Kolattukudy, P.E., and Rogers, L., 1978, Synthesis of multimethyl-branched fatty acids by avian and mammalian fatty acid synthetase and its regulation by malonyl-CoA decarboxylase in the uropygial gland, Arch. Biochem. Biophys., 186:152.

Buschke, W., 1933, Die Hautdrüssenorgane (Hardersche Drüsen, Inguinaldrüsen, Praputialdrüsen, Analdrüsen, Kaudaldrüsen, Kieferdrüsen) der Laboratoriumsnagetiere und die Frage ihrer Abhangigkeit von den Geschlechtsdrüsen, Z. Zellforsch., 18:217.

Chiquoine, A.D., 1958, The identification and electron microscopy of myoepithelial cells in the Harderian gland, Anat. Rec., 132:569.

Chistensen, F., and Dam, H., 1953, A sexual dimorphism of the Harderian glands in hamsters, Acta Physiol. Scand., 27:333.

Cohn, S.A., 1955, Histochemical observations on the Harderian gland of the albino mouse, J. Histochem. Cytochem., 3:342.

Hanahan, D.J., Brockerhoff, H., and Barron, E.J., 1960, The site of attack of phospholipase (lecithinase) A on lecithin: A reevaluation. Position of fatty acids on lecithins and triglycerides, J. Biol. Chem., 235:1917.

Harder, J.J., 1694, Glandula nova lachrymalis una cum ductu excretorio in cervis & damis, Acta eruditorium lipsiae, 49-52.

Hoffman, R.A., 1971, Infuence of some endocrine glands, hormones and blinding on the histology and porphyrins of the Harderian glands of golden hamsters, Am. J. Anat., 132:463.

Hsu, R.Y., Wasson, G., and Porter, J.W., 1965, The purification and properties of the fatty acid synthetase of pigeon liver, J. Biol. Chem., 240:3736.

Jacob, J., 1978, Chemical composition of the preen gland secretions from some ciconiiform birds, Lipids, 13:274.

Kalckar, H.M., 1965, Galactose metabolism and cell "sociology," Science, 150:305.

Kasama, K., Rainey, W.T., Jr., and Snyder, F., 1973, Chemical identification and enzymatic synthesis of a newly discovered lipid class—hydroxyalkylglycerols, Arch. Biochem. Biophys., 154:648.

Kasama, K., Uezumi, N., and Itoh, K., 1970, Characterization and identification of glyceryl ether diesters in Harderian gland tumor of mice, Biochim. Biophys. Acta, 202:56.

Kennedy, G.Y., 1970, Harderoporphyrin: A new porphyrin from the Harderian glands of the rat, Comp. Biochem. Physiol., 36:21.

Lin, W.-L., and Nadakavukaren, M.J., 1981, Harderian gland lipids of male and female golden hamsters, Comp. Biochem. Physiol., 70B:627.

Lynen, F., 1969, Yeast fatty acid synthetase, in: "Methods in Enzymology," vol. 14, pp. 17-33, J.M. Lowenstein (ed.), New York: Academic Press.

Murawski, U., and Jost, 1974, Unsaturated wax esters in the Harderian gland of the rat, Chem. Phys. Lipids, 13:155.

Nicolaides, N., and Apon, J.M.B., 1976, Further studies of the saturated methyl branched fatty acids of vernix caseosa lipid, Lipids, 11:781.

Otsuru, O., Otsuka, H., Kasama, T., Seyama, Y., Sakai, T., and Yohro, T., 1983, The characterization of 2,3-alkanediol diacyl esters obtained from the Harderian glands of Mongolian gerbil (Meriones unguiculatus), J. Biochem., 94:2049.

Payne, A.P., 1977, Pheromonal effects of Harderian gland homogenates on aggressive behaviour in the hamster, J. Endocrinol., 73:191.

Pevet, P., Balemans, M.G.M., Legerstee, W.C., and Vivien-Roels, B., 1980, Circadian rhythmicity of the activity of hydroxyindole-O-methyl transferase (HIOMT) in the formation of melatonin and 5-methoxytryptophol in the pineal, retina, and Harderian gland of the golden hamster, J. Neural Transm., 49:229.

Rock, C., Fitzgerald, V., Rainey, W.T., Jr., and Snyder, F., 1976, Mass spectral identification of 2-(O-acyl)hydroxy fatty acid esters in the white portion of the rabbit Harderian gland, Chem. Phys. Lipids, 17:207.

Rock, C.O., and Snyder, F., 1975, Metabolic interrelation of hydroxy-substituted ether-linked glycerolipids in the pink portion of the rabbit Harderian gland, Arch. Biochem. Biophys., 171:631.

Sakai, T., 1981, The mammalian Harderian gland: Morphology, biochemistry, function and phylogeny, Arch. Histol. Japon., 44:299.

Scaife, J.R., Wahle, K.W.J., and Garton, G.A., 1978, Utilization of methylmalonate for the synthesis of branched-chain fatty acids by preparations of chicken and sheep adipose tissue, Biochem. J., 176:799.

Seyama, Y., Kawaguchi, A., Okuda, S., and Yamakawa, T., 1978, New assay method for fatty acid synthetase with mass fragmentography, J. Biochem., 84:1309.

Seyama, Y., Ohashi, K., Imamura, T., Kasama, T., and Otsuka, H., 1983, Banched chain fatty acids in phospholipids of guinea pig Harderian gland, J. Biochem., 94:1231.

Seyama, Y., Otsuka, H., Kawaguchi, A., and Yamakawa, T., 1981, Fatty acid synthetase from the Harderian gland of guinea pig: biosynthesis of methyl-branched fatty acids, J. Biochem., 90:789.

Smith, A., Calder, A.G., Lough, A.K., and Duncan, W.R.H., 1979, Identification of methyl-branched fatty acids from the triacylglycerols of subcutaneous adipose tissue of lambs, Lipids, 14:953.

Tashiro, S., Smith, C.C., Badger, F., and Kezur, E., 1940, Chromodacryorrhea, a new criterion for biological assay of acetylcholine, Proc. Soc. Exp. Biol. Med., 44:658.

Tattrie, N.H., 1959, Positional distribution of saturated and unsaturated fatty acids on egg lecithin, J. Lipid Res., 1:60.

Thiessen, D.D., and Rice, M., 1976, Mammalian scent marking and social behavior, Psychol. Bull., 83:505.

Valeri, V., Lopes, R.A., Migliorini, R.M., and Camargo, A.C.M., 1973, Lipids in the Harderian gland of the guinea pig (Cavia Porcellus), Ann. Histochim., 18:301.

Watanabe, M., 1980, An autoradiographic biochemical and morphological study of the Harderian gland of mouse, J. Morphol., 163:349.

Wetterberg, L., Geller, E., and Yuwiler, A., 1970a, Harderian gland: an extraretinal photoreceptor influencing the pineal gland in neonatal rats? Science, 167:884.

Wetterberg, L., Yuwiler, A., Ulrich, R., Geller, E., and Wallace, R., 1970b, Harderian gland: influencing on pineal hydroxyindol-O-methyltransferase activity in neonatal rats, Science, 170:194.

White, D.A., 1973, The phospholipid composition of mammalian tissues, in: "Form and Function of Phospholipids," G.B. Ansell, R.M.C. Dawson, and J.N. Hawthorne (eds.), p. 441, Elsevier, Amsterdam.

Wooley, G.W., and Worley, J., 1954, Sexual dimorphism in the Harderian gland of the hamster (Cricetus auratus), Anat. Rec., 118:416.

Yamazaki, T., Seyama, Y., Otsuka, H., Ogawa, H., and Yamakawa, T., 1981, Identification of alkyldiacylglycerols containing methyl branched chains in the Harderian gland of guinea pig. J. Biochem., 89:683.

SURFACE CHANGES OBSERVED ON NEURAL TISSUES DURING DEVELOPMENT OF AVIAN EMBRYOS

Annette M.C. Rapin

Department of Biochemistry
Biocenter of the University of Basel
Klingelbergstr. 70
CH - 4056 Basel, Switzerland

ABSTRACT

Neural crest (NCr) cells are precursors of autonomic ganglia and of a number of other neural and non-neural tissues. The way in which they differentiate depends on their migratory pathway from their origin along the closing neural tube to their final location, and the cells interact during this time with a number of other cells and extracellular material. In spite of the differences in their migratory pathways, quail NCr cells from all regions of the neural tube were marked in the same way by fluorescent lectins, and the cholinergic and adrenergic ganglia derived from these NCr cells had at first the same lectin binding pattern as the parent cells. Further development of the ganglia, in older embryos or during prolonged culture of explants from young ganglia, led to an increase in lectin binding. Noteworthy was the appearance of receptors for the GalNAc-specific SBA (soybean agglutinin), which marked the fibers exclusively and not the cell bodies. Might these carbodydrate-containing molecules on the surface of the fibers possibly play a role during interaction between fibers and their target organs?

Development of cells is influenced not only by intracellular contacts, but also by the cells' extracellular environment. We found that adrenals from quail embryos release into their culture medium factors which favor, fairly specifically, survival and development of quail sympathetic ganglia in culture. This is an intriguing relationship, since the adrenal is one of the target organs of autonomic nerves, and since adrenomedullary cells are derived from neurons of the sympathetic chain.

INTRODUCTION

Cell Sociology

This paper is dedicated, with gratitude and affection, to Dr. Herman M. Kalckar, in honor of his 75th birthday. The years which I spent in his laboratory were very valuable to me, from a human as well as from a scientific point of view.

It was Dr. Kalckar who first made me realize the importance of "cell sociology" (Kalckar, 1965), and the work which I am now doing in the laboratory of Dr. Max M. Burger in Basel is focused on the interaction of cells with their environment, i.e., with other cells, or with their extracellular milieu.

Environmental interactions are particularly relevant during the development of an organism: cells migrate, they meet each other, and some of them stick preferentially to each other to form an organ. One still does not understand completely how these processes are initiated and regulated. Evidence from different systems indicates that cell surface molecules may play a decisive role for preferential cell recognition and adhesion (see for instance, Burger, 1974; Moscona, 1968; Thiery et al., 1977; Roseman, 1970).

Most interesting is the development of the nervous system, where connections are often made over long distances. How does a growing nerve fiber find its own proper target organ, and how does it interact with it?

Framework of Our Studies

For studies on the development of the nervous system of vertebrates, it is interesting to go back to one of its precursor structures, the neural crest. This is a transitory structure whose cells, at first undetermined, appear on the closing neural tube of the embryo, and then migrate along definite pathways to differentiate into a variety of neural (autonomic and sensory ganglia), neural associated (glia and Schann cells), and some non-neural tissues (some of the cranial bones and cartilage, calcitonin-secreting cells, pigment cells). For reviews about neural crest in avian embryos, see Le Douarin (1980) and Noden (1978).

NCr* (neural crest) cells are undifferentiated, and there is still

---

*Abbreviations: cAMP, cyclic adenosine monophosphate; Con A, concanavalin A; D, age of embryo in days; EGF, epidermal growth factor; FCS, fetal calf serum (Gibco); FGF, fibroblast growth factor; (continued)

considerable controversy as to the point at which they become committed to differentiate into a particular tissue. It is known, however, that their migration pathways depend on their point of origin on the crest, along the neural tube, and that the milieu (cellular and extracellular) with which they interact during their migration and at their ultimate location must play some inductive role in their final differentiation (Le Douarin, 1980; Noden, 1978; Weston, 1963). Since surface structures are thought to play a role in such interactions, it was of interest to characterize them at different points during the developmental process. Our studies have focussed on (a) NCr cells before, or in the earliest stage of migration, and (b) autonomic ganglia, as derivatives of the NCr, right after their formation, and during later stages of their development.

EXPERIMENTAL SYSTEM

Neural Crest and Autonomic Ganglia of Quail Embryos

Study of the surface of NCr cells was initiated in collaboration with Nicole Le Douarin and Catherine Ziller, in Paris (for initial reports on this work, see Rapin et al., 1980; Rapin and Burger, 1983). Since the migratory pathway of NCr cells depends on their location along the neural tube, one can ask whether differences in the surfaces of the cells at different locations along the tube might possibly play a role in the determination of the cells' migratory pathway.

N. Le Douarin and her associates originally made transplantations of NCr segments between quail and chick embryos, and since the cells of these two species have a different nuclear structure, it was possible to follow the NCr cells during their migration, and to establish their pathways (Le Douarin, 1973; Le Douarin and Teillet, 1974). This was the reason for our choosing bird embryos for our studies on the neural crest and autonomic ganglia. An advantage of the avian system is that fertilized eggs can be incubated for different lengths of time, giving embryos at definite stages of development (total egg incubation time until hatching: chick, 21 days; quail, 16 days).

Our preliminary characterization of the cell surface has been

---

FITC, fluorescein isothiocyanate; GalNAc, N-acetylgalactosamine; MEM, minimal essential medium, Earle's (Gibco); NCr, neural crest; NGF, nerve growth factor; PdGF, platelet derived growth factor; RCA_{60}, Ricinus communis agglutinin of MW 60'000 = ricin; RCA_{120}, Ricinus communis agglutinin of MW 120'000; WGA, wheat germ agglutinin; SBA, soybean agglutinin.

Table 1: Lectins Used for Marking Cultures

Lectins	Sugar Specificity
Con A (Concanavalin A)	Glucose, mannose
RCA_{60} (Toxin from Ricinus communis; MW 60'000)	N-Acetylgalactosamine, galactose
RCA_{120} (Ricinus communis agglutinin; MW 120'000)	Galactose
SBA (Soybean agglutinin)	N-Acetylgalactosamine
WGA (Wheat germ agglutinin)	N-Acetylglucosamine, sialic acid

directed toward carbohydrates, since other systems have shown that surface glycoproteins and glycolipids may well play a role in cellular interactions (Burger, 1974; Roseman, 1970). This laboratory has been studying for many years the binding of lectins to different types of cells (Burger, 1973). We therefore chose to use fluorescent lectins with different sugar specificities to mark cultures made from explants of the NCr or of autonomic ganglia from quail or chick embryos.

Neural Crest

The neural crest is a transitory structure. In chick and quail embryos NCr cells leave the neural tube between the 1st and the 2nd day of incubation; cephalic NCr cells remain located in the head region, whereas cells from the cervical and truncal regions migrate into the somites, or some of them between somites and neural tube, and then on further to their definite locations after another day. The pathways and the ultimate destiny of NCr cells from different regions have now been well established, essentially thanks to the isotope-labeling experiments of Weston (1963) and the chick-quail transplantation experiments of Le Douarin and associates (Le Douarin, 1977). As far as neural derivatives of the NCr are concerned, it is now known that cephalic NCr will give rise to cholinergic ganglia, whereas NCr cells from the trunk region can produce either adrenergic or cholinergic ganglia, depending on where they started from and which pathway they took. See Le Douarin (1980) for a map showing the correspondence between location of the NCr along the neural tube and the autonomic ganglia produced by the cells which started from these different levels.

Might the pathways chosen by these NCr cells be due in part to differences on their outer membrane which might guide them to one or another tissue?

As a first attempt to answer this question, we made cultures from NCr segments free of neural tube, excised from different levels of the neural tube and then marked these cultures with fluorescent lectins. No topographical differences were found: NCr cells from all levels of the neural tube were marked in the same way. Table 2 and Figure 1 show that Con A, WGA, and RCA_{120} were bound to the cell membranes, but not the GalNAc-specific lectins SBA and RCA_{60}.

Autonomic Ganglia

It is quite laborious to follow the NCr cells during their migration (need for isotopic or transplantation experiments, see above), and we therefore looked instead at the autonomic ganglia which are derived from these cells, resorting again to the technique of marking of explant cultures by fluorescent lectins.

Development in vivo. Since we are primarily interested in development, we examined ganglia from embryos of different ages, comparing adrenergic (sympathetic chain) and cholinergic (ciliary, Remak) ganglia, and these with the neural crest.

No differences were found between sympathetic and parasympathetic ganglia, and the results given here and in the rest of the paper are from the sympathetic chain of quail embryos.

Young ganglia (5-6th embryonal day for the quail) had the same lectin specificity as did NCr cells, with no detectable binding of the GalNAc-specific lectins. There is no doubt that NCr cells undergo some modifications during their migration, their interactions with other tissues, and the beginning of their differentiation into neural

Table 2: Appearance of Lectin Receptors During Development

	In Vivo Development: Lectin marking as function of age in D of embryo; culture time 1-2 days for:			In Vitro Development: Lectin marking as funcion of culture time in days, for 6 D sympathetic ganglia.	
	NCr	Ganglia		Culture Time	
Lectins	2 D	6 D	9 D	2 Days	9 Days
Con A	++	++	++(+)	++	+++
WGA	++	++	++(+)	++	+++
RCA_{60}	-	-	++	-	+(+)
RCA_{120}	++	++	++(+)	+(+)	++(+)
SBA	-	-	++	-	++(+) fibers only

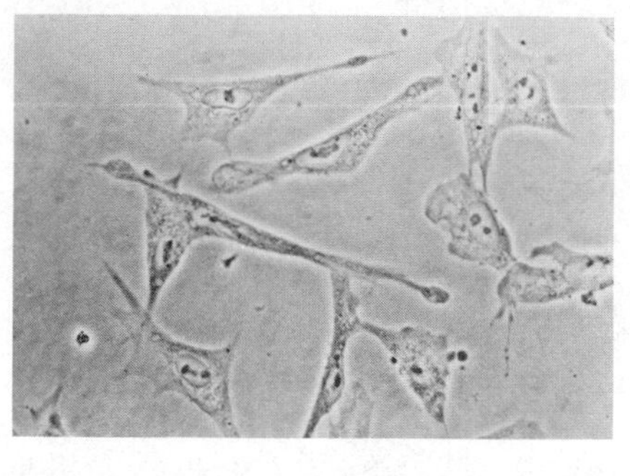

a1

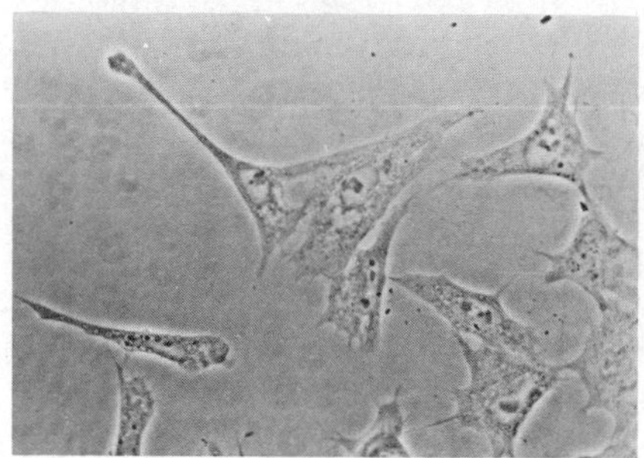

b1

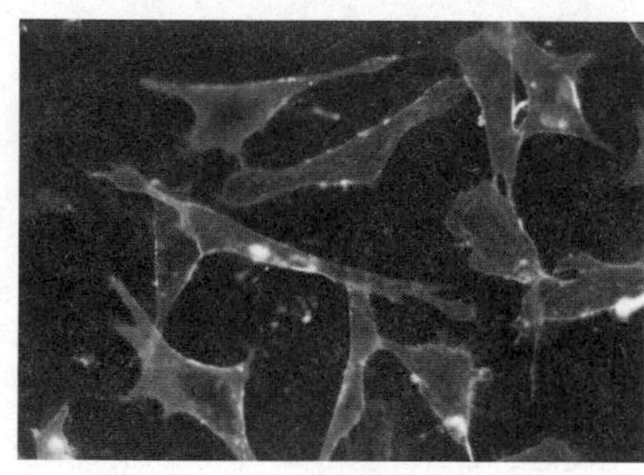

a2

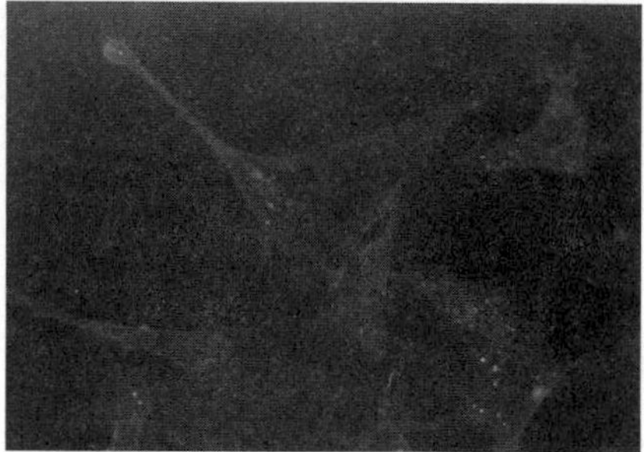

b2

Fig. 1. Binding of FITC-Con A (a1, a2) and FITC-SBA (b1, b2) to 1 day cultures of migrating cephalic NCr cells from quail embryos. Pairs of pictures represent same field, seen with phase contrast or under fluorescent illumination.

cells. These modifications were not, however, reflected in the lectin pattern which we observed.

Further differentiation did, however, change this pattern: there was a gradual increase of binding of all lectins as the embryo grew older, and striking was the appearance of receptor sites for SBA and RCA_{60} in the ganglia when they had developed a more extensive and intricate pattern of fibers. See Table 2.

Development in vitro. The observations described above were made from two day cultures of the ganglia explants, a time which allows for restoration of the fibers cut during the dissection, but not for very extensive further development. We were interested in seeing how the neural cells would develop in culture, and to that effect we cultured explants from young ganglia for a prolonged length of time. Here again, development was accompanied by an increase in lectin marking the appearance of SBA and RCA_{60} sites. Particularly striking here was that fibers were more marked than the cell bodies, and Figure 2 shows that, except for some irregular binding to extracellular material, the GalNAc-specific SBA was bound to the fibers exclusively.

It is essentially with its neurites that a nerve cell communicates with other cells, and with them that it finds its way to its target organ and makes contact with it. It is thus tempting to speculate that this increase in surface carbohydrates on the nerve fibers, during their development, might be of some relevance for these interaction phenomena. This speculation is strengthened by the observation of the increase in lectin binding in both in vivo and in vitro development; but it will still be necessary to isolate and characterize these fiber lectin receptors and to see whether they can interact specifically with some of the target cells of the sympathetic ganglia. It would also be interesting to know whether the GalNAc containing molecules are newly synthetized on the neurite membranes, or whether they are cryptic substances which become evident due to modulation of the membranes during extension and maturation of the fibers.

CELL COMMUNICATION AT A DISTANCE

Trophic Factors in Cell Culture

The growth and development of cells is conditioned not only by contact with other cells, but by their physical and chemical environment. It has been shown, for instance, that fibronectin, a membrane glycoprotein, is secreted by many different cells and that it, along with other extracellular material, plays a role for adhesion of cells as well as for guidance during their migration (see for instance, Aplin and Hughes, 1981; Yamada, 1983), and Sieber-Blum et al (1981) have shown that it favors adrenergic differentiation of neural crest cells in culture.

When cells are grown in culture, one attempts to have them develop in a way that will be as close as possible to development in the whole organism. Tissue culture has many idiosyncrasies and it still holds a number of mysteries. One of these concerns serum, which is an almost essential requirement in culture medium; serum contains a number of "growth factors," but the nature and mode of action are not yet known for all of them. Most tumor cells can be cultured in the absence of serum, and they have been found to secrete a number of growth factors. Normal cells also secrete certain factors which favor their own growth, or which can be carried, e.g., by the blood stream, to other tissues; examples would be EGF (epithelial growth factor), FGF (fibroblast growth factor), PdGF (platelet derived growth factor). Most growth factors are

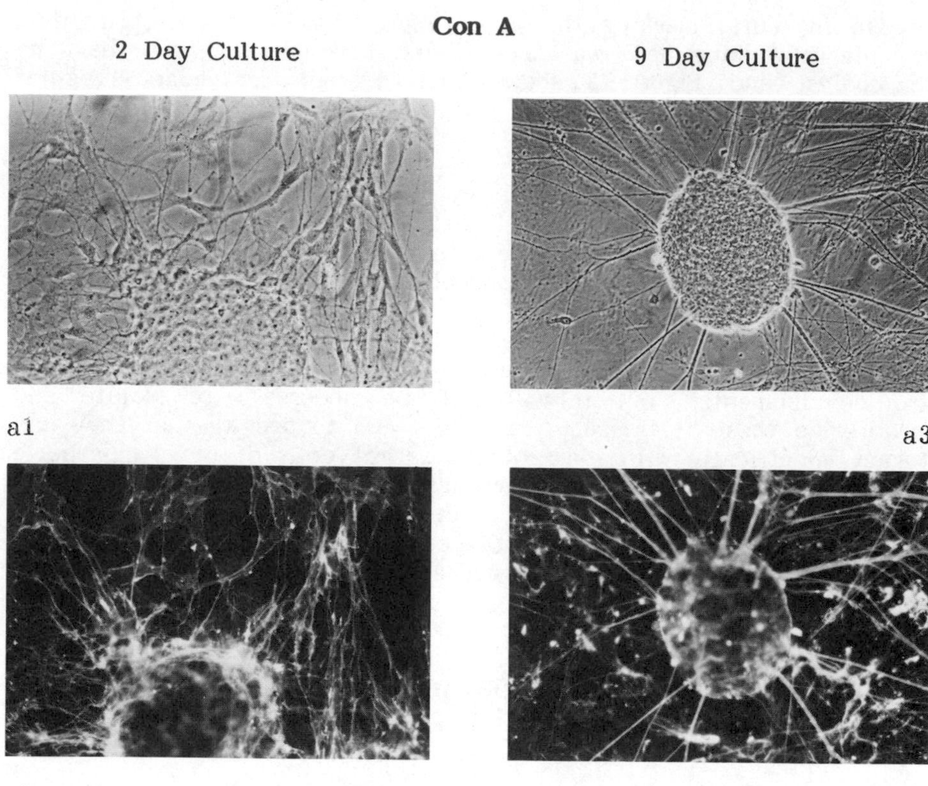

Fig. 2. Development in culture of explants from sympathetic ganglia of 6 D quail embryos. Binding of FITC-Con A (a1-a2) (this page), and FITC-SBA (b1-b4) (opposite page). Pairs of pictures represent the same field, seen with phase contrast or with fluorescent illumination.

polypeptides (hormones) or small proteins which stimulate cells after binding to specific membrane receptors. There are many cases where several growth factors work synergistically, and they also often work concomitantly with other messengers, such as cAMP. Several growth factors play a role during development, and it could be shown with the culture system that different factors stimulate cells or tissues of different embryonic ages. New growth factors are constantly being found, and much work is being done to try to understand their mode of action. For reviews about growth factors, see Antoniades and Owen (1982), and Sato (1979).

In the nervous system the nerve growth factor, NGF, has been particularly well characterized. This protein stimulates fiber outgrowth from the sympathetic neurons and from dorsal root sensory

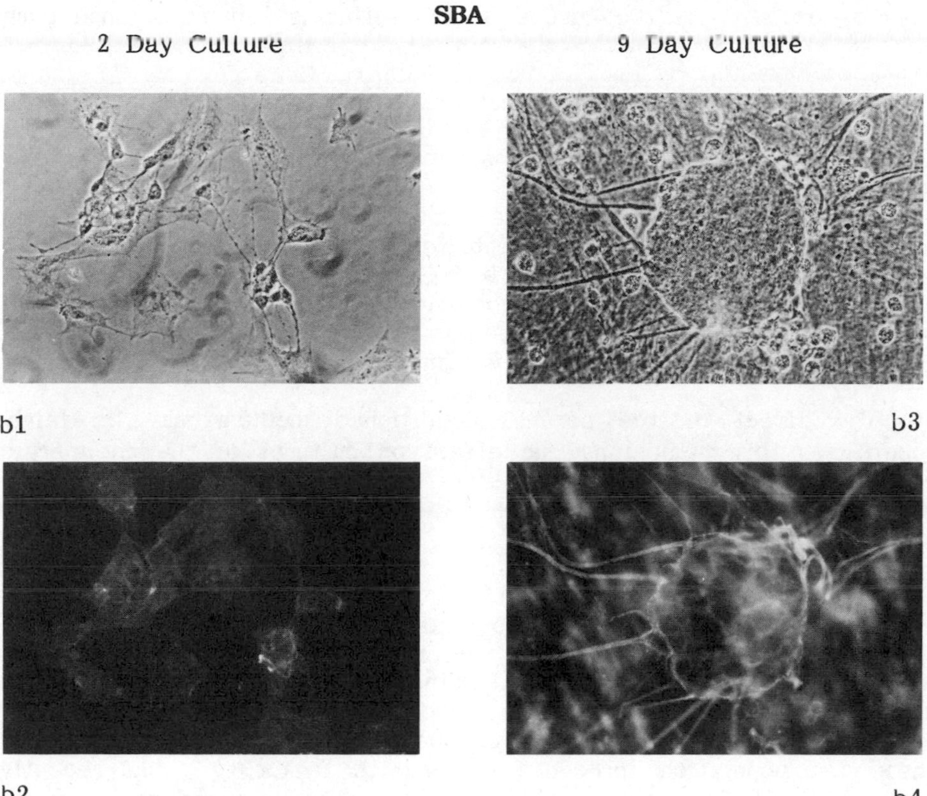

Fig. 2. (continued)

neurons, and it is essential for their survival in most culture conditions. Several other trophic factors for nerve cells have been described or are still being discovered. Some of them are secreted by neural or satellite (e.g., glia) cells, but many neurotrophic factors are produced by non-neuronal cells (one of the best sources of NGF is, for instance, the mouse salivary gland). It is often the target cells of the nerves which produce neurotrophic factors, and these are transported back to the cell body along the axon; this has been particularly well demonstrated for NGF by Thoenen and collaborators (Hendry et al., 1974).

In our own cultures of sympathetic ganglia from quail embryos we found that survival and development, with fiber outgrowth, were very much increased when the sympathetic explants were cocultured with explants from the adrenal gland of the same or of older embryos. We later observed that instead of being stimulated by the proximity of this target organ, the cultures of sympathetic ganglia could also be favored by "adrenal conditioned medium," i.e., a cell-free medium that had been collected after growth in it for several days of explants of the adrenal gland. This is illustrated in Figure 3.

The adrenals apparently secrete into the medium some factor(s) which favors development in culture of the sympathetic ganglia, and we are now in the process of characterizing this material. We know so far that the conditioned medium may contain some NGF, but experiments with antibodies have shown that this is not the major source of its activity. It is, however, possible that this growth factor works in conjunction with NGF, as do some other factors. The stimulatory effect of the conditioned medium is essentially lost by dialysis; it is destroyed by heating to 100°, but it resists 20 min. heating at 60°.

The effect of the adrenal conditioned medium may be fairly specific as this medium had no effect on cultures of the cholinergic ciliary ganglia or on the sensory dorsal root ganglia; it did, however, have a slight effect on explants from the cholinergic part of the Remak ganglia.

The adrenal is not merely a target organ of sympathetic nerves, but its medullary cells are derived from the sympathetic, and one can speculate that this might be a reason for the favorable interaction of these two tissues in culture. Since in the avian adrenal, which was used to make the conditioned medium, the cortical and medullary cells are mixed, we have not yet been able to establish which of these two populations produces the growth factor. It has recently been shown (Fauquet et al., 1983) that glucocorticoids favor adrenergic differentiation in cultures of NCr cells from quail embryos; these steroids had, however, but a moderate effect in our system. Ziegler and collaborators (1983) found that an extract from rat adrenocortical cells stimulated outgrowth of fibers from rat adrenocortical cells and from pheochromocytoma (PC 12) cells, a tumor counterpart of adrenomedullary cells, from the rat. Our conditioned medium could not, however, reproduce this effect on PC 12 cells; and on the other hand, culture medium conditioned by growth of PC 12 cells had no favorable effect on development of our quail sympathetic ganglia in culture.

CONCLUDING REMARKS

We are left at the end of this survey with a number of open questions, and with the way open for further investigations. This brief report on some of my investigations does not come to firm conclusions, but I hope it shows that cell sociology has been and still is a main concern in my research. I now hope to find out more about the interaction of nerve fibers with their targets (attempt to characterize a possible "fiber antigen"), and about how growth of fibers is stimulated by a soluble factor produced by one of their target organs.

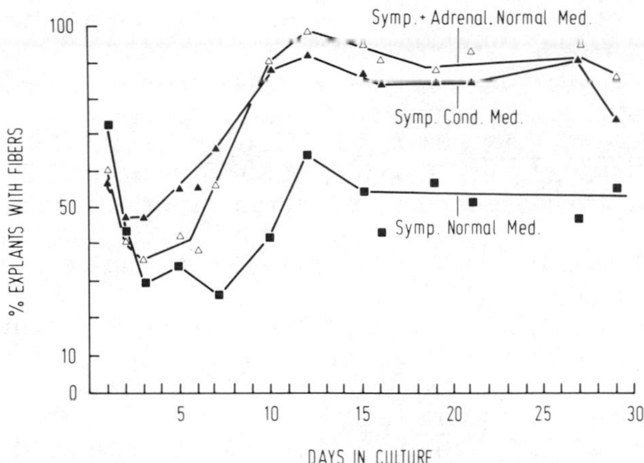

Fig. 3. Fiber outgrowth in cultures of explants of sympathetic ganglia from 6 D quail embryos. Effect of culture in adrenal-conditioned medium ▲—▲, compared with coculture with adrenal explants in normal medium (MEM + 10% FCS) Δ—Δ.

ACKNOWLEDGEMENTS

This work has been performed in the laboratory of Dr. Max M. Burger at the Biocenter in Basel. I am very grateful to him for his patience, his counsel, and his support. It was Dr. Nicole Le Douarin (Nogent-sur-Marne) who first suggested that we bind lectins to neural crest cells, and I thank her for initiating this project and for the friendly collaboration between our laboratories in Basel and Paris. Particular thanks go to Dr. Catherine Ziller (Nogent-sur-Marne) for introducing to me dissection on culture of tissues from quail embryos, and for many helpful discussions since then. The excellent technical assistance of Mrs. Ulla Finne, Ms. Silvia Mathis, and Mrs. Heidi Sommer Taatjes is gratefully acknowledged. This research was supported by grants No. 3.513-0.79 as well as No. 3.269-0.82 from the Swiss National Fonds, to Max M. Burger.

REFERENCES

Antoniades, H.M., and Owen, A.J., 1982, Growth factors and regulation of cell growth, Ann. Rev. Med., 33:445.

Aplin, J.D., and Hughes, R.C., 1981, Cell adhesion on model substrata: threshold effects and receptor modulation, J. Cell Sci., 50:89.

Burger, M.M., 1973, Surface changes in transformed cells detected by lectins, Fed. Proc. Fed. Am. Soc. Exp. Biol., 32:91.

Burger, M.M., 1974, The surface membrane and cell-cell interactions, in: "Third Study Program of the Neurosciences," F. Schmitt and F. Worden (eds.), Cambridge, MIT Press, 68:773.

Fauquet M., Smith, J., and Le Douarin, N.M., 1983, Effects of steroid hormones on neuronal differentiation in cultures of neural crest, Neurosci. Lett., 14:(Suppl)S111.

Hendry, I.A., Stöckel, K., Thoenen, H., and Iversen, L.L., 1974, The retrograde axonal transport of nerve growth factor, Brain Res., 68:103.

Kalckar, H.M., 1965, Galactose metabolism and cell "sociology," Science, 150:305.

Le Douarin, N.M., 1973, A biological cell labeling technique and its use in experimental embryology, Dev. Biol., 30:217.

Le Douarin, N.M., 1980, The ontogeny of the neural crest in avian embryo chimaeras, Nature, 286:663.

Le Douarin, N.M., and Teillet, M.A., 1974, Experimental analysis of the migration and differentiation of neuroblasts of the autonomic nervous system and of neuroectodermal mesenchymal derivaties, using a biological cell marking technique, Dev. Biol., 41:162.

Le Douarin, N.M., Teillet, M.A., and Le Lievre, C., 1977, Influence of the tissue environment on the differentiation of neural crest cells, in: "Cell and Tissue Interactions," J.W. Lash and M.M. Burger (eds.), New York, Raven Press.

Le Douarin, N.M., Smith, J., Teillet, M.A., Le Lievre, C.S., and Ziller, C., 1980, The neural crest and its developmental analysis in avian embryo chimaeras, TINS, 3:39.

Moscona, A.A., 1968, Cell aggregation: properties of specific cell-ligands and their role in the formation of multicellular systems, Dev. Biol., 18:250.

Noden, D.M., 1978, Interactions directing the migration and cyto-differentiation of avian neural crest cells, in: "Receptors and Recognition," Ser. B, vol. 4: "Specificity of Embryological Interactions," D.R. Garrod (ed.), London, Chapman and Hall.

Rapin, A.M.C., and Burger, M.M., 1983, Development of autonomic ganglia from avian embryos. Cell surface properties and effect of growth factors, Neurosci. Lett., 14:(Suppl.)296.

Rapin, A.M.C., Burger, M.M., Ziller, C., and Le Douarin, N.M., 1980, differentiation in the neural crest of avian embryos, in: "Protides of Biological Fluids," H. Peeters (ed.), Oxford and New York, Pergamon Press, 27:589.

Roseman, S., 1970, The synthesis of complex carbohydrates by multiglycosyltransferase systems and their potential function in intercellular adhesion, Chem. Phys. Lipids, 5:270.

Sato, G.H., and Ross, R. (eds.), 1979, "Hormones and Cell Culture," Books A and B, New York, Cold Spring Harbor.

Sieber-Blum, M., Sieber, F., and Yamada, K.M., 1981, Cellular fibronectin promotes adrenergic differentiation of quail neural crest cells in vitro, Exp. Cell Res., 133:285.

Thiery, J.P., Brackenbury, R., Rutishauser, U., and Edelman, G.M., 1977, Adhesion among neural cells of the chick embryo, in: "Progress in Clinical and Biological Research," 15:199, "Cellular

Neurobiology," Z. Hall, R. Kelly, and C.F. Fox (eds.), New York, Alan R. Liss.

Weston, J.A., 1963, A radioautographic analysis of the migration and localization of trunk neural crest cells in the chick, Dev. Biol., 6:279.

Yamada, K.M., 1983, Cell surface interactions with extracellular materials, Ann. Rev. Biochem., 52:761.

Ziegler, W., Hofmann, H.D., and Unsicker, K., 1983, Rat adrenal non-chromaffin cells contain a neurite out-growth-promoting factor immunologically different from nerve growth factor, Dev. Brain Res., 7:353.

ENERGY-REQUIRING REGULATION OF HEXOSE TRANSPORT, AS STUDIED IN FIBROBLAST CULTURES OF A METABOLIC MUTANT

Herman M. Kalckar* and Donna B. Ullrey

Chemistry Department
Boston University
College of Liberal Arts
684 Commonwealth Avenue
Boston, Massachusetts 02215

PROLOGUE

I am indeed thankful to my friend Ed Haber for his initiative to come up with a full monograph based on the symposium on selected aspects of membrane biology, what some of us used to call "Ektobiology."

My interest in transport systems was reawakened by reading a manuscript by Harold Amos and his coworker which later appeared in print in 1972. In this article as well as a succeeding one (Martineau et al., 1972; Christopher et al., 1976a) a metabolically mediated type of regulation of the hexose transport system was described. At that time I had become interested in the literature on the up-regulation of the hexose transport system in oncogenically transformed fibroblast cultures through my friend next door at the MGH, Kurt Isselbacher (cf. Isselbacher, 1972). The mutual friendship with Dr. Amos and Dr. Christopher resulted a few years later in a very fortunate event for me: Bill Christopher joined our group at MGH. Added to this, Dr. Isselbacher, upon his return from Dr. Michael Stoker's laboratory in London, brought new ideas with him and various cell lines, of which the NIL hamster fibroblasts became one of our favorite cell lines. This cell line and another fibroblast line which I shall mention shortly, constituted our living "organism" when

---

*Most of the experimental work discussed here was supported by the National Science Foundation, Grants PCM 8021552 and PCM 8302034 (to HMK).

my excellent coworker Donna B. Ullrey and I moved to Boston University in 1979. We tried here to disclose further the complex parameters which dictate the regulation of the hexose transport system in cultured mammalian fibroblasts.

In this endeavor we received great stimulation from Dr. Jacques Pouyssegur, Universite de Nice in France, who offered us various mutants of a hamster lung fibroblast line, some hyperglycolytic, some hypoglycolytic. You might soon understand why we chose to concentrate our efforts on the hypoglycolytic strain and its differentiated type of transport regulation.

The transport or uptake regulation which will be subject to discussion is a mediated type of down-regulation of the hexose uptake system observed in cultured fibroblasts. Since the word "down-regulation" is a long word and I have always tried to avoid abbreviations, I have decided to use the brief and expressive word "curb."

INTRODUCTION

It is well known that chick embryo fibroblasts (CEF) and NIL hamster cultures, deprived for 6 to 24 hours of glucose in their maintenance medium, show a greatly enhanced hexose transport activity, when monitored for uptake of hexose, using radioactive glucose or glucose analogues (Martineau et al., 1972; Christopher et al., 1976a,b,c). Moreover, this transport enhancement could be ascribed to an increase in the population of functional hexose transport carriers (Kletzien and Perdue, 1975; Christopher et al., 1976a; Christopher, 1977). This type of up-regulation was found to be preserved in plasma membrane preparations from CEF cultures starved of glucose (Yamada et al., 1983).

The hexose transport system in the NIL fibroblasts is subject to a mediated curb by exposure to the following aldohexoses: D-glucose, D-glucosamine, D-mannose, and D-galactose (cf. Ullrey et al., 1975; Ullrey and Kalckar, 1981). Fructose was unable to mediate a transport curb, presumably because it does not possess detectable affinity for the hexose carrier system (Ullrey and Kalckar, 1981; cf. also Lipmann and Lee, 1978). In practically all the mammalian cell lines the aldohexoses can be metabolically interconverted; in CEF, galactose seems to be an exception (cf. Lipmann and Lee, 1978, Table 5).

The affinities of the aldohexoses and their analogues for the hexose carriers remained at a K_m of the order of 4 to 5 mM, in fed as well as in starved cells (Kletzien and Perdue, 1975) and for galactose uptake in NIL fibroblasts, it was of the same order of magnitude (Christopher, 1977). After a shift from glucose to a glucose-free medium (or a fructose medium) the V_{max} increased gradually and after 20 hours of maintenance on an aldose-free medium, the V_{max} usually climbed five- to tenfold (Kletzien and Perdue, 1975; Christopher, 1977). A combination of glucose-feeding and exposure of NIL cultures to cycloheximide (CHx) consistently brought about a further decrease in the hexose carrier population, corresponding to a V_{max} of less than 30 percent of the fed state (Christopher et al., 1976b,c; Christopher, 1977).

Conversely, the increased hexose carrier population was found to be subject to a gradual but marked down-regulation if the fructose medium (or sugar-free medium) was supplemented with glucose. However, this mediated transport curb was found to be more complex than anticipated, since it did not ensue if CHx was present at the onset of glucose-refeeding (Christopher et al., 1976b,c)*.

RESULTS

The mediated curb seems to depend on oxidative energy metabolism, since inhibitors like 2,4 dinitrophenol (DNP), oligomycin or malonate released the hexose transport curb imposed by glucose or D-glucosamine (Kalckar et al., 1979; Ullrey and Kalckar, 1981). The above mentioned additional losses of carriers seen, if fed-NIL cultures were exposed to small amounts of CHx of the order of 2 μg/ml (cf. Christopher et al., 1976b,c), could be strikingly mitigated or even counteracted, if DNP or the other inhibitors, such as malonate, were added at the onset of the CHx addition (Kalckar et al., 1979; Ullrey and Kalckar, 1981, Table 5; see also Kalckar, 1983). This is illustrated in Table 1 (cf. Kalckar, 1983).

It is apparent from Table 1 that DNP was able to release most of the severe uptake-curb imposed on the glucose-fed cells by CHx. Upon replacement of glucose by fructose, CHx addition partly prevented the development of the "uncurbed" state, and a significant further release of the curb ensued if DNP was also present (Kalckar et al., 1979; Kalckar, 1983).

These involved phenomena are difficult to interpret without taking a number of other cell functions into account. Christopher has emphasized the potential importance of protein turnover in terms of a vigorous recycling of the hexose carrier system with a possible "involution" of the system (Christopher and Morgan, 1981). Intracellular cathepsin B levels were found to be high in glucose-fed NIL fibroblasts and low in glucose-starved cells or fructose-fed cells (Christopher, 1979; Christopher and Morgan, 1981).

---

*Variations on the levels of cyclic AMP did not affect the transport curb (Christopher et al., 1975)

Table 1: DNP Counteracts Glucose-Mediated Uptake Curb and the Additional Curb Imposed by CHx

Culture Medium	U-^{14}C Galactose Uptake, nmol/mg Protein/10 min	
	Glucose Feeding	Fructose Feeding
No additions	1.58	3.63
DNP 0.2 mM	3.02	3.55
CHx 2 μg/ml	0.23	1.44
DNP + CHx	2.03	1.97

NIL fibroblasts were cultured for 20 hr at 37° C in Eagle's medium containing 22 mM glucose or fructose, with or without DNP or CHx as indicated; U-^{14}C galactose, 0.1 mM for 10 min. (modified from Kalckar et al., 1979, Proc. Natl. Acad. Sci. USA, 76:6454).

At another occasion, I have tried to elaborate on the importance of Christopher's recycling hypothesis, by invoking the possibility of endocytosis of the hexose transport carriers in the glucose-fed state (Kalckar, 1983). It was furthermore found that endocytosis and translocation of hexose carriers in adipose fat cells were arrested by DNP (Kono et al., 1981; cf. Kalckar, 1983). Transposing these concepts to our discussion of the regulatory events in the NIL cultures, one might characterize the state of the glucose-carrying transporters as being more oriented towards the intracellular milieu with its cathepsins, as compared with the prevailing states when the carriers are empty. The glucose-mediated transport curb and the so-called "involuted" state (and its tendency to endocytosis) may be closely associated events (Christopher and Morgan, 1981). Perhaps we have managed to get the beginning of a working hypothesis which we can try to prove or disprove (Kalckar, 1983).

As mentioned earlier, in CEF and NIL cultures the glucose effects did not seem to be specific for glucose, since mannose and D-glucosamine also brought about a marked transport curb (Kalckar et al., 1979; Ullrey and Kalckar, 1981). Our more recent work on a metabolic mutant has, however, forced us to revise some of our views on the nature of the mediated curb.

In ordinary avian and mammalian fibroblast lines, glucose is not only able to serve as a ligand for the hexose transport system, but it can also serve as an effective metabolite in either of the two aerobic pathways: the energy generating (i.e., the di- and tricarboxylate pathway) and the pentose-shunt.

Isolation of a particular hamster lung fibroblast line 023 described by Franchi et al. (1978) deserve some comments. The 023 line which is tumorigenic, possesses a hexose transport system susceptible to a marked glucose-mediated curb (Franchi et al., 1978; Ullrey et al., 1982). Glucose starvation brought about an enhancement of the hexose transport system; this enhancement developed practically undisturbed, even in the presence of 10 μg per ml of CHx (Franchi et al., 1978).

From the 023 fibroblast line, Pouyssegur et al. (1980) were able to select and isolate a number of metabolic mutants, hyperglycolytic as well as hypoglycolytic. The mutant, called DS-7, is unable to generate any lactic acid from glucose and was found to be grossly defective in the enzyme phosphoglucose isomerase (Pouyssegur et al., 1980). We call it the PGI mutant (or "PGI⁻") in the present text.

A metabolic scheme trying to illustrate the site of the block in the pathway should also help to clarify a number of intriguing new features (see Fig. 1). Gluconeogenesis from amino acids is beyond the capacity of the PGI mutant whereas fructo-neogenesis is not (cf. Fig. 1). Conversely fructose, fed to PGI⁻ cells, was unable to sustain more than 10 to 15 percent of the UDP-hexose levels observed in glucose-fed cells (Kalckar and Ullrey, 1984a). Mannose was the only aldohexose which we found able to generate large

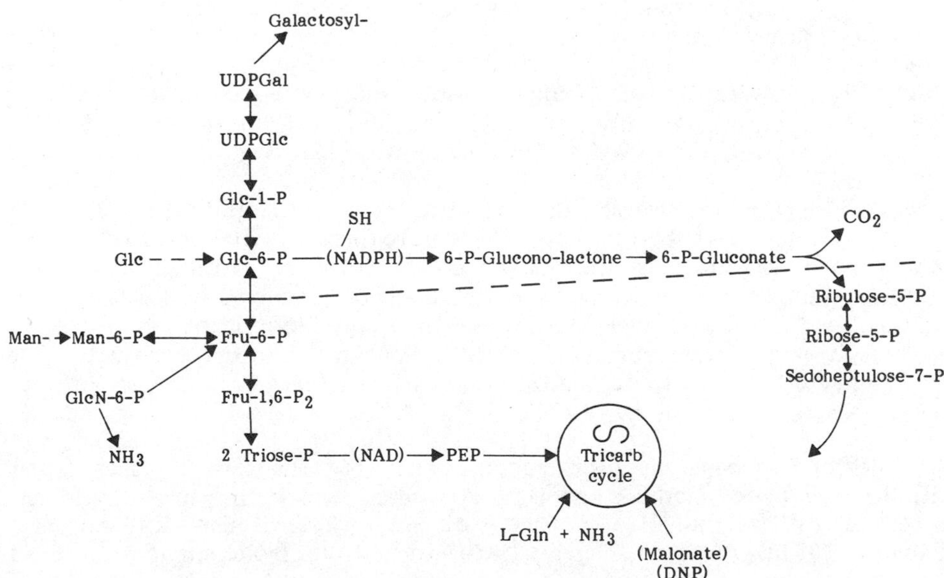

Fig. 1. Schism in carbohydrate metabolism of the PGI mutant (modified from Kalckar and Ullrey, Fed. Proc., 43:2243, 1984).

amounts of lactic acid (Ullrey and Kalckar, 1982; Kalckar and Ullrey, 1984a).

Surprisingly enough, mannose or D-glucosamine which were able to mediate a curb of the hexose transport system in the parental line 023, were not able to elicit a curb in the PGI mutant (Ullrey et al., 1982; Ullrey and Kalckar, 1982; Kalckar and Ullrey, 1984a). As far as the mediated curb of the hexose transport or uptake system of the mutant is concerned, only glucose was active (see Table 2).

We examined the regulation pattern in the PGI mutant in a simplified maintenance medium in which L-glutamine was eliminated. It is known that L-Gln in near-confluent cultures is able to serve as a source of pyruvate and lactate (for references cf. Ullrey and Kalckar, 1982). The following regulatory change in metabolism and uptake patterns could be observed in the PGI^- cultures deprived of L-glutamine (ibid, 1982).

i. Glucose was unable to mediate a marked curb of the hexose uptake or transport system.

ii. Mannose remained unable to elicit a curb.

iii. Glucose and mannose restored the ability to mediate a curb.

iv. Lactic acid formation by mannose was practically extinguished by glucose (competition with glucose-6-hexokinase?).

v. Galactose and mannose were also able to mediate a curb, and this constellation did not interfere with lactate formation from mannose.

The latter feature is due to the fact that galactose is phosphorylated by a specific kinase, galactokinase. In contrast, phosphorylation of glucose and mannose (as well as D-glucosamine and 2-deoxyglucose) is catalyzed by a common kinase, hexokinase (cf. Leloir, 1951). Moreover, it is known that Glc-6-P which tends to accumulate in the glucose-fed PGI mutant*, interferes with the phosphorylation of the glucose derivatives (cf. Pouyssegur et al., 1980; Kalckar and Ullrey, 1984).

Table 3 summarizes the requirements for the two hexoses (item iii) in order to mediate an uptake curb, which in turn could be released by malonate (cf. Ullrey et al., 1982; also Kalckar and Ullrey, 1984a). Regarding lactate formation from mannose, it is

---

*UDPHexose accumulated in the glucose-fed mutant, yet no accumulation of glycogen was detectable (Kalckar and Ullrey, 1984b)

Table 2: Catabolic Down-Regulation of 3-O-MeGlc Transport by Glucose and Glucosamine in the PGI Mutant and Its Parental Strain

Conditioning Medium	3-O-[^{14}C]MeGlc Transported, pmol/mg cell protein per 20 sec DS-7 (PGI$^-$)	023 (Parental)
No hexose	13.0	17.8
No hexose + malonate	23.9	20.3
Glucose	3.6	4.6
Glucose + malonate	15.3	12.0
Glucosamine	10.7	5.7
Glucosamine + malonate	10.7	17.5
Fructose	15.2	18.2
Fructose + malonate	21.7	20.8

Concentration in medium: glucose, 22 mM; glucosamine, 5 mM; fructose, 22 mM; malonate, 25 mM; DME medium, no pyruvate, but with 4 mM L-glutamine. Dialyzed fetal calf serum, 10%. Transport was measured at 23° C (from Ullrey et al, 1982, Proc. Natl. Acad. Sci., 79:3777.)

noteworthy that 2 mM glucose was able to exert more than 80 percent inhibition of glycolysis of 20 mM mannose (see last column and footnote in Table 3). However, the mediated transport curb was not affected by the presence or absence of glycolysis. This is also illustrated by the fact that replacement of glucose by galactose which is phosphorylated by the specific galactokinase permits lactate formation from mannose as well as a transport curb (Kalckar and Ullrey, 1984a).

The requirements for two hexoses in order to mediate the uptake curb and the independent features regarding lactic acid formation are summarized in Table 3. The partial release of the composite uptake curb by malonate must be ascribed to the interference of the oxidative metabolism, generated by mannose (Ullrey and Kalckar, 1982).

Returning to the model proposed by Christopher, an involution of hexose-filled carriers would expose the carriers to intracellular cathepsins with varying degrees of inactivation by proteolysis (Christopher and Morgan, 1981). The state of involution may be the typical conformation of the "fed state" of the hexose carriers, a

Table 3: Hexose Transport Curb and Lactate Formation in the PGI Mutant, Maintained in Medium Devoid of L-Glutamine and Pyruvate

Substrate*	[^{14}C]Galactose uptake, nmol/mg of protein†	Lactate Formed μ mol/mg of cell protein per 20 hr
Mannose	3.45	33.70
Glucose	3.39	1.60
Mannose and glucose	1.38	6.83
Mannose, glucose, and malonate	2.68	11.84

The concentrations of substrates were: 20 mM mannose, 2 mM glucose, and 25 mM malonate (from Kalckar & Ullrey, 1984a).
*Over 20 hr at 37° C.
†10 min. at 37° C.

state which presumably could be released by inhibitors of oxidative energy metabolism (Kalckar et al., 1979). As mentioned, in normal fibroblasts, glucose serves not only as a ligand for the carriers but also as a metabolite in both oxidative pathways; mannose or D-glucosamine can fully replace glucose, as for instance in the parental strain 023 (Ullrey and Kalckar, 1982).

In the PGI mutant the situation is more circumscribed. In the absence of L-glutamine, mannose as well as glucose are needed to bring about the uptake or transport curb (Ullrey and Kalckar, 1982; Kalckar and Ullrey, 1984a). In the mutant, glucose is unable to take care of energy metabolism (Pouyssegur et al., 1980; Ullrey et al., 1982). This can be effectively generated from mannose (Ullrey and Kalckar 1982; Kalckar and Ullrey, 1984a).

We have tried to assess the importance of the Glc-6-P dehydrogenase pentose shunt for the mediated transport curb by testing human fibroblast cultures from subjects with a defective Glc-6-P dehydrogenase. However, a comparison with transport regulations in fibroblasts from normal subjects has remained elusive, since both types of cultures showed only marginal regulatory responses (Kalckar and Ullrey, unpublished observations [1983]).

Gay and Amos (1983) have, however, observed an important reciprocal relationship between the mediated regulation of the hexose transport system in CEF and the cellular levels of phosphoribosyl pyrophosphate (PRPP). Glucose-fed CEF accumulated high levels of PRPP, whereas fructose-fed or xylose-fed CEF showed low cellular levels of PRPP; the transport rates were as usual in the opposite direction (Gay and Amos, 1983). Inhibition of oxidative phosphorylation, which might interfere with accumulation of PRPP, also released the glucose-mediated curb (Gay and Amos, 1983).

Interpreting these relationships within the above mentioned framework, i.e., maintaining the PGI mutant in medium devoid of L-glutamine, one would tend to assign glucose as the generator of ribose-5-P, whereas mannose would serve the role of keeping the ATP and hence the PRPP levels high.

Both glucose and mannose are ligands of the hexose transport system of the mutant. Quite apart from the mediated curb, which requires both hexoses, would there be other criteria for conformational changes, elicited by one or the other hexose?

Franchi et al. (1978) have reported that the enhanced hexose uptake which dominates glucose-starved fibroblasts of the parental line, 023, is sensitive to N-ethyl maleimide (NEM) in the following way. Low concentrations of NEM (0.1 mM or less) brings about an elimination of enhanced hexose uptake, but sparing the "basal" activity, i.e., the activity found in glucose-fed 023 cultures. The latter basal activity was found to be insensitive to the NEM treatment (Franchi et al., 1978).

Perhaps this approach tailored to the PGI mutant, especially under conditions as described in Table 3, might offer an independent way to differentiate between the "uncurbed" state which would be sensitive to NEM and the one which would remain insensitive. Or phrased more directly, would mannose or glucose confer sensitivity to NEM, assuming that the presence of both hexoses would confer insensitivity to NEM as well as imposing a curbed transport? Alternatively, would both hexoses also be required for bringing about a conformation which is insensitive to NEM?

The release of the glucose-mediated transport curb of hexoses by inhibitors of oxidative energy metabolism (or oxidative phosphorylation) was found to be markedly expressed in fibroblast lines from hamsters as well as mice and most recently from chick embryos (Kalckar et al., 1979; Ullrey, unpublished observation, 1982; Ullrey et al., 1982; Gay and Amos, 1983).

REFERENCES

Christopher, C.W., Johnson, W.C., and Ullrey, D., 1975, Comparisons of galactose metabolism in sugar-fed and sugar-deprived hamster cells in the presence or absence of prostaglandin E_1. J. Gen. Physiol., 66:20A.

Christopher, C.W., Kohlbacher, M.S., and Amos, H., 1976a, Transport of sugars in chick-embryo fibroblasts. Evidence for a low affinity system and a high affinity system for glucose transport, Biochem. J., 158:439.

Christopher, C.W., Ullrey, D., Colby, W., and Kalckar, H.M., 1976b, Paradoxical effects of cycloheximide and cytochalasin B on hamster cell cultures, Proc. Natl. Acad. Sci. USA, 73:2429.

Christopher, C.W., Colby, W., and Ullrey, D., 1976c, Derepression and carrier turnover, evidence for two distinct mechanisms of hexose transport regulation in animal cells, J. Cell. Physiol., 89:683.

Christopher, C.W., 1977, Hexose transport regulation in cultured hamster cells, J. Supramol. Struc., 6:485.

Christopher, C.W., 1979, Initiation of intracellular protein degradation by limited proteolysis, in: "Limited Proteolysis in Microorganisms," G.N. Cohen and H. Holzer (eds.), DHEW Publication No. (NIH)79-1591, pp. 37-42.

Christopher, C.W., and Morgan, R.A., 1981, Are lysosomes involved in hexose transport regulation? Turnover of hexose carriers and the activity of thiol cathepsins are arrested by cyanate and ammonia, Proc. Natl. Acad. Sci. USA, 78:4416.

Franchi, A., Silvestre, P., and Pouyssegur, J., 1978, 'Carrier activation' and glucose transport in Chinese hamster fibroblasts, Biochem. Biophys. Res. Comm., 85:1526.

Gay, R.J., and Amos, H., 1983, Purines as 'hyperrepressors' of glucose transport. A role of phosphoribosyl diphosphate, Biochem. J., 214:133.

Isselbacher, K.J., 1972, Increased uptake of amino acids and 2-deoxy-D-glucose by virus-transformed cells in culture, Proc. Natl. Acad. Sci. USA, 69:585.

Kalckar, H.M., Christopher, C.W., and Ullrey, D., 1979, Uncouplers of oxidative phosphorylation promote derepression of the hexose transport system in cultures of hamster cells, Proc. Natl. Acad. Sci. USA, 76:6453.

Kalckar, H.M., 1983, Regulation of hexose transport-carrier activity; another confrontation with cellular recycling, in: "An Era in New York Biochemistry," M. Pullman (ed.), Trans. N.Y. Acad. Sci., 41:83.

Kalckar, H.M., and Ullrey, D.B., 1984a, Further clues concerning the vectors essential to regulation of hexose transport as studied in fibroblast cultures from a metabolic mutant, Proc. Natl. Acad. Sci. USA, 81:1126.

Kalckar, H.M., and Ullrey, D.B., 1984b, Hexose uptake regulation mediated through aerobic pathways. Schism in a fibroblast mutant, Fed. Proc., 43:2242.

Kletzien, R.F., and Perdue, J.F., 1975, Induction of sugar transport in chick embryo fibroblasts by hexose starvation, J. Biol. Chem., 250:593.

Lipmann, F., and Lee, S.G., 1978, A glucose binding transport factor isolated from normal and malignantly transformed chicken fibroblasts, in: "Microenvironments and Metabolic Compartmentation," pp. 263-281.

Leloir, L.F., 1951, The metabolism of hexosephosphate, in:

"Phosphorus Metabolism," vol. I, McElroy and Glass (eds.), The Johns Hopkins Press.

Martineau, R.M., Kohlbacher, M.S., and Amos, H., 1972, Enhancement of hexose entry into chick fibroblasts by starvation: differential effects of galactose and glucose, Proc. Natl. Acad. Sci. USA, 69:3407.

Pouyssegur, J., Franchi, A., Salomon, M.C., and Silvestre, P., 1980, Isolation of a Chinese hamster fibroblast mutant defective in hexose transport and aerobic glycolysis: its use to dissect the malignant phenotype," Proc. Natl. Acad. Sci. USA, 69:2698.

Ullrey, D., Gammon, M.T., and Kalckar, H.M., 1975, Uptake patterns and transport enhancements in cultures of hamster cells deprived of carbohydrates, Arch. Biochem. Biophys., 167:410.

Ullrey, D.B., and Kalckar, H.M., 1981, The nature of regulation of hexose transport in cultured mammalian fibroblasts: aerobic "repressive" control by D-glucosamine, Arch. Biochem. Biophys., 209:168.

Ullrey, D.B., Franchi, A., Pouyssegur, J., and Kalckar, H.M., 1982, Down-regulation of the hexose transport system: metabolic basis studied with a fibroblast mutant lacking phosphoglucose isomerase, Proc. Natl. Acad. Sci. USA, 79:3777.

Ullrey, D.B., and Kalckar, H.M., 1982, Schism and complementation of hexose mediated transport regulation as illustrated in a fibroblast mutant lacking phosphoglucose isomerase, Biochem. Biophys. Res. Comm., 107:1532.

Yamada, K., Tillotson, L.G., and Isselbacher, K.J., 1983, Regulation of hexose carriers in chicken embryo fibroblasts. Effect of glucose starvation and role of protein synthesis, J. Biol. Chem., 258:9786.

HUMAN HISTOCOMPATIBILITY ANTIGENS: GENES AND PROTEINS

Jack L. Strominger*

Department of Biochemistry and Molecular Biology
Harvard University
Cambridge, Massachusetts 02138

Class I (HLA-A,B,C) and Class II (HLA-DR) human histocompatibility antigens encoded on chromosome 6 in the Major Histocompatibility Complex (Fig. 1) have remarkably similar structures, despite the differences in function. Each is a heterodimer composed of four extracellular domains (two of which are conserved Ig-like domains and one of which is polymorphic), located in two chains (α or heavy and β or light), in addition to the transmembrane and intracytoplasmic regions (Fig. 2). From a gross structural standpoint the antigens differ in a few ways. In the class I antigens three domains are in the heavy chain while only one domain is in the light chain (β_2-microglobulin). However, in the class II antigens each chain has two domains. Moreover, in the class II antigens both chains pierce the membrane and have small intracytoplasmic regions while in the class I antigens only the heavy chain does so.

Both groups of molecules are involved in immune defense. Their discovery as histocompatibility antigens, that is their involvement in transplantation rejection, derives from the extreme population polymorphism which is related to their function and therefore transplant rejection is a by product of an essential function of these molecules. Class I molecules act as restricting elements in the elimination of virus infected cells while class II antigens are involved

---

*I worked with Herman Kalckar for about a year at the National Institutes of Health during 1953. It was a very exciting and formative time in my career, and I will always be grateful to him for the part he played in it. I remember Herman best from this period from a knock on my door during a Washington snowstorm. It was Herman on skis to inform me that he was on his way to the lab to change the bottles on our homemade fraction collector.

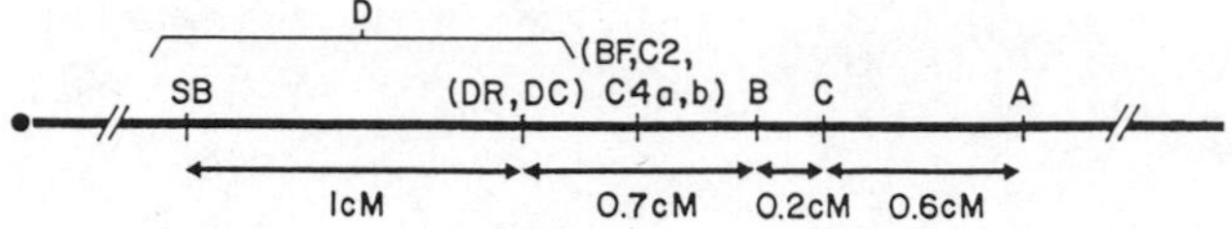

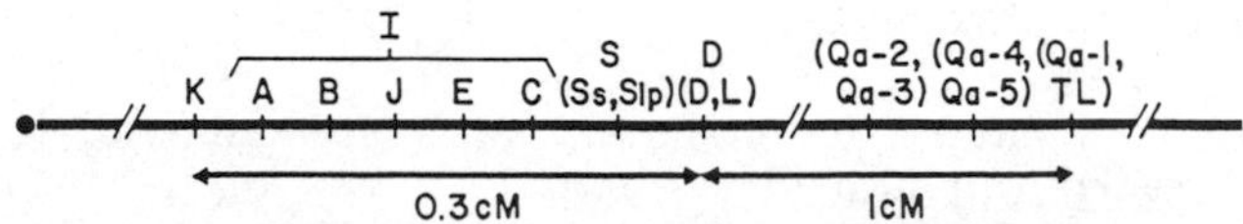

Fig. 1. Structure of the HLA complex on chromosome 6 and its murine homologue on chromosome 17 based on genetics and serology. It should be noted that the order within each of the following groups is unknown: Ss and Slp; L and D; Qa1 and Qa2; DR and DC; C2, BF, C4a and C4b.

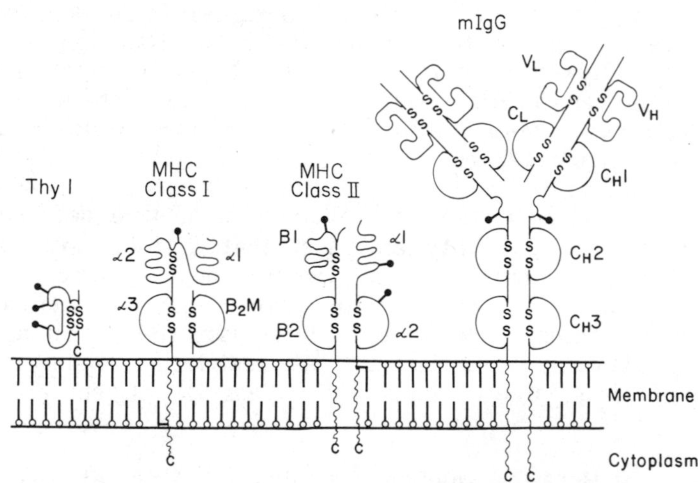

Fig. 2. Models of membrane proteins with homology to immunoglobulin. The class I and class II MHC antigens each contain two immunoglobulin-like domains adjacent to the membrane.

in cell-cell interaction in the generation of antibodies (macrophage, T helper cell and B lymphocyte). In view of their function it is not surprising that they are closely related structurally to immunoglobulins. An even simpler related molecule is thy-1, an antigen found on mouse thymocytes originally; more recently a homologue has apparently been found in the squid. These molecules are quite old in evolution and must have all descended from the same ancestral gene.

The number of sets of class II ("HLA-DR") antigens which are expressed in homozygous human cells is not yet entirely clear, but it is clear that the number of sets is considerably larger than is expressed in the mouse (Ia antigens). In the first place there are three families of antigens known in man, now called DR, DC, and SB, while only two families are known in the mouse, I-A and I-E (homologous to DC and DR, respectively). From a protein standpoint DR, DC, and SB have all been separated. The DR family has been shown to be composed of at least three subsets which share a common heavy chain but have distinct "DR-like" light chains.

From the study of the genes which encode these proteins even more complexity is observed (Fig. 3). In the DR family the single heavy chain and the three light chain genes have been detected. At least two heavy chain and two light chain genes occur in the DC family, but there is evidence of the possibility of greater complexity in both the heavy and light chains. For example, there could be as many as five heavy chain genes in this family if all of the detected crosshybridizing sequences turn out to be complete genes. In the SB family there are at least two heavy chain genes and at least three light chain genes. Some of the observed complexity could be due to the occurrence of nonexpressed pseudogenes. In any event the human HLA-D region appears to be more complex than the corresponding murine Ia region (Fig. 4), at least as so far observed, and appears to have been created by a large gene expansion. If so, this would represent the only example presently known in which the complexity of a genetic region in man is considerably greater than that in the mouse. However, an alternate possibility, though it may seem less likely, that the mouse has undergone gene deletion over evolutionary time, is not entirely excluded. The fact that the human class II region is much larger than that in the mouse was already known from recombinant frequencies (Fig. 1), although the data had not been so interpreted.

The class I genes of both man and mouse are extremely complex. In the range of 20 to 30 genes have been detected in this region in both species, and only a small number of these (about five) can be ascribed to the classical HLA-A,B,C or H-2K,D,L molecules. The nature and functions of the remainder and what portion of them are pseudogenes is presently unknown. An outline of the structures of class I and class II genes, as well as of an immunoglobulin heavy

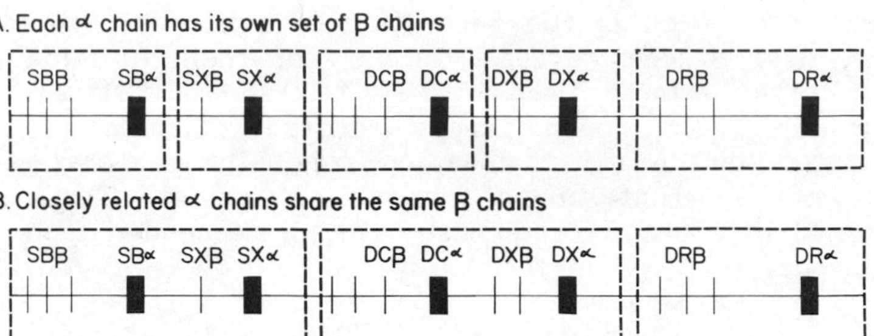

Fig. 3. Organization of the class II gene region of the human MHC. This cartoon, illustrating the extent of complexity found, is not intended to be accurate with respect to the precise number of genes or their relative locations, since this information is not yet available.

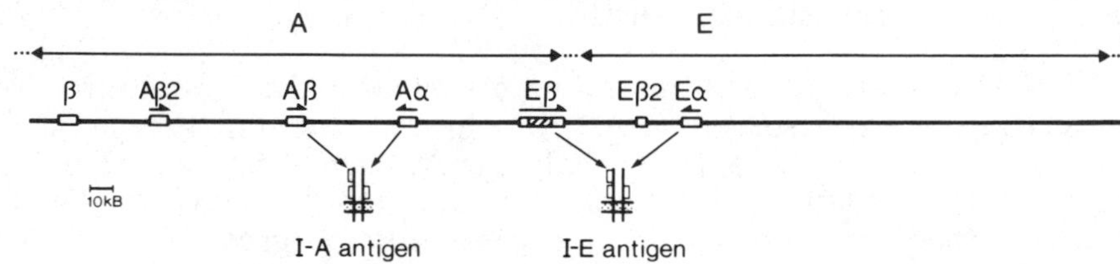

Fig. 4. Organization of the class II gene region of the murine MHC.

chain constant regions gene, is shown in Figure 5.

Several practical applications of the DNA probes generated in the course of these studies have become apparent. In the first place polymorphisms can be detected at the DNA level (by Southern blots) just as they can be detected serologically. For example, the polymorphism of the heavy chain of the DC antigens is readily detected by this method. Other polymorphisms are harder to define with the probes available, although they can be detected. It is anticipated that as better probes become available, typing of individuals will be more accurately accomplished at the DNA level than is presently possible serologically. In addition, the probes have been used in situ hybridization experiments to localize the genes on human chromosomes rather precisely. Moreover, they have been used to study a chromosomal rearrangement of chromosome 6 involving reciprocal translocations of chromosome 6 with chromosome 14. The precise break point in this translocation could be defined with the use of these probes, and it is likely that much more accurate diagnosis of genetic abnormalities in man will be possible in the future. These probes are also being used to study possible

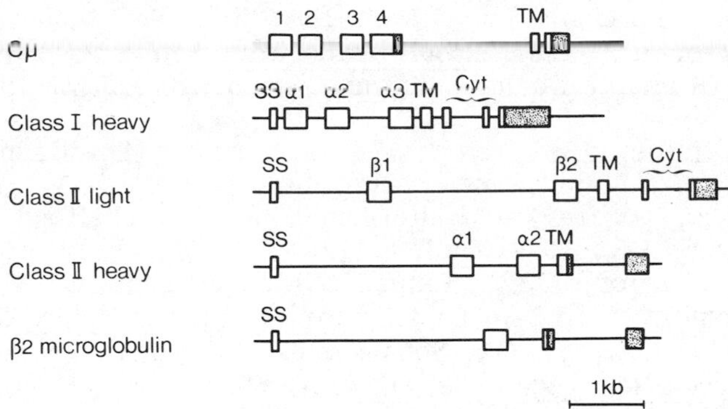

Fig. 5. Genomic structures of five related genes (SS, signal sequence; TM, transmembrane region; cyt, cytoplasmic region; exons representing extracellular domains are so numbered). Shaded boxes denote 3' untranslated regions. Cμ represents the constant region of a Cμ heavy chain gene.

polymorphisms in HLA-linked autoimmune diseases (e.g., multiple sclerosis, juvenile onset diabetes, rheumatoid arthritis).

This short account is based mainly on the following recent references from our laboratory. References to the work of others can be found in the papers cited.

ACKNOWLEDGEMENT

With the collaboration of Robert deMars (University of Wisconsin), Daniel Cohen and Jean Dausset (Hopital St. Louis, Paris), John Seidman, Avi Ben-Nun, and Cynthia Morton (Harvard Medical School) Ronald Germain, Lani Kirsch and Glenn Evans (N.I.H.), Jiri Novotny (Massachusetts General Hospital), Claude Mawas (Centre d'Immunologie, Marseille) and Walter Nance (Medical College of Virginia). Supported by research grants from N.I.H. (AI-10736, AM-13230, and AM-30241) and the Kroc Foundation.

REFERENCES

Arnot, D., Auffray, C., Boss, J., Grossberger, D., Kappes, D., Korman, A., Kuo, J., Lillie, J., Okada, K., Roux-Dosseto, M., Schamboeck, A., and Strominger, J.L., 1983, Proceedings of the 5th International Congress of Immunology, Academic Press, Tokyo, Japan.

Auffray, C., Ben-Nun, A., Roux-Dosseto, M., Germain, R.N., Seidman, J.G., and Strominger, J.L., 1983, EMBO J., 2:121.

Auffray, C., Korman, A.J., Roux-Dosseto, M., Bono, R., and Strominger, J.L., 1982, Proc. Natl. Acad. Sci. USA, 79:6337,

Auffray, C., Kuo, J., DeMars, R., and Strominger, J.L., 1983, Nature, 304:174.

Auffray, C., Lillie, J.W., Arnot, D., Grossberger, D., Kappes, D., and Strominger, J.L., in press, Nature.

Kaufman, J.F., Auffray, C., Korman, A.J., Shackelford, D.A., and Strominger, J.L., 1984, Cell, 36:1.

Kirsch, I.R., Morton, C.C., Nance, W.E., Evans, G.A., Korman, A.J., and Strominger, J.L., in press, Proc. Natl. Acad. Sci. USA.

Korman, A.J., Auffray, C., Schamboeck, A., and Strominger, J.L., 1982, Proc. Natl. Acad. Sci. USA, 79:6013.

Korman, A.J., Knudsen, P.J., Kaufman, J.F., and Strominger, J.L., 1982, Proc. Natl. Acad. Sci. USA, 79:1844.

Morton, C.C., Brown, J., Nance, W.E., Woods, D.E., Kirsch, I.R., Evans, G.A., Korman, A.J., and Strominger, J.L., in press, in: "Immunogenetics—Its Application to Clinical Medicine," (Proceedings of the Conference on Immunogenetics, August 17-19, 1983, Tokyo), T. Sasazuki and T. Tada (eds.), Academic Press, Tokyo, Japan.

Roux-Dosseto, M., Auffray, C., Lillie, J.W., Korman, A.J., and Strominger, J.L., 1983, in "Gene Expression," UCLA Symp. Mol. Cell. Biol., vol. 8, D. Hamer and M. Rosenberg (eds.), Alan R. Liss, Inc., New York.

Roux-Dosseto, M., Auffray, C., Lillie, J.W., Boss, J., Cohen, D., DeMars, R., Mawas, C., Seidman, J.G., and Strominger, J.L., 1983, Proc. Natl. Acad. Sci. USA, 80:6036.

Schamboeck, A., Korman, A.J., Kamb, A., and Strominger, J.L., 1983, Nuc. Acids Res., 11:8663.

Shackelford, D.A., Kaufman, J.F., Korman, A.J., and Strominger, J.L., 1982, Immunol. Rev., 66:133.

WHEN ECTO AND ENDOBIOLOGY MERGE: MONITORING ISCHEMIC CELL DEATH

Edgar Haber

Harvard Medical School
Massachusetts General Hospital
Boston, Massachusetts 02114

The theme of this symposium is ectobiology, an examination of the structure and function of the cell membrane. It is, of course, the membrane that defines the cell and differentiates it from extracellular fluid. When the integrity of the membrane has been breached and molecules may diffuse nonselectively between intra- and extracellular fluid, the cell ceases to exist. It is now believed that the first irreversible change that occurs in mammalian cells when they are deprived of oxygen and nutrients is the formation of rather large holes in the membrane.

The diffusion of macromolecules from extracellular to intracellular fluid might be a tool for identifying those cells that have lost their membrane integrity. An antibody, which is both a macromolecule and also capable of high affinity binding to some intracellular component might be a suitable marker for identifying cells that possess membrane defects.

An organ of interest for study is the heart, since its death (and thereby the death of the organism that it serves) is generally meditated by a loss of blood supply. Myosin is the principal protein of the cardiac cell. Its covalent structure is unique to the heart (Finck, 1965), allowing the development of antibodies that differentiate between cardiac and either skeletal- or smooth-muscle myosins. In the intact organism, cardiac myosin is protected from extracellular fluid by the cell's plasma membrane. When cell death occurs and the membrane breaks down, myosin is exposed to extracellular fluid. It is then available to react with labeled antibodies or antibody fragments.

In order to demonstrate myosin antibody binding to cells with

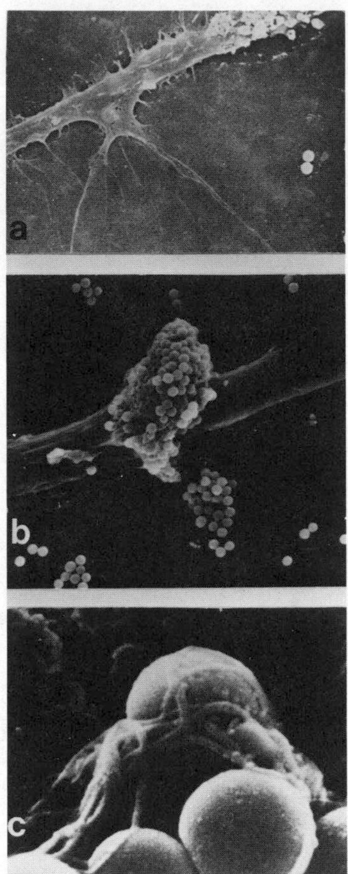

Fig. 1. (a) Scanning electron micrograph of an intact myocyte demonstrating a lack of Co-AMM binding. (b) Co-AMM attachment across a tear in the membrane of a necrotic myocyte. (c) Higher magnification (x100,000) showing binding of Co-AMM to exposed myofibrils of necrotic myocyte. (From Khaw et al., Science 217:1052, 1982. Reprinted with permission of the American Association for the Advancement of Science)

compromised membranes, mouse myosites were studied in culture (Khaw et al., 1982). The hearts of neonatal CD mice were treated with 0.25% Trypsin-EDTA with vigorous stirring until the cells were dissociated. After stabilization in culture for 3 to 4 days the beating myosites were then incubated either in a high glucose culture medium or one that was free of glucose. This was immediately followed by the addition of flourescine labeled polystyrene microspheres of 1 μ diameter either covalently coupled to antimyosin or to a monoclonal antibody of the same isotype and of irrelevant specificity (anti-alprenolol). The cells were then incubated at 37° C for an additional 24 h. After separating unbound spheres and dissociating the cells from the plastic, they were examined by scanning electron micrography.

Figure 1 shows electron micrographs of an intact myocyte from the high glucose incubation (a) showing very few adherent spheres and necrotic myocytes from the low glucose incubation (b) with a dense accumulation of antimyosin coupled spheres over a hole in the cell membrane through which intracellular contents appear to be herniating. In Figure 1c, at much higher magnification, the binding of antimyosin coupled spheres to cardiac myofibrils exposed through a hole in the membrane of a necrotic myocyte is demonstrated. This provides a graphic demonstration of the loss of membrane integrity in cells with compromised nutrition and the ability of antibodies to myosin mark the membrane defect.

For quantitative studies, cells could be sorted in in a fluorescence-activated cell sorter to separate populations labeled to varying degrees with flourescine labeled microspheres. Because of the high degree of flourescense provided by the spheres and its quantized distribution, it was easy to separate cells that were labeled with 0-5 spheres with those that bound larger numbers. Control monoclonal antibody coupled to flourescent spheres-labeled myocytes grown in an HG medium-normal O_2 atmosphere showed mainly nonspecific adsorption of beads (0-5). When myocytes were grown under identical conditions but exposed to antimyosin labeled spheres, the presence of damaged myocytes that had adsorbed greater than 25 beads per cell became apparent as well as a population of either minimally compromised or intact myocytes which had adsorbed smaller numbers of beads (0-5). The ratio of high to low fluorescence events could be employed as an indicator of injury. Maximum damage was seen in cells grown in a glucose free medium in a nitrogen atmosphere and minimal damage in an high glucose medium with a normal O_2 atmosphere.

Figure 2 shows FACS III dot-plot analyses (where the abscissa is cell size, the ordinate intensity of fluorescence, and each point a fluorescent event). The spheres labeled with control antibody show only a very minor population of labeled cells in contrast to the specific antibody labeled spheres where a major part of the population has a high flourescense. A further increase in flourescent cells occurs in a glucose-free medium under nitrogen atmosphere, an intervention aimed at promoting cell death by simulating an ischemic environment. The ratio of high to low fluorescence events increased approximately four-fold under these conditions. Thus it appears that antimyosin coupled flourescent spheres can specifically discriminate necrotic from live myocytes and that nonspecific adsorption of spheres coated with another, irrelevant antibody is minimal. Cells

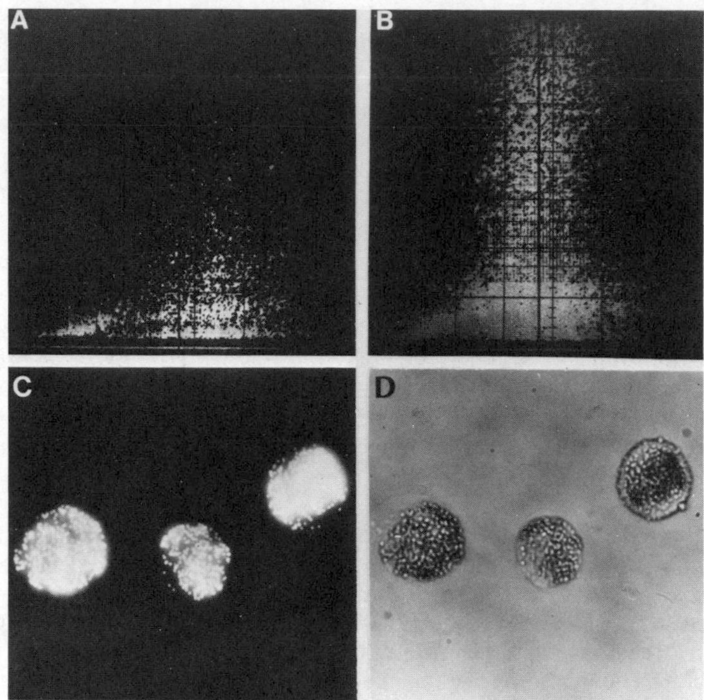

Fig. 2. (a) and (b) The dot-plot analyses (abscissa, cell size; ordinate, fluorescence intensity; each point is a fluorescence event) obtained on the fluorescence-activated cell sorter utilizing microspheres bound to an irrelevant mononclonal antibody (frame a) in a hyperoxic high-glucose medium, demonstrating a minimal amount of nonspecific adsorption. In contrast, the corresponding dot plots using myosin specific microspheres (frame b) show marked fluorescence of necrotic myocytes (10^5 cells were analyzed). (c) A fluorescence micrograph of sorted necrotic myocytes with a large number of bound myosin-specific microspheres and (d) the corresponding light micrograph. (From Khaw et al., Science 217:1052, 1982. Reprinted with permission of the American Association for the Advancement of Science)

that were heavily labeled (greater than 20 beads) were nonviable and did not survive in culture. In contrast, those that carried only a few microspheres (0-5) continued to contract and grow in culture following their retrieval from the cell sorter.

In order to test this concept in an in-vivo model, infarction was produced in dogs by ligation of the left anterior descending coronary artery (Khaw et al., 1976). Four hours later ^{125}I-labeled, myosin-specific antibody or antibody fragments were injected intravenously.

At varying times subsequent to the injections, animals were killed, the hearts perfused with triphenyltetrazolium chloride, and the myocardium was examined (Khaw et al., 1978). Figure 3 (left) shows a section of a heart treated in this way. The light area represents a largely subendocardial infarction. The central panel of the figure is a tracing of the slice superimposed on an autoradiograph. The exposed area corresponds to the infarct as revealed by the triphenyltetrazolium stain. The right panel shows specific radioactivity in the area of the infarct relative to myocardium on the posterior wall. It is apparent that the major concentration of radioactivity is in the subendocardial region, with lesser concentration in the subepicardial region that had been subjected only to spotty necrosis as indicated in the triphenyltetrazolium stain. Microautoradiographs showed that individual necrotic myocytes could be identified and differentiated from adjacent viable cells (Khaw et al., 1979). In order to demonstrate that antibody concentration was specific and not simply the result of passive diffusion of a macromolecule into infarcted cells, specific antibody labeled with ^{131}I and nonimmune globulin labeled with ^{125}I were injected simultaneously into the coronary arteries (Khaw et al., 1978). At the center of the infarct, the antibody has concentrated 34-fold in relation to normal myocardium, while the nonimmune globulin is only 7-fold in excess. In normal tissue, as expected, concentrations are equal. When compared to a marker of relative flow (the distribution of radioactive microspheres that had been injected into the left atrium), it was clear that the concentration of labeled antibody was inversely related

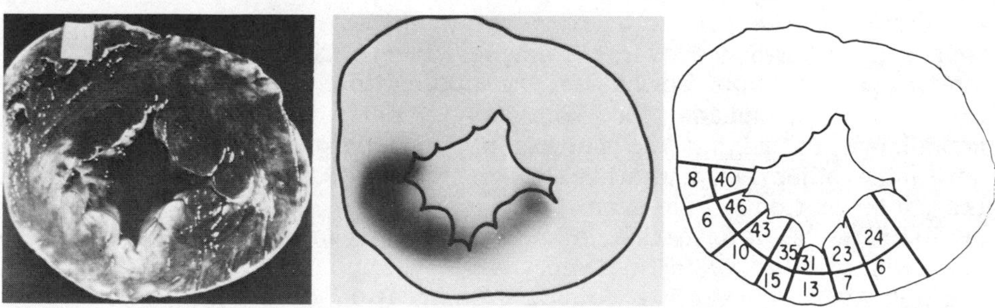

Fig. 3. Histochemically delineated myocardial infarction in a ventricular slice (left) seen as white or lighter-colored regions, and normal myocardium seen as darker regions. The corresponding macroautoradiograph is shown in the center with the outline of the ventricular slice. The right panel shows relative antibody uptake in the indicated areas. Ratios of antibody uptake were determined as specific radioactivity in relation to normal posterior ventricular myocardium. (From Haber & Khaw, Proc. N. Engl. Cardiovasc. Soc. II:38, 1979. Reprinted with permission of the American Heart Association, Inc.)

to relative blood flow (Khaw et al., 1976, 1978). There seems to be sufficient collateral circulation in this ischemic model to provide delivery of antibody, even to the regions of minimal blood flow.

It would be very desirable to apply this method to the detection, localization and quantification of myocardial infarcts in the living subject. Optimal imaging with a gamma or positron camera requires the labeling of antibodies with radionuclides of appropriate half-life and affinity. We have accomplished this by covalently linking diethylene triamine pentaacetic acid (DTPA) to antibody or antibody fragment and then binding cationic radionuclides by chelation (Khaw et al., 1980). Successful images have been obtained using the following radionuclides linked to antibody: ^{111}In-DTPA-Ab Fab (Khaw et al., 1978), ^{68}Ga-DTPA Fab (Khaw et al., 1980), and ^{99m}Tc-DTPA-AM-Fab (Khaw et al., 1983).

Heterogeneous affinity column-purified antibody, though useful, cannot be standardized because of the variability of the immune response. Quantities of antibody are necessarily limited. The application of monoclonal antibodies readily overcomes these difficulties. Monoclonal antibodies to human myosin have been obtained (Khaw et al., 1984). These antibodies cross reacted with canine cardiac myosin and thus could be tested in the canine model. Specific localization in the region of infarction was demonstrated by scintigraphy following intravenous injection of 99m-technetium labeled antibody. We were now prepared to embark on clinical studies.

A major problem in the evaluation of patients with coronary occlusion is the differentiation between irreversibly damaged and ischemic, though recoverable myocardium. Current techniques for limiting infarct size emphasize the application of thrombolytic agents or mechanical means for establishing early reflow. A significant impediment in judging the success of this approach is the inability to determine objectively whether or not myocardium has been spared by the intervention. An even more pressing clinical issue is the availability of information allowing for a reasoned decision to be made as to whether or not coronary bypass surgery or percutaneous angioplasty is appropriate following restoration of coronary flow. It is generally agreed that reperfusion of large regions of necrotic myocardium is of no value and may well be detrimental (Kloner et al., 1983).

Available imaging methods lack specificity in the evaluation of cardiac necrosis. Tc-99m labeled pyrophosphate may bind to damaged but recoverable myocardium, and thus overestimate the volume of damage (Bonte et al., 1974; Coleman et al., 1976; Willerson et al., 1975; Zaret et al., 1976). Thallium-201 is simply a marker for blood flow and thus cannot differentiate among necrotic, severely ischemic or scarred myocardium (Okada, in press). Examination of cardiac wall motion shortly after corornary occlusion is of little value, since

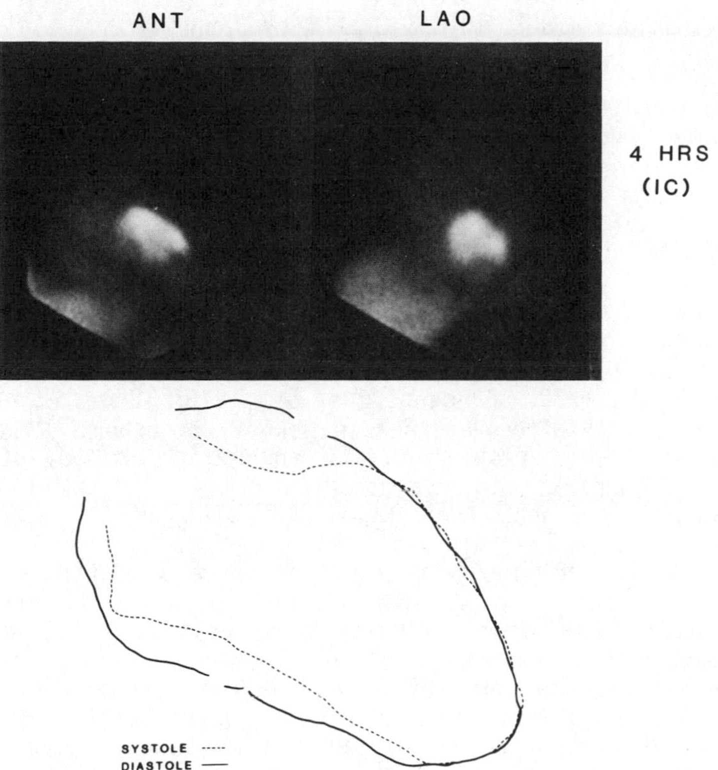

Fig. 4. (a) Antimyosin images in the anterior position (top left panel) and LAO 45 degree position (top right panel) of a patient with acute anterior infarction (SH). Antimyosin was administered by the intracoronary route four hours prior to imaging. There is antimyosin concentration in the anterolateral wall and upper septum. (b) End-diastolic (solid lines) and end-systolic (dotted lines) outlines of the contrast ventriculogram performed at 10 days. Akinesis of the anterolateral wall of the left ventricle corresponds to the zone of antimyosin uptake.

potentially recoverable myocardium may not resume function for some time after an ischemic insult (Braunwald and Kloner, 1982).

We examined 32 patients with myocardial infarction and unstable angina by scintigraphy following either intravenous or intracoronary injection of the Fab fragments of antimyosin antibodies, either monoclonal or polyclonal, labeled with technetium-99m. The results were compared with regional wall motion studies by ventriculography in 27 patients or by gated blood pool scanning in five at 10 to 14 days following the onset of symptoms. Discrepancy between wall motion abnormality and antimyosin uptake occurred in 5 of the 32 patients studied. There was but one instance of antimyosin Fab uptake without akinesis in a patient with anterior myocardial infarction diagnosed by electrocardiogram and elevated serum creatine phosphokinase. Four patients with small inferior infarctions showed akinesis without evident Fab uptake. None of the patients with unstable angina pectoris and no evidence of infarction showed antimyosin uptake. Figure 4 shows scintigraphic and wall motion studies in a typical patient.

The use of antibodies for imaging in man is a relatively new technique. One concern was the posssibility of acute allergic reactions to the foreign protein in previously sensitized individuals (Epenetos et al., 1982; Goldenberg et al., 1978; Mach et al., 1980; Belitsky et al., 1978) or late serum sickness in others. Evidence of hypersensitivity was not observed. The amount of antibody utilized is small (150-300 μg) in contrast reported doses of 1 to 150 mg in tumor imaging experiments (Belitsky et al., 1978).

Thus imaging with antimyosin Fab specifically identifies necrotic myocardium on the basis of loss of cell membrane integrity at a relatively early time after the onset of symptoms. It is a change in the ectobiology of the cardiac cell that allows its identification. This method appears to be safe in initial clinical use but its greatest potential lies in future development of quantification of infarct volume by such methods as single photon tomography or nuclear magnetic resonance.

REFERENCES

Belitsky, P., Ghose, T., Aquino, J., Tai, J., and MacDonald, A.S., 1978, Radionuclide imaging of metastasis in renal cell carcinoma patients by 131I-labeled antitumor antibod, Radiology, 126:515.

Bonte, F.J., Parkey, T.W., Graham, K.D., Moore, J., and Stokely, E.M., 1974, A new method for radionuclide imaging of myocardial infarcts, Radiology, 110:473.

Braunwald, E., and Kloner, R.A., 1982, The stunned myocardium: prolonged, postischemic ventricular dysfunction, Circulation, 66:1146.

Coleman, E.R., Klein, M.S., Roberts, R., and Sobel, B.E., 1976, Improved detection of myocardial infarction with technetium-99m stannous pyrophosphate and serum MB creatine phosphokinase, Am. J. Cardiol., 37:732.

Epenetos, A.A., Britton, K.E., Mather, S., Shepherd, J., Granowska, M., Taylor-Papadimitrious, J., Nimmon, C.C., Durbin, H., Hawkins, L.R., Malpas, J.S., and Bodmer, W.F., 1982, Targeting of iodine-123-labeled tumor-associated monoclonal antibodies to ovarian, breast, and gastrointestinal tumors, Lancet, 2:999.

Finck, H., 1965, Immunochemical studies with myosin. III. Immunochemical comparison of myosins from chicken skeletal, heart and smooth muscle, <u>Biochim. Biophys. Acta</u>, 111:231-238.

Goldenberg, D.M., Deland, F., Kim, E., Bennett, S., Primus, F.J., van Nagell, J.R., Estes, N., DeSimone, P., and Rayburn, P., 1978, Use of radiolabeled antibodies to carcinoembryonic antigen for the detection and localization of diverse cancers by external photoscanning, <u>N. Engl. J. Med.</u>, 298:1384.

Khaw, B.A., Beller, G.A., Haber, E., and Smith, T.W., 1976, Localization of cardiac myosin-specific antibody in experimental myocardial infarction, <u>J. Clin. Invest.</u>, 58:439.

Khaw, B.A., Fallon, J.T., Beller G.A., and Haber, E., 1979, Specificity of localization of myosin-specific antibody fragments in experimental myocardial infarction: histologic, histochemical, autoradiographic and scintigraphic studies, <u>Circulation</u>, 60:1527.

Khaw, B.A., Fallon, J.T., Strauss H.W., and Haber, E., 1980, Myocardial infarct imaging of antibodies to canine cardiac myosin with Indium-111-diethylenetriamine pentaacetic acid, <u>Science</u>, 209:295.

Khaw, B.A., Gold, H.K., Leinbach, R., Fallon, J.T., Strauss, W., Pohost G.M., and Haber, E., 1978, Early imaging of experimental myocardial infarction by intracoronary administration of ^{131}I-labeled anticardiac myosin $(Fab')_2$ fragments, <u>Circulation</u>, 58:1137.

Khaw, B.A., Mattis, J.A., Melincoff, G., Strauss, H.W., Gold, H.K., and Haber, E., 1984, Monoclonal antibody to cardiac myosin: imaging of experimental myocardial infarction, <u>Hybridoma</u>, 3:11.

Khaw, B.A., Scott, J., Fallon, J.T., Cahill, S.L., Haber, E., and Homcy, C., 1982, Myocardial injury: Quantitation by cell sorting initiated with antimyosin fluorescent spheres, <u>Science</u>, 217:1050.

Khaw, B.A., Strauss, H.W., Carvalho, A., Locke, E., Gold, H.K., and Haber, E, 1983, Concerning the labeling of DTPA-coupled proteins with Tc-99m, <u>J. Nucl. Med.</u>, 24:545.

Kloner, R.A., Ellis, S.G., Lange, R., and Braunwald, E., 1983, Studies of experimental coronary artery reperfusion. Effects on infarct size, myocardial function, biochemistry, ultrastructure and microvascular damage, <u>Circulation</u>, 68:8.

Mach, J.P., Carrel, S., Farni, M., Ritschard, J., Donath, A., and Alberto, P., 1980, Tumor localization of radiolabeled antibodies against carcinoembryonic antigen in patients with carcinomas, <u>N. Engl. J. Med.</u>, 303:5.

Okada, R.D., in press, Kinetics of thallium-201 in reperfused canine myocardium, <u>J. Am. Col. Cardiol.</u>

Willerson, J.T., Parkey, R.W., Bonte, F.J., Meyer, S.L., and Stokely, E.M., 1975, Acute subendocardial myocardial infarction in patients. Its detection by technetium-99m stannous pyrophosphate myocardial scintigrams, <u>Circulation</u>, 51:436.

Zaret, B.L., DiCola, V.C., Donabedian, R.K., Puri, S., Wolfsen, S., Freedman, G.S., and Cohen, L.S., 1976, Dual radionuclide study of myocardial infarction: Relationships between myocardial uptake of potassium-43, technetium-99m stannous pyrophosphate, regional myocardial blood flow and creatine phosphokinase depletion, <u>Circulation</u>, 53:422.

EUKARYOTIC TRANSPORT CONTROL: THE TURNOVER YEARS

C. William Christopher

The Biological Science Center and Biology Department
Boston University
Boston, Massachusetts 02215

ABSTRACT

The years 1974 to 1979 marked a transition period for Herman Kalckar. While at the Massachusetts General Hospital, he moved out of the Biochemical Research Unit into the Huntington Laboratories and, subsequently, to the Chemistry Department at Boston University. It was during this five-year period of changes in personnel and sites of research activity, a turnover period, that the "Carrier Turnover Model" for the regulation of transport in animal cells was developed. The model was designed to explain how a balance between carrier synthesis, opposed by carrier degradation, is regulated by some, as yet unknown, process of oxidative glucose metabolism. Although the process is not linked to glycolysis, it is known to require an "energized state" of the cell. The energized state is, however, not dependent on the intracellular supply of ATP. Instead, the energized state appears to be crucial to the process of carrier degradation which apparently takes place through the action of lysosomal proteinases.

NEVER INTO RETIREMENT

A renewed interest in the mechanism of nutrient transport regulation in animal cells was just beginning when Herman M. Kalckar officially retired from his academic appointments and when I began work with him. In 1972, a paper by Harold Amos and his associates (Martineau et al., 1972) had just appeared in the Proceedings of the National Academy of Sciences. The paper, which caught Herman's interest (and was communicated by him), described how, compared with continuous exposure to glucose, prolonged sugar starvation

Fig. 1. Herman M. Kalckar at the Huntington Labs in June, 1975.

resulted in striking increases in glucose uptake by cultured chicken embryo fibroblasts (CEF). At about the same time, Kurt Isselbacher contributed a paper that described the transport of hexoses and amino acids, comparing the relatively low rates of transport by untransformed baby hamster kidney (BHK) cells with enhanced transport by polyoma virus-transformed BHK (pyBHK) cells (Isselbacher, 1972). The results of Amos' starvation experiments using untransformed CEF and the observations of Isselbacher using BHK and pyBHK stirred Herman's imagination and he asked, "Are sugar starvation effects on hexose transport the same as the transformation effects?" (Kalckar and Ullrey, 1973). Thus was launched a long series of experiments that, since Herman's "retirement," has spanned more than 10 years, resulted in more than 20 published works, and spawned a plethora of unpublished results. I am fortunate to have been part of that story, and I fondly remember the work environment and the lunch meetings where strategies of immediate importance were mixed with often hilarious, impromptu chapters out of the unpublished "History of Biochemistry According to HMK." Those precious moments were mercurial and, unfortunately, will never be recorded. Even though the times (at a personal level and financial support level) were not easy, those "munch our lunch" meetings were always held in a relaxed atmosphere with Herman and his unmatched sense of humor in subtle command (Fig. 1).

TURNOVER YEARS

Besides the scientific work of the early 1970's, other events were transpiring that ultimately shaped my association with Herman

Kalckar. The Biochemical Research Unit at the Massachusetts Genral Hospital was being dissolved as Herman "retired" and stepped down as its chief. Paul Zamecnik, then Chief of the John Collins Warren Laboratories of the Huntington Memorial Hopsital at MGH (The Huntington Labs), invited Herman to join his group and soon Herman, with a remnant of his Biochemical Research staff, moved two floors down the Jackson Tower to the Huntington Labs. This was late 1973 and marked the first of what I refer to as Herman's turnover events.

At this time and across town at Harvard Medical School, Harold Amos, Mary Kohlbacher and I were working with the sugar-starved CEF and were finishing up a detailed kinetic analysis of sugar transport. The paper that resulted from this work (Christopher et al., 1976a) contained descriptions of our observations that the increase in hexose transport in sugar starved cells was primarily due to increases in the V_{max} for D-glucose, 2-deoxy-D-glucose, and 3-O-methyl-D-glucose (Table 1). This strongly suggested that transport increases were due to increases in the number of available carriers (rather than a change in carrier affinity for the sugars). This conclusion agreed with other observations and was of major importance in the development of the Carrier Turnover Model of transport control. In the course of the experiments it was found that cells could be treated with N-ethylmaleimide which inactivated hexokinases and allowed the direct measure of transport using otherwise metabolizable substrates if 5 to 10 sec. assay times were

Table 1: Transport Kinetics in Cultured Chicken Embryo Fibroblasts

Culture Condition†	Transported Ligand*					
	D-Glc		2-dGlc		3-0-meGlc	
	K_m	V_{max}	K_m	V_{max}	K_m	V_{max}
+ Glucose	1	2.5	1.5	6.5	2.5	3.5
- Glucose	1	10.	1.5	20.	3.5	16.5
(fold increase)	-	4	-	3	1.4	5

*Confluent cells were washed, treated with 0.5 mM N-ethylmaleimide for 5 min. at 37° to inactivate hexokinases and then assayed for the transport of D-glucose (D-Glc), 2-deoxy-D-glucose (2-dGlc) and 3-0-methyl-D-glucose (3-0-meGlc) in 10 s at room temperature (adapted from Christopher et al., 1976a).

†Eagle's MEM $\pm$ D-Glc for 24 hours.

used. This procedure was to be used later in the Huntington Labs to inactivate galactokinase and thus allow the measure of D-galactose transport in cultured hamster cells (Christopher, 1977). Another operational procedure that was put to use later was the observation that cells treated for 18 to 24 hours with low concentrations of cycloheximide lost significant transport activity if the cells were also maintained in the presence of D-glucose. After joining Herman Kalckar in 1973, the experiments that we performed in the Huntington Labs (described below) led one to the next and contributed to the general interest in transport regulation that was prominent at that time. Such was Herman's perception of that interest, that he organized a highly successful international symposium sponsored by the Fogarty International Center (NIH) where Herman was a resident scholar in the spring of 1976. The results of that symposium filled an entire volume of the Journal of Cellular Physiology that appeared a mere eight months after the meetings; a remarkable testimony to Herman's ability to get a worthwhile job done and done well.

In 1979, Paul Zamecnik also "retired" marking the termination of activity in what, for years, had been known as the Huntington Labs and the beginning of new positions for all of us. Paul went to continue his own research at the Worcester Foundation and the rest of us went our own separate (or almost separate) ways. The Chemistry Department at Boston University welcomed Dr. Herman Kalckar and appointed him Distinguished Professor of Chemistry; I joined in the Biology Department at BU a few hundred yards away from Herman and his laboratory.

The time since the "turnover" from MGH to BU has been punctuated with ups and downs much of which has been influenced by factors such as cutbacks in funding. Yet, through it all, Herman has maintained an even disposition, has persevered and has forged into a new area using animal cell mutants in his effort to identify the elusive metabolic factor (or factors) that controls transport in normal cells. With the technical help of his faithful and gifted associated, Donna Ullrey, Herman now labors in a kind of solitude that he breaks with excursions to visit places and friends as near as MGH and as far as mainland China. It is fitting that, in testimony to his influence, so many of his past and present associates can contribute to this recognition of his 75th birthday.

THE TURNOVER MODEL

Cultured animal cells that have been maintained in medium containing glucose and have reached a state of population confluence have low, basal rates of hexose and amino acid transport. Martineau et al. (1972) first showed that maintenance of chicken cells in medium devoid of D-glucose resulted in a gradual increase in hexose uptake

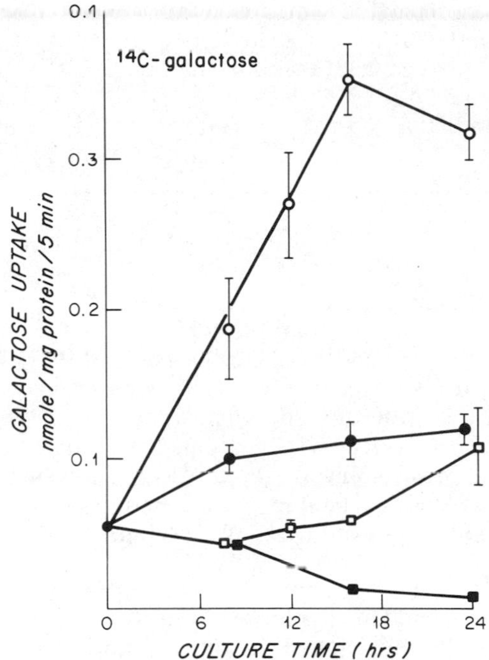

Fig. 2. Confluent Nil hamster cells were washed and changed to glucose-free Dulbecco's modified Eagle's minimal essential medium (D-MEM) with (●) and without (o) 10 μg cycloheximide per ml medium or D-MEM containing 4.5 mg D-glucose/ml with (■) and without (□) 10 μg cycloheximide/ml. At times indicated, cultures were washed with Dulbecco's phosphate-buffered saline (D-PBS) and assayed for the uptake of 1.5 x 10^{-5} M D-[^{14}C]galactose at room temperature for 5 min. Values are the mean ± S.D. nmol galactose taken up per mg cell protein to triplicate cultures (where not indicated by bar, S.D. was less than the size of the symbol). (From Christopher et al., 1976b.)

that, in 18 to 24 hours, could reach rates 20- to 50-fold higher than the basal rate. The increased rate of uptake was established to be a transport phenomenon by Kletzien and Perdue (1974, 1975) who also confirmed that the increase required protein and RNA synthesis because inhibitors of either process blocked the starvation-induced increases (see also, Martineau et al., 1972, and Amos et al., 1977). In 1976 we confirmed the findings using mammalian cells and discovered an interesting corollary. As represented in Figure 2, confluent Nil hamster cells (that will manifest 5- to 10-fold starvation-induced hexose transport enhancements) will also lose over 90 percent (10- to 20-fold decreases) of their hexose transport capacity in 24 hours if maintained in glucose-containing medium to

Table 2: Relative Activities of Hexose Carriers and Cathepsin B in Glucose-Fed and Glucose-Starved Hamster Cells

Culture Condition (24 hours)	Galactose Transport*	Cathepsin B Activity†
+ Glucose	1.00	1.00
- Glucose	5.67	0.06

*Galactose transport (nmol/mg/10 min) for glucose-fed cells was set at 1.00 relative value and the increase in the rate of transport for glucose-starved cells set relative to the glucose-fed cells.
†Cathepsin B activity of lysosomes taken from glucose-fed and glucose-starved cells was measured as the release of 2-naphthylamine from Benzoyl-D,L-arginine-2-naphthylamine in 15 min. at 40 ° and pH 6.

which has also been added cycloheximide (or puromycin) (Christopher et al., 1976b). As mentioned above, this phenomenon had been suggested in the results from earlier work with the CEF (Christopher et al., 1976a). Direct measurement of galactose transport in cells rendered galactokinaseless by brief treatment of the cells with N-ethylmaleimide confirmed that the changes (both the increases and the decreases) in transport rates were changes in the number of available, functional carriers (Christopher, 1977). Our findings have been confirmed and extended in important ways by others (Amos et al., 1977; Franchi et al., 1978; Gay and Hilf, 1980; Inui et al., 1980; Yamada et al., 1983), and most recently, Yamada et al. (1983) proved conclusively that rates of hexose uptake, that change as a function of culture conditions, are indeed due to changes in the number of functional carriers in membranes. It was suggested as early as 1977 that perhaps the loss of carriers was caused by proteolysis and that the basal steady state of transport in animal cells was the net result of a balance between carrier synthesis and carrier degradation. In contemplating which, if any, intracellular proteinases might be involved in such a mechanism, we learned that acid proteinases, particularly sulfhydryl-requiring (thiol) cathepsins of the lysosome can be irreversibly inactivated by changes in physiological conditions such as subtle changes in lysosomal pH (Ohkuma and Poole, 1978; Ohkuma et al., 1982; Starkey and Barret, 1973).

When I measured the activity of cathepsin B in extracts obtained from isolated lysosomes, it was found that, indeed, when cells were starved for glucose (and had high rates of transport) their cathepsin B activity was decreased (Table 2). There was, moreover, a consistent pattern of reciprocal changes in cathepsin B activity an transport activity (Christopher, 1979). Later it was found that a faithful reciprocal relationship between carrier activity and cathepsin B activity persisted when lysosomotropic probes were used to block the cathepsins in situ (Christopher and Morgan, 1981). The evidence pointed strongly in the direction of a lysosomal involvement in the control of hexose and amino acid transport in cultured animal cells. Furthermore, one of the most striking effects of viral transformation of animal cells is often the major increase in hexose transport (Isselbacher, 1972; Weber, 1973; reviewed by Hatanaka, 1974). It turns out that, at least in Nil hamster cells and mouse 3T3 cells, transformation by oncogenic viruses is accompanied by decreases in cathepsin B activity (Lockwood and Shier, 1977; Morgan et al., 1981). Again there is a reciprocal relationship: high transport, low cathepsin B. In addition to transformation-induced increases in transport, it has been known for years that freshly isolated cells (i.e., culture transfers) have high rates of transport that gradually decrease as the cells grow in density to confluence at which time the transport rate is basal and remains that way until the cultures are perturbed again. We compared cathepsin B (and also B-glucuronidase) activity with transport of both amino acids and hexoses as a function of culture time from transfer. Table 3 shows the results of these experiments; again, there is a reciprocal relationship between transport and lysosomal hydrolase activity.

Sometime in this period of study, it was discovered that losses of carrier activity induced by blocking protein synthesis could be prevented when a variety of reagents were included in medium for glucose-fed cells along with cycloheximide (Table 4). Among the most interesting blockers were fluoride, 2,4-dinitrophenol (DNP) and oligomycin. Fluoride is itself an inhibitor of protein synthesis (among other properties) and glucose-fed cells treated with both fluoride and

Table 3: Hexose and Amino Acid Transport versus Lysosomal Enzyme Activities as a Function of Culture Time

Transport or Enzyme Activity	Culture Time		Fold Change
	t=8 h	t=7 days	
D-galactose uptake (nmol/mg/5 min)	1.8	0.08	25 (decrease)
α-Aminoisobutyric acid (nmol/mg/5 min)	1.5	0.46	3 (decrease)
Cathepsin B (pH 6) (nmol/mg/min)	0.02	2.85	143 (increase)
β-glucuronidase (pH 4.5) (nmol/mg/min)	0.15	4.45	30 (increase)

Table 4: Effect of Metabolic Inhibitors on the Cycloheximide-Induced Loss of Hexose Carrier Activity in Nil Hamster Cells

Additions to Culture Medium (present for 24 hours)	Relative Galactose Uptake*	
	Glucose-Fed Cells	Glucose-Fed Plus Cycloheximide
No additions	1.00	0.08
Ammonium chloride (50 mM)	0.81	0.90
Cytochalasin B (21 μM)	3.27	1.18
Sodium fluoride (5 mM)†	1.02	1.05
2,4-dinitrophenol (0.2 mM)	1.91	1.28
Oligomycin (10 μg/ml)	1.64	1.04
Carbamoyl phosphate (10 mM)	2.27	0.92
Sodium cyanate (10 mM)	0.69	0.81

*Galactose uptake was assayed as described in the legend of Figure 2.

†Protein synthesis (incorporation of ^{3}H-leucine) was 92% inhibited by 5 mM NaF.

Table 5: Effects of 2,4-dinitrophenol and Oligomycin on Galactose Uptake

Culture Medium (24 hours)	Galactose Uptake (nmol/mg protein/10 min)	
No additions	1.58	(1.00)*
2,4-dinitrophenol (0.2 mM)	3.02	(1.91)
Cycloheximide (2 μg/ml)	0.23	(0.15)
Cycloheximide and 2,4-dinitrophenol	2.03	(1.28)
No additions	1.23	(1.00)
Oligomycin (10 μg/ml	2.06	(1.65)
Cycloheximide (2 μg/ml)	0.14	(0.11)
Cycloheximide and oligomycin	1.30	(1.04)

*Relative uptake values in parentheses (adapted from Kalckar et al., 1979).

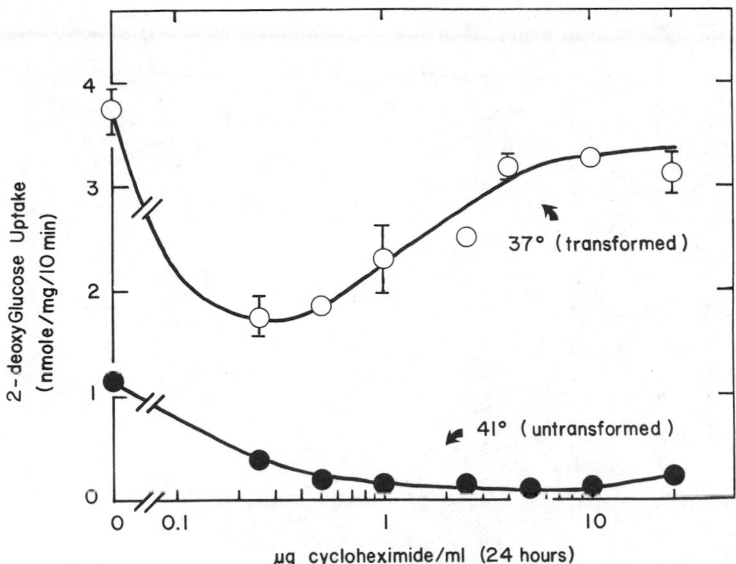

Fig. 3. Cultured chicken embryo fibroblasts infected with the temperature sensitive mutant of Rous sarcoma virus (RSBts68) were grown to confluency at the nonpermissive temperature (41° C) and then half the cultures were maintained at 37° C (half at 41° C) for 48 hr. The cultures were then washed and refed for 24 hr with D-MEM (4.5 mg glucose/ml) containing amounts of cycloheximide as indicated. Cultures were then washed and allowed to take up 0.1 M 2-deoxy-D-[^{14}C]glucose for 10 min at room temperature.

cycloheximide were doubly blocked, yet loss of carrier activity did not occur. Because fluoride also interferes with glycolysis and perhaps the level of ATP, two other inhibitors of ATP utilization (or synthesis) were tested for their effects on loss of carrier activity that consistently occurs when the synthesis of new carriers is prevented by protein synthesis inhibitors. Both DNP and oligomycin block the loss of carrier activity (Table 5) and, interestingly, allow for a substantial enhancement of transport carrier in control cultures, maintained in the absence of cycloheximide (Table 5). Was this the result of an ATP-dependent process? Perhaps it is, but if it is, the effects we saw were not the result of depleted pools of ATP. Starved cells (Rapaport et al., 1980) and DNP-treated cells maintain surprisingly high levels of ATP, the latter apparently the result of the compensatory increase in glycolytic ATP production. If not the ATP supply, the apparent energy-dependent carrier inactivation process must relate to the energized state of the cell (Kalckar et al., 1979).

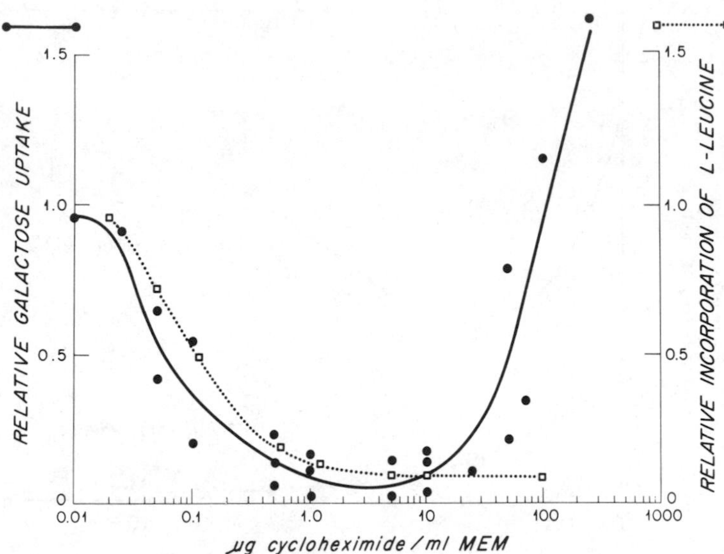

Fig. 4. Nil hamster cells were cultured as described in the legend of Figure 2, washed and refed with medium containing glucose and the various amounts of cycloheximide indicated. After 18 to 24 hr, the cells were washed and allowed to take up galactose for 5 or 10 min. Results are nmol of galactose taken up per mg of protein per assay time (●) and are expressed relative to uptake by cells cultured without cycloheximide. Protein synthesis (the incorporation of [^{3}H]leucine into ethanol-insoluble material, □) was for 2 hr at both the beginning and the end of the culture time while the cells were in media containing cycloheximide. This figure is compiled from several experiments; individual points are averages of duplicate or triplicate samples. (From Christopher et al., 1976b.)

The cycloheximide effects on transport are even more perplexing when analyzing results of the drug on transport rates in chicken cells infected with temperature sensitive mutants of the Rous sarcoma virus. Figure 3 shows a titration curve superimposed on a temperature-dependent change in 2-deoxyglucose transport. Cells phenotypically transformed at the permissive temperature, 37° C, have about 3-fold higher rates of transport than parallel cultures at the nonpermissive temperature. Something in the phenotypically transformed cells is much more sensitive to the effects of cycloheximide than in the phenotypically normal cells. In fact, there is a protection against the loss of carrier activity when the cycloheximide concentrations had also been observed to occur in Nil hamster cells (Fig. 4). Amenta et al. (1978) have referred to certain properties of

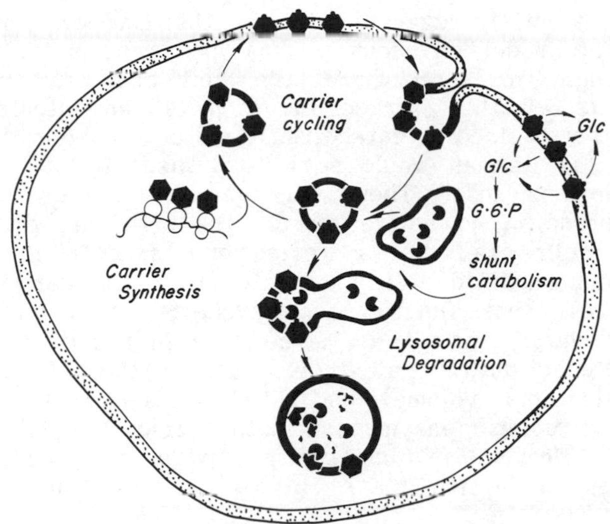

Fig. 5. Schematic model of opposing forces in the carrier turnover model. On the left, ribosomes synthesize new carriers (hexagonal symbols) which make their way first to an intracellular pool and then to the functional position in the plasma membrane. On the right, carriers are internalized as vesicles containing carriers by endocytosis and are either recycled to the plasma membrane or are intercepted by lysosomes. Lysosomal membranes and vesicle membranes fuse to expose the carrier molecule to the intralysosomal hydrolases. Inactivation and degradation of carriers removes potentially functional carriers from the pool as long as lysosomes remain intact and catabolites originating out of the hexosemonophosphate shunt pathway maintain the lysosome (and/or its enzymes) in an active state.

cycloheximide that strongly suggest lysosomal proteinases are blocked by cycloheximide.

These experiments, taken together, suggest that hexose carriers (and perhaps A-system amino acid carriers; Nishino et al., 1978; Christopher et al., 1980) are constantly being synthesized and subject to degradation. When an energy dependent system, perhaps an ATPase proton pump used in the lysosome for maintaining low intralysosomal pH (Ohkuma et al., 1982) is perturbed, the cells' degradative mechanism is incapacitated. Likewise, any destruction or inhibition of the lysosomal proteinases allows for an increase in the number of carriers by sparing their destruction. This sparing of the carriers suggests a means for recycling the internalized carriers to a functional position in the plasma membrane, and, coupled with the degradation potential, suggested to me the Carrier Turnover Model (Fig. 5). The model envisions plasma membrane associated carriers cycling through the interior of the cell back to the functional membrane site at the extracellular/intracellular interface. Under conditions where glucose catabolism (and/or anabolism) is optimal, carriers are (as Herman would say) "ambushed" by the lysosome and its hydrolytic enzymes. These carriers do not then return to the plasma membrane in an active form. The high rate of transport by transformed cells may be a reflection of deficient degradative enzymes or an impaired ability to internalize the carriers (or both). What regulates these functions, however, is not known; although, clearly, the overall control lies somewhere in the glucose amphibolic pathways. Recent nutritional studies such as those by Amos and his colleagues (in this volume) and studies using mutants that are defective in glycolytic enzymes and other glucose catabolic pathways (Kalckar and Ullrey, this volume) may provide clues (or an answer) to the mechanism of action that controls nutrient transport in animal cells.

REFERENCES

Amenta, J.S., Sargus, M.J., and Baccino, F.M., 1978, Inhibition of basal degradation in rat embryo fibroblasts by cycloheximide: Correlation with activities of lysosomal proteases, J. Cell. Physiol., 97:267.

Amos, H., Musliner, T.A., and Asdourian, H., 1977, Regulation of glucose carriers in chick fibroblasts, J. Supramol. Struct., 7:499.

Christopher, C.W., 1977, Hexose transport regulation in cultured hamster cells, J. Supramol. Struct., 6:485.

Christopher, C.W., 1979, Proteolysis as a potential factor in hexose transport regulation in cultured animal cells, in: "Limited Proteolysis in Microorganisms," G.N. Cohen and H. Holzer (eds.), D.H.E.W. Publication No. (NIH) 79-1591, U.S. Government Printing Office, Washington, D.C., p. 37.

Christopher, C.W., Kohlbacher, M.S., and Amos, H., 1976a, Transport of sugars in chick-embryo fibroblasts. Evidence for a low-affinity system and a high-affinity system for glucose transport, Biochem. J., 158:439.

Christopher, C.W., Ullrey, D., Colby, W., and Kalckar, H.M., 1976b, Paradoxical effects of cycloheximide and cytochalasin B on hamster cell hexose uptake, Proc. Natl. Acad. Sci. USA, 73:2429.

Christopher, C.W., Nishino, H., Schiller, R.M., Isselbacher, K.J., and Kalckar, H.M., 1980, Catabolic control of the enhanced alanine-preferring system for amino acid transport in glucose-starved hamster cells requires protein synthesis, Proc. Natl. Acad. Sci. USA, 76:1878.

Christopher, C.W., and Morgan, R.A., 1981, Are lysosomes involved in hexose transport regulation? Turnover of hexose carriers and the activity of thiol cathepsins are arrested by cyanate and ammonia, Proc. Natl. Acad. Sci. USA, 78:4416.

Franchi, A., Silvestre, P., and Pouyssegur, J., 1978, "Carrier activation" and glucose transport in Chinese hamster fibroblasts, Biochem. Biophys. Res. Commun., 85:1526.

Gay, R.J., and Hilf, R., 1980, Density-dependent and adaptive regulation of glucose transport in primary cell cultures of the R3230AC rat mammary adenocarcinoma, J. Cell. Physiol., 102:155.

Hatanaka, M., 1974, Transport of sugars in tumor cell membranes, Biochim. Biophys. Acta, 355:77.

Inui, K.-I., Tillotson, L.G., and Isselbacher, K.J., 1980, Hexose and amino acid transport by chicken embryo fibroblasts infected with temperature properties of whole cells and membrane vesicles, Biochem. Biophys. Acta, 598:616.

Isselbacher, K.J., 1972, Increased uptake of amino acids and 2-deoxy-D-glucose by virus transformed cells in culture, Proc. Natl. Acad. Sci. USA, 69:585.

Kalckar, H.M., and Ullrey, D., 1973, Two distinct types of enhancement of galactose uptake into hamster cells; tumor-virus transformation and hexose starvation, Proc. Natl. Acad. Sci. USA, 70:2502.

Kalckar, H.M., Christopher, C.W., and Ullrey, D., 1979, Uncouplers of oxidative phosphorylation promote derepression of the hexose transport system in cultures of hamster cells, Proc. Natl. Acad. Sci. USA, 76:6453.

Kletzien, R.F., and Perdue, J.F., 1974, Sugar transport in chick embryo fibroblasts. I. A functional change in the plasma membrane associated with the rate of cell growth, J. Biol. Chem., 249:3366.

Kletzien, R.F., and Perdue, J.F., 1975, Induction of sugar transport in chick embryo fibroblasts by hexose starvation, J. Biol. Chem., 250:593.

Lockwood, T.D., and Shier, W.T., 1977, Regulation of acid proteases during growth, quiescence and starvation in normal and transformed cells, Nature, 267:252.

Martineau, R., Kohlbacher, M., Shaw, S., and Amos, H., 1972, Enhancement of hexose entry into chick fibroblasts by starvation: differential effect on galactose and glucose, Proc. Natl. Acad. Sci. USA, 69:3407.

Morgan, R.A., Inge, K.L., and Christopher, C.W., 1981, Localization and characterization of N-ethylmaleimide sensitive inhibitor(s) of thiol cathepsin activity from cultured Nil and polyoma virus-transformed Nil hamster cells, J. Cell. Physiol., 108:55.

Nishino, H., Christopher, C.W., Schiller, R.M., Gammon, M.T., Ullrey, D., and Isselbacher, K.J., 1978, Sodium-dependent amino acid transport by cultured hamster cells: Membrane vesicles retain transport changes due to glucose starvation and cycloheximide, Proc. Natl. Acad. Sci. USA, 75:5048.

Ohkuma, S., and Poole, B., 1978, Fluorescence probe measurement of the intralysosomal pH in living cells and the perturbation of pH by various agents, Proc. Natl. Acad. Sci. USA, 75:3327.

Ohkuma, S., Moriyama, Y., and Takano, T., 1982, Identification and characterization of a proton pump on lysosomes by fluorescein isothiocyanate-dextran fluorescence, Proc. Natl. Acad. Sci. USA, 79:2758.

Rapaport, E., Christopher, C.W., Ullrey, D., and Kalckar, H.M., 1980, Selective high metabolic lability of uridine triphosphate in response to glucosamine feeding of untransformed and olyoma virus-transformed hamster fibroblasts, J. Cell. Physiol., 104:253.

Robins, E., and Hirsch, H.E., 1968, Glycosidases in the nervous system: II. Localization of β-galactosidase, β-glucuronidase and β-glucosidase in individual nerve cell bodies, J. Biol. Chem., 243:4253.

Starkey, P.M., and Barret, A.J., 1973, Human cathepsin B. Inhibition by α_2-macroglobulin and other serum proteins, Biochem. J., 131:823.

Weber, M.J., 1973, Hexose transport in normal and Rous sarcoma virus-transformed cells, J. Biol. Chem., 248:2978.

Yamada, K., Tillotson, L.G., and Isselbacher, K.J., 1983, Regulation of hexose carriers in chicken embryo fibroblasts. Effect of glucose starvation and role of protein synthesis, J. Biol. Chem., 258:9786.

INDEX